ALTERNATIVE
ENERGY SOURCES II

ALTERNATIVE ENERGY SOURCES II

Edited by T. Nejat Veziroğlu

ALTERNATIVE ENERGY SOURCES II

VOLUME 1
Solar Energy 1

Edited by

T. Nejat Veziroğlu
Clean Energy Research Institute, University of Miami

●

HEMISPHERE PUBLISHING CORPORATION

Washington   New York   London

DISTRIBUTION OUTSIDE THE UNITED STATES

McGRAW-HILL INTERNATIONAL BOOK COMPANY

Auckland   Bogotá   Guatemala   Hamburg   Johannesburg
Lisbon   London   Madrid   Mexico   Montreal
New Delhi   Panama   Paris   San Juan   São Paulo
Singapore   Sydney   Tokyo   Toronto

Proceedings of the Second Miami International Conference on Alternative Energy Sources, December 10–13, 1979, Miami Beach, Florida, presented by the Clean Energy Research Institute, University of Miami, Coral Gables, Florida, in cooperation with the International Association for Hydrogen Energy, the International Atomic Energy Agency, the International Solar Energy Society, Florida International University, the Florida Solar Energy Center, and the Mechanical Engineering Department, University of Miami.

EDITOR
T. Nejat Veziroğlu
Clean Energy Research Institute
University of Miami
Coral Gables, Florida, U.S.A.

EDITORIAL BOARD
Laxman G. Phadke
Northeastern University
Tahlequah, Oklahoma, U.S.A.

Harold J. Plass, Jr.
University of Miami
Coral Gables, Florida, U.S.A.

John W. Sheffield
University of Missouri
Rolla, Missouri, U.S.A.

EDITORIAL ASSISTANTS
Kazim Akyuzlu
Mechanical Engineering Department
University of Miami
Coral Gables, Florida, U.S.A.

Halim Gurgenci
Mechanical Engineering Department
University of Miami
Coral Gables, Florida, U.S.A.

Aykut Mentes
Mechanical Engineering Department
University of Miami
Coral Gables, Florida, U.S.A.

ALTERNATIVE ENERGY SOURCES II

1 2 3 4 5 6 7 8 9 0   B R B R   8 9 8 7 6 5 4 3 2 1

Library of Congress Cataloging in Publication Data

Miami International Conference on Alternative Energy
    Sources, 2d, Miami Beach, Fla., 1979.
    Alternative energy sources II.

    Proceedings of the Second Miami International Conference on Alternative Energy Sources, held Dec. 10–13, 1979 in Miami Beach, Fla. and presented by the Clean Energy Research Institute, University of Miami, Coral Gables, Fla. and others.
    Proceedings of the first conference (1977) published under title: Alternative energy sources.
    Includes bibliographical references and index.
    CONTENTS: v. 1-3. Solar energy.—v. 4. Indirect solar energy.—v. 5. Geothermal power/energy programs.
    1. Renewable energy sources—Congresses. I. Veziroğlu, T. Nejat. II. Miami, University of, Coral Gables, Fla. Clean Energy Research Institute, III. Title.
TJ163.15.M49   1979   333.79   80-25788
ISBN 0-89116-208-9 (set)
ISBN 0-89116-213-5 (v. 1)

Contents

CONCENTRATING COLLECTORS

ENERGY STORAGE

Preface

The world population is growing fast. It is now about 4.5 billion and is projected to double every 35 years. Peoples of the world are demanding more and more energy, since they aspire to raise their standard of living, and since the standard of living is directly proportional to the energy consumed. Nations consuming more energy per capita have better living standards than the others. Because of this natural desire to improve living standards, world energy consumption is increasing much faster than the world population. In fact, it is doubling every 10 to 12 years.

Today most of the world's energy demand is met by fossil fuels, mainly petroleum and natural gas. But their production is not keeping up with the demand. By now we know that world production of fossil fuels will start to decrease in the next 20 to 30 years. If we do not start introducing alternative energy sources to meet the growing demand, a proportional reduction in the living standards would result. Fortunately, there are many options before us: solar energy, in its direct and indirect forms, nuclear breeders, thermonuclear power, geothermal energy, synthetic fluid fuels, and hydrogen as an energy carrier to complement the nonfossil energy sources. However, before these energy alternatives can be utilized, in most cases it is necessary to conduct extensive research and development work.

The second Miami International Conference on Alternative Energy Sources, held December 10-13, 1979, two years after the successful first conference, provided a forum where the world's leading energy scientists, economists, and planners, from some 40 countries, met and presented their latest research findings in 42 technical sessions. The papers presented covered the technological advances for the utilization of energy alternatives as well as conservation, environment, economics, planning, strategies, and policy matters.

The papers recommended by the session chairpersons and co-chairpersons, together with the invited lecture and keynote address (published in the introduction, Vol. 1) are arranged in the nine volumes in 43 parts, by subject. The index is found at the end of Vol. 9. The reader should be advised that it was difficult to specifically classify some of the papers when there was an overlap in the subject matter. In such cases, we tried to make the best possible choice.

Alternative Energy Sources II is a valuable reference collection for engineers, architects, scientists, economists, planners, and decision makers in their efforts to find solutions to the important and growing problem of our times—energy.

T. Nejat Veziroğlu

Acknowledgments

We gratefully acknowledge support of the International Association for Hydrogen Energy, the International Atomic Energy Agency, the International Solar Energy Society, Florida International University, the Florida Solar Energy Center, and the Mechanical Engineering Department of the University of Miami.

We also extend sincere appreciation to the invited lecturer, Robert Tanenhaus, International Energy Agency, Paris, France; to the keynote speaker, James E. Funk, University of Kentucky, Lexington, Kentucky, U.S.A.; and to the banquet speaker, the Honorable Mike McCormack, U.S. Representative from the state of Washington, U.S.A.

Special thanks are due our authors, lecturers, and panelists, who provided the substance of the conference as published in this nine-volume compendium.

And last, but not least, we extend our gratitude to the session chairpersons and co-chairpersons for organizing and executing the technical sessions and for helping in selection of the papers published. In acknowledgment, we list these session officials on the following pages.

The Conference Committee

Conference Committee and Staff

CONFERENCE COMMITTEE

John O'M. Bockris
Texas A & M University

Hatice S. Cullingford
Los Alamos Scientific Laboratory

James E. Funk
University of Kentucky

Howard P. Harrenstien
University of Miami

Herbert Hoffman
Oak Ridge National Laboratory

Henry Piper
Oak Ridge National Laboratory

Harold J. Plass, Jr. (Co-Chairman)
University of Miami

John W. Sheffield (Co-Chairman)
University of Missouri-Rolla

T. Nejat Veziroğlu (Chairman)
University of Miami

CONFERENCE STAFF

Coordinators
Carol Pascalis Vogt
Lucille Walter

Special Assistants
Barbara Berman
Mary Connors
Helen Hooper
Marlene Pernas
Terri Yonon

Graduate Assistants
Kazim Akyuzlu
Halim Gurgenci
Ilker Gurkan
Aykut Mentes
Ranga Samudrala

Session Officials

FIRST PLENARY SESSION   CONFERENCE OPENING

Chairperson: J. Catz, University of Miami, Coral Gables, Florida, USA

Co-Chairperson: J. W. Sheffield, University of Missouri, Rolla, Missouri, USA

SESSION 1A   SOLAR COLLECTORS I

Chairperson: E. A. Farber, University of Florida, Gainesville, Florida, USA

Co-Chairperson: H. H. Robertson, University of Miami, Coral Gables, Florida, USA

SESSION 1B   PHOTOVOLTAICS I

Chairperson: J. C. Belote, The BDM Corporation, McLean, Virginia, USA

Co-Chairperson: J. G. Hirschberg, University of Miami, Coral Gables, Florida, USA

SESSION 1C   NUCLEAR ENERGY I

Chairperson: H. S. Cullingford, Los Alamos Scientific Laboratory, Los Alamos, New Mexico, USA

Co-Chairperson: J. Alexander, University of Miami, Coral Gables, Florida, USA

SESSION 1D   GEOTHERMAL ENERGY I

Chairperson: M. Inan, Technical University of Istanbul, Istanbul, Turkey

Co-Chairperson: I. Gurkan, University of Miami, Coral Gables, Florida, USA

SESSION 1E   UNUSUAL ENERGY CONVERSION

Chairperson: H. Piper, Oak Ridge National Laboratory, Oak Ridge, Tennessee, USA

Co-Chairperson: A. A. M. Sayigh, University of Riyadh, Riyadh, Saudi Arabia

SESSION 1F   ECONOMICS AND POLICY I

Chairperson: D. J. Santini, Argonne National Laboratory, Argonne, Illinois, USA

Co-Chairperson: T. Khalil, University of Miami, Coral Gables, Florida, USA

SESSION 2A   SOLAR COLLECTORS II

Chairperson: Y. Bayazitoglu, Rice University, Houston, Texas, USA

Co-Chairperson: I. Gurkan, University of Miami, Coral Gables, Florida, USA

SESSION 2B   PHOTOVOLTAICS II

Chairperson: A. C. Pal, The George Washington University, Washington, D.C., USA

Co-Chairperson: L. Phadke, Northeastern Oklahoma State University, Tahlequah, Oklahoma, USA

SESSION 2C   NUCLEAR ENERGY II

Chairperson: J. Baublitz, U.S. Department of Energy, Washington, D.C., USA

Co-Chairperson: D. Jopling, Florida Power and Light Company, Miami, Florida, USA

SESSION 2D   COAL TECHNOLOGY I

Chairperson: H. T. Gencsoy, West Virginia University, Morgantown, West Virginia, USA

Co-Chairperson: L. J. Vogt, Public Service Indiana, Plainfield, Indiana, USA

SESSION 2E   HYDROGEN ENERGY I

Chairperson: F. J. Salzano, Brookhaven National Laboratory, Upton, New York, USA

Co-Chairperson: J. W. Sheffield, University of Missouri, Rolla, Missouri, USA

SESSION 2F   ECONOMICS AND POLICY II

Chairperson: J. O'M. Bockris, Texas A&M University, College Station, Texas, USA

Co-Chairperson: L. E. Poteat, University of Miami, Coral Gables, Florida, USA

SESSION 3A   **SOLAR COLLECTORS III**

Chairperson: J. T. Pytlinski, New Mexico State University, Las Cruces, New Mexico, USA

Co-Chairperson: H. Wiseman, University of Miami, Coral Gables, Florida, USA

SESSION 3B   **SOLAR HEATING AND COOLING I**

Chairperson: A. A. M. Sayigh, University of Riyadh, Riyadh, Saudi Arabia

Co-Chairperson: H. Samudrala, University of Miami, Coral Gables, USA

SESSION 3C   **NUCLEAR ENERGY III**

Chairperson: R. T. Perry, University of Wisconsin, Madison, Wisconsin, USA

Co-Chairperson: H. Gurgenci, University of Miami, Coral Gables, Florida, USA

SESSION 3D   **GEOTHERMAL ENERGY II**

Chairperson: P. R. Franke, Los Alamos Scientific Laboratory, Los Alamos, New Mexico, USA

Co-Chairperson: A. Mentes, University of Miami, Coral Gables, Florida, USA

SESSION 3E   **HYDROGEN ENERGY II**

Chairperson: T. Ohta, Yokohama National University, Yokohama, Japan

Co-Chairperson: A. Mitsui, University of Miami, Coral Gables, Florida, USA

SESSION 3F   **ECONOMICS AND POLICY III**

Chairperson: L. V. Stover, Solar Reactor Corporation, Miami, Florida, USA

Co-Chairperson: J. Brown, Florida International University, Miami, Florida, USA

SESSION 4A   HEAT STORAGE AND TRANSFER I

Chairperson: J. D. Powell, Scripps Institution of Oceanography, La
                    Jolla, California, USA
Co-Chairperson: H. J. Plass, Jr., University of Miami, Coral Gables,
                    Florida, USA

SESSION 4B   SOLAR HEATING AND COOLING II

Chairperson: M. Lokmanhekim, Lawrence Berkeley Laboratory,
                    Berkeley, California, USA
Co-Chairperson: A. Bowen, University of Miami, Coral Gables, Florida,
                    USA

SESSION 4C   RURAL SOLAR APPLICATIONS

Chairperson: H. Harrenstien, University of Miami, Coral Gables,
                    Florida, USA
Co-Chairperson: A. A. El-Bassuoni, University of Miami, Coral Gables,
                    Florida, USA

SESSION 4D   COAL TECHNOLOGY II

Chairperson: P. A. Kittle, ARCO Chemical Company, Glenolden,
                    Pennsylvania, USA
Co-Chairperson: L. J. Vogt, Public Service Indiana, Plainfield, Indiana,
                    USA

SESSION 4E   HYDROGEN ENERGY III

Chairperson: E. Snape, Ergenics, Inc., Wyckoff, New Jersey, USA
Co-Chairperson: R. Tomonto, University of Miami, Coral Gables,
                    Florida, USA

SESSION 4F   ECONOMICS AND POLICY IV

Chairperson: R. C. Forrester, III, Oak Ridge National Laboratory,
                    Oak Ridge, Tennessee, USA
Co-Chairperson: R. Zuckerman, University of Miami, Coral Gables,
                    Florida, USA

SESSION 5A   HEAT STORAGE AND TRANSFER II

Chairperson: A. J. Parker, Jr., Mueller Associates, Inc., Baltimore,
                    Maryland, USA

Co-Chairperson: K. Akyuzlu, University of Miami, Coral Gables,
                       Florida, USA

SESSION 5B   SOLAR HEATING AND COOLING III

Chairperson: A. R. Martinez, National Research Council, Caracas,
                       Venezuela
Co-Chairperson: G. Kvajic, University of Miami, Coral Gables, Florida,
                       USA

SESSION 5C   OCEAN ENERGY

Chairperson: C. C. Richard, U.S. Naval Academy, Annapolis,
                       Maryland, USA
Co-Chairperson: I. Gurkan, University of Miami, Coral Gables, Florida,
                       USA

SESSION 5D   WIND ENERGY I

Chairperson: D. L. Miller, The Pennsylvania State University,
                       Middletown, Pennsylvania, USA
Co-Chairperson: H. Gurgenci, University of Miami, Coral Gables,
                       Florida, USA

SESSION 5E   HYDROGEN ENERGY IV

Chairperson: R. E. Billings, Billings Energy Corporation, Provo,
                       Utah, USA
Co-Chairperson: J. F. Steel, Jr., University of Miami, Coral Gables,
                       Florida, USA

SESSION 5F   ENERGY POLICIES AND PUBLIC
                       UNDERSTANDING

Moderator: J. Shacter, Union Carbide, Oak Ridge, Tennessee, USA
Co-Moderator: L. E. Poteat, University of Miami, Coral Gables,
                       Florida, USA

SESSION 6A   HEAT TRANSFER AND STORAGE III

Chairperson: S. I. Guceri, University of Delaware, Newark,
                       Delaware, USA
Co-Chairperson: R. Samudrala, University of Miami, Coral Gables,
                       Florida, USA

SESSION 6B   **SOLAR ENERGY ECONOMICS**

Chairperson: C. Krischner, University of New Mexico, Albuquerque, New Mexico, USA

Co-Chairperson: A. Mertol, Lawrence Berkeley Laboratory, Berkeley, California, USA

SESSION 6C   **INDUSTRIAL SOLAR APPLICATIONS**

Chairperson: I. Sakr, National Research Centre, Cairo, Egypt

Co-Chairperson: A. Mentes, University of Miami, Coral Gables, Florida, USA

SESSION 6D   **WIND ENERGY II**

Chairperson: R. L. Moment, Rockwell International, Golden, Colorado, USA

Co-Chairperson: I. Gurkan, University of Miami, Coral Gables, Florida, USA

SESSION 6E   **SYNTHETIC FUELS**

Chairperson: R. F. McAlevy, III, Stevens Institute of Technology, Hoboken, New Jersey, USA

Co-Chairperson: K. Wong, University of Miami, Coral Gables, Florida, USA

SESSION 6F   **ECONOMICS AND POLICY V**

Chairperson: M. Avriel, Technion—Israel Institute of Technology, Haifa, Israel

Co-Chairperson: M. I. Mantell, University of Miami, Coral Gables, Florida, USA

SESSION 7A   **SOLAR POWER SYSTEMS**

Chairperson: D. Gidaspow, Illinois, Institute of Technology, Chicago, Illinois, USA

Co-Chairperson: A. Mentes, University of Miami, Coral Gables, Florida, USA

SESSION 7B   **NATIONAL GOALS FOR SOLAR ENERGY: ECONOMIC AND SOCIAL IMPLICATIONS**

Chairperson: R. H. Bezdek, U.S. Department of Energy, Washington, D.C., USA

Co-Chairperson: H. Harrenstien, University of Miami, Coral Gables,
                   Florida, USA

SESSION 7C   HYDRO AND TIDAL POWER

Chairperson: F. E. Naef, Lockheed Corporation, Washington, D.C.,
                   USA
Co-Chairperson: R. Samudrala, University of Miami, Coral Gables,
                   Florida, USA

SESSION 7D   BIOCONVERSION

Chairperson: S. Farooq, University of Miami, Coral Gables, Florida,
                   USA
Co-Chairperson: E. Daly, University of Miami, Coral Gables, Florida,
                   USA

SESSION 7E   ENERGY CONSERVATION

Chairperson: T. A. King, Mueller Associates, Inc., Baltimore,
                   Maryland, USA
Co-Chairperson: J. H. Parker, Florida International University, Miami,
                   Florida, USA

SESSION 7F   WORKSHOP-ADDRESSING THE PUBLIC ON
                   ENERGY

Chairperson: J. Shacter, Union Carbide, Oak Ridge, Tennessee, USA
Co-Chairperson: L. G. Phadke, Northeastern Oklahoma State
                   University, Tahlequah, Oklahoma, USA

INTRODUCTION

International Energy Problems

ROBERT TANENHAUS
International Energy Agency
Organization for Economic Cooperation and Development
Paris, France

INTRODUCTION

The United States-inspired International Energy Agency (IEA) in Paris, France, coordinates the development and implementation of national energy policies among most (twenty) of the Western (developed, market economy) countries, including the United States. The international energy problems the member countries are trying to address together are described below.

OIL "GAP"

Increasing energy demand and decreasing self-sufficiency are expected to cause the IEA countries' reliance on imported fuels to increase by over 40% (43%) between 1977 and 1990. As today, oil will make up the overwhelming majority (over 80%) of the imports. There probably will not be enough oil to serve the demand without stronger national and international policy measures.

The IEA estimates that the so-called "demand gap" could be about one-third. In comparison, the "gap" caused by the Iranian situation was about 4%. Even more than presently, therefore, a widening "gap" would apply stronger upward pressure on the real price of imported oil and result in greater uncertainty over the reliability of oil supply. The result would be the increasing drain of energy costs on the economics of member countries, the increasing likelihood of energy shortages disrupting their economies and the increasing influence of oil dependence on their political decisions.

1979: POST-IRANIAN REVOLUTION

In addition, an overview of trends in the OPEC policies, since the Iranian revolution suggests that they are converging toward more rapidly rising prices and constant or decreasing production, corresponding more to domestic revenue and energy requirements.

As long as the market favors the producers, in economic terms, the OPEC countries will be able to turn to higher prices rather than higher exports and to maintain a tight market and miximize revenue, possibly without full consideration of the jeopardy to other countries' growth and to the international financial system. The degree to which they may do so would be influenced by, besides economic factors, political and military considerations.

In the first three quarters of 1979, for the IEA countries and the majority of

developings countries[1] without sufficient oil for domestic use, net oil imports
are estimated to cost at least $109 billion and $18 billion, respectively, including $36 billion in the United States[2]. The price hikes are estimated to
drain off about 8% of expected GNP growth in the period for both sets of coun-
tires, including the United States.

SOLUTIONS

The situation is likely to prevail until the countries, including the United
States, greatly reduce their dependence on imported oil through both conservation and alternative energy sources. Because of the OPEC countries' ability
to maintain a tight market, neither approach can assure oil supply or price
relief in the near future. But together the two approaches can reduce the use
of imported oil fastest and thereby recoupe the IEA countries' diminished economic and political independence the quickest.

As a beginning, in early March of 1979, the countries agreed to try to reduce
expected oil imports by 5% within the next year, that is, to hold net oil imports steady, and, since then, many countries fixed 1980 targets for themselves.

There is no one fuel, solution or country that, by itself, can alter the prospect before us. The IEA countries, both individually and together, must take
a broad and complementary array of actions if the member countries are to achieve
the maximum self-sufficiency with the least societal cost over time.

The United States is especially important as both an energy producer and consumer, comprising about half of the IEA totals in both roles.

This is why the United States and the other Western countries, both individually
and together through organizations like the IEA, are working to accelerate their
indiependence from imported oil. For example, in the supply area, they must increase the use of domestic fuels and trade in these fuels with each other. In
conservation, they must increase the efficiency of energy use by people and
machines. Ultimately, they must develop new energy technologies. Since most
of the rest of the world is also drawing on the same energy resources, the IEA
and the member countries arc reaching out to work together with these countries
as appropriate.

IEA OBJECTIVES

The last annual review showed that all member countries have made some progress
towards increasing energy supplies to varying degrees. Methods include: substitution for oil, increased domestic production, supply diversification, holding down demand through conservation, and/or a combination of all these measures.

However, it also showed that the countries' efforts undertaken and planned at
that time were probably inadequate to prevent a serious imbalance in oil supplies and demand as early as the late 1980's. Events in early 1979 showed
that these efforts were inadequate for today's world with a rapidly changing
political situation.

In the past, the IEA countries had pledged themselves to try to prevent a crisis
in the medium-term by holding their total oil imports to a target of 26 million

1   Does not include OECD member countries
2   F.O.B. only

barrels/day (1300 mtoe) in 1985.

In early March 1979, the countries agreed to attempt to reduce expected oil imports by 5% within the next year, that is, to hold net oil imports steady.
Since that time, many countries have fixed targets for themselves to secure the
1985 objective.

However, an even more vigorous pursuit and re-inforcement of the IEA's Twelve
Principles, and of national and international policy measures are necessary
to assure that oil and other forms of energy will be available in sufficient
quantities and at reasonable prices.

The IEA's Principles for Energy Policy are noted below.

1. Reduce oil imports by conservation, supply expansion and oil substitution.

2. Reduce conflicts between environmental concerns and energy requirements.

3. Allow domestic energy prices sufficient to bring about conservation and
    supply creation.

4. Slow energy demand growth relative to economic growth by conservation and
    substitution.

5. Replace oil in electricity generation and industry.

6. Promote international trade in coal.

7. Reserve natural gas to premium users.

8. Steadily expand nuclear generating capacity.

9. Emphasize R&D, increasing international collaborative projects.

10. Establish a favorable investment climate, establish priority for exploration.

11. Plan alternative programs should conservation and supply goals not be fully
    attained.

12. Co-operate in evaluating world energy situation, R&D and technical require-
    ments with developing countries.

CONSERVATION

As mentioned before, conservation is a necessary, but not a sufficient, response
to the IEA countries. It provides three important advantages to oil-importing
countries. It saves money and reduces the impact of oil costs and price hikes
on their economies. It reduces their dependence on imported oil and, therefore,
reduces their sensitivity to OPEC supply changes and to OPEC influence on their
policies. Conservation and alternative fuels together can reduce imported oil
faster than relying on either one alone.

ALTERNATIVE SOURCES

Therefore, let us consider the prospects for existing forms of fuels as alternatives for declining oil. These fuels include: coal, natural gas, nuclear
power, hydroelectric, geothermal and various forms of solar energy.

Let us look at what measures will be necessary to achieve these prospects. First
of all, exploration and development of oil in IEA countries should continue in
order to slow the depletion rate, although any unanticipated additions will not
significantly affect the situation just described.

COAL

The IEA countries increasingly are looking to coal to fill much of the gap.
Use of coal is expected to double by 2000 (to 1170 Mtoe), based on its favorable
competitive economics. According to the IEA's recent study, Steam Coal:Prospects
to 2000, an accelerated international effort might add over 50% more by that
time (330 Mtoe). However, coal's share of total energy requirements would only
have increased 3% from 19% in 1977 to 22% in 2000. Total coal use would then
be 1500 Mtoe, or 30.5 million barrels/day. At this level, world coal reserves
last more than an additional 200 years.

To achieve this accelerated use, each producer or consumer country will have to
remedy several of the following constraints:

- antiquated infrastructures,

- cumbersome permit processes,

- erroneous belief that all energy prices will be the same,

- erroneous belief, given recent oil prices and technological advances in coal,
  that coal cannot be both acceptably "clean" and economic,

- legislation performing poorly in achieving both environmental and energy
  objectives,

- an obsolete energy-environmental balance.

This redress can be accomplished by promoting a more favorable climate for investment and development of coal projects; encouraging substitution of oil by
coal in power generation, process heat and some applications of space heating;
and implementing effective information programs aimed at the general public,
investors and traders on energy policy issues, fuel choices and relative costs,
including environmental trade-offs.

To establish a more favorable investment and development climate the IEA coun-
tires recently concluded an agreement on the international trade of steam coal.
The agreement offers a model of cooperation between international producer and
consumer nations.

NATURAL GAS

Natural gas is important, because of its relatively easy substitution for oil
and its environmental "cleanliness". Although the production of natural gas by
some IEA countries, such as the United States, the United Kingdom and Norway
may continue to increase in the near future, in the IEA countries as a whole,
it is projected to continue to decline at a increasing rate. This is expected
to reach over 1% per year by the late 1980's.

Gas imports by pipeline and LNG are expected to increase dramatically, more
than replacing the lost production. This gas is expected to come increasingly
from non-IEA sources, especially after 1985 as North Sea production begins to
decline. However, even with increasing imports, gas' share of requirements is
expected to fall from 20% in 1976 to 17% in 1990.

To achieve this much the IEA countries need to encourage exploration and development of gas in their countries by appropriate pricing policies. In this
context price decontrol in the United States is particularly welcome by the
other IEA countries; concentrate gas use on premium requirements; and develop
the infrastructure necessary for expanding the availability of gas.

In addition, for liquid natural gas (LNG) trade to grow beyond its present small contribution to Japan, Spain and the United States to the projected levels pricing, terminal siting and safety issues will have to be resolved and development efforts will have to be intensified.

NUCLEAR

After the oil crisis in 1974, many IEA countries accelerated their nuclear programs to replace imported oil in electricity generation. This approach was expected to "save" a lot of oil, since electricity generation has a low energy efficiency compared with direct use. However, IEA countries have lowered their original projections as electricity demand has decreased and the acceleration has aggravated the problems that need to be solved.

Nuclear power can still make a significant contribution to total energy requirements rising from 3% in 1976 to about 10% in 1990. It can do so, if the IEA countries make special efforts, both nationally and through the promotion of broader international cooperation; to resolve the problems as speedily as possible; consistent with safety, environmental and security standards satisfactory to the concerned countries; and the need to prevent the proliferation of nuclear weapons.

HYDRO AND GEOTHERMAL POWER

Hydro and geothermal generated electricity is projected to grow at about the same rate as total requirements through 1990. The share of requirements is expected to remain at about the same level (6%). Most significant large and small scale sites have been identified and their productive capacity determined, except for some deep geothermal sources. The task ahead is to weigh their scenic and ecological value if left unharnessed against the potential power and the value of these attributes after harnessing. The appropriate sites would then need to be rapidly developed to provide the projected amount of power by 1990. At that time, these sources would be nearly completely developed.

SOLAR AND RENEWABLE SOURCES

Solar technologies are increasingly becoming practical and competitive solutions. Sometimes they can also be "kinder" to the natural and social environments than highly polluting conventional energy systems. However, in the IEA countries, they are primarily in the research and development stage. Only recently, as more and more technologies are maturing, the work increasingly is focusing on how to bring them to commercialization. Although it is difficult to judge with the limited information available, under present plans the IEA countries are not likely to rely on renewable energy sources for more than 3% of their primary energy requirements in 1990, omitting large-scale hydro. The share may be slightly higher in some countries. The principal constraints are costs and institutional barriers. Many countries have some financial incentives, but although some countries are actively trying to remove institutional barriers, many still remain.

CONCLUSION

There is no one fuel, solution or country that, by itself, can alter the prospect before us. The IEA countries, both individually and together, must take a braod and complementary array of actions if the member countries are to achieve the maximum self-sufficiency with the least societal cost over time. The United States is especially important as both an energy producer and consumer, comprising about half of the IEA totals in both roles. This is why the other IEA countries have exhorted the United States to action and warmly welcomed the United States' responses to the energy problem, although not necessarily agreeing with all of the specific measures.

Since most of the rest of the world is also drawing on the same energy resources, the IEA and the member countries are reaching out to work together with these countries as appropriate. This is why the IEA is studying and recommending ways to increase its member countries' independence from imported oil by the domestic development of each major fuel and trade in these fuels among member countries; by conservation, without significantly dampening the countries' economies; by increasing the initial efficiency of energy use by people and machines and reusing waste energy and ultimately; by the development of new energy technologies among member and non-member countries.

Getting from Here to There

JAMES E. FUNK
University of Kentucky
Lexington, Kentucky 40502, USA

Good morning, Ladies and Gentlemen. The title of this address, "Getting from Here to There" is taken, with permission and gratitude, from a book by the same name written by Prof. W.W. Rostow of the University of Texas. Prof. Rostow is an economist and in his book he describes the broad economic trends which he sees affecting the world in the near to mid-term future. Many of his observations are important to the scientific and technical community and I would like to share some of these with you this morning. They are especially important for this conference, which deals with alternative energy sources.

Let's begin by talking about what it's going to be like "then and there". We all know what it's like "here and now", and its quite clear that we have some very serious problems, but it's going to be better when we get "there" in the hopefully not too distant future. By then we will have accomplished the following:

1.) Birth rates in the developing world will have dropped from 40/1000 to 20 or fewer.

2.) The level and productivity in agriculture will have greatly improved.

3.) We will have solved the scientific and technical problems in creating a new economic energy source, based on renewable or essentially infinite sources.

4.) We will have faced and begun to overcome, by expanded production, substitution, economy and recycling, a growing list of raw material shortages.

5.) We will have proved we know how to control air, water, and other forms of pollution and that we have the will to allocate the resources necessary for its control.

6.) We will have restructured our patterns of investment, including the scale and directions of research and development to accomplish the above results, and to provide the technologies in agriculture, energy and raw materials required in the coming century.

7.) We will have learned how to reconcile high and steady rates of growth with price stability.

8.) Per capita income in the developing nations will have doubled and their competence in a much widened range of technologies will have greatly improved.

The accomplishment of these eight conditions will require the best efforts of science and technology, as well as all other components of society. One of these conditions, I am sure you have noticed, requires the creation of a new economic energy source, based on renewable or essentially infinite resources.

The importance of this task has also been emphasized by Geoller and Weinberg, who have said "...man must develop an alternative energy source. Moreover, the incentive to keep the price of prime energy as low as possible is immense. In the Age of Substitutability energy is the ultimate raw material. The living standard will almost surely depend primarily on the cost of prime energy. We, therefore, urge moving as vigorously as possible, not only to develop satisfactory inexhaustible energy sources--the breeder, fusion, solar and geothermal power--but to keep the program sufficiently broad so that we can determine, perhaps within 50 years, the cheapest inexhaustible energy source."

Now that is a tall order, especially when the emphasis is placed on the price of energy production. In developing energy conversion processes we're always working against thermodynamics and, more specifically, we're trying to bend the second law. Boss Kettering, the great engineer and scientist from General Motors, once said that the second law of thermodynamics simply stated that you can't push something going faster than you are. In energy conversion processes, however, the second law shows itself in conversion efficiency and cost. • We trade off efficiency or yield against capital cost, and both efficiency and capital cost determine the final production cost, or the price we must pay to use the energy in its converted and presumably more conventient form. • A good design requires processes and equipment which are efficient in the second law sense to accomplish high efficiency energy conversion. Such processes and equipment, however, are usually expensive. The design choices to be made, therefore, fundamentally involve second law effects and capital costs. In the various alternative energy source concepts being presented at this conference, we are attempting to exploit a circumstance of nature to develop energy or power which we can use for our own benefit. Whether it is coal conversion, fusion, or ocean thermal gradients, a careful consideration of the factors involved in implementation of the technology will reveal the interplay and connection between costs and thermodynamics. What I am suggesting is that the interplay between second law effects and economics is much more important in energy conversion than in, say, a chemical plant producing a high-priced commodity. In order to develop economic energy sources in the future, we are going to have to pay much more attention to the important relationship between costs and the second law than we have in the past.

Hydrocarbon fuels are very attractive thermodynamically as well as in other ways, and the importance of these fuels to the industrialized and developing nations hardly needs to be reiterated here. The worldwide pattern of consumption versus economically recoverable reserves is somewhat askew as a result of the price and great convenience in use of the fluid hydrocarbons. Forty-five percent of worldwide energy consumption is in the form of petroleum whereas only 16% of the reserves is in this form. Twenty-one percent of the consumption is natural gas, which is only 8% of the reserve and coal, which makes up 65% of the reserve only contributes 32% to consumption. Coal-fueled the Industrial Revolution beginning in the late eighteenth century but gave way in the recent past to cheap and convenient oil and natural gas. One of the workers in our laboratory has a sign on her new, modern and very powerful analytical instrument saying "There's no fuel, like an old fuel." There may be as much truth as humor in this, since we seem to be approaching a time when coal will again play a major role as a primary energy source. Coal conversion to liquids and gases seems likely to be one of the means by which we deal with the near term problem. In the longer term, however, additional primary energy sources will have to be found.

Both the breeder reactor and fusion offer essentially infinite energy sources. With fusion there are scientific and technical problems to be solved

before the economic questions can be faced. Fission power from the breeder reactor, and in fact, from nuclear power in general is questionable in the long
term due to unresolved institutional and political questions.

Solar energy in all its forms -- direct, wind, temperature gradients in
the sea, the biomass -- is coming slowly and is expensive. However, energy
from the sun is as close to an infinite source as we're likely to get for some
time, and strong efforts to move solar into our energy picture in an economic
way are certainly warranted.

Conservation and wise use of our energy resources must be an important
element of our energy strategy. Conservation alone, however, will not get us
from here to there. We must have a concerted effort to expand energy production
in all forms. This is especially important to the developing nations of the
world, because a sufficient supply of reasonably priced energy is a prerequisite
for an improving standard of living. Expanded production in the industrialized
nations will contribute in a substantial way to improvement in the less developed
areas of the world. It is sometimes stated that the development of alternative
energy technologies is proceeding too slowly, and that they will contribute
in a small way to production in the coming decade. Consider, however, the observation of the petroleum economist Walter Levy: "A 2% perceived shortage of
oil may result in a 20% increase in price". Thus, even a small contribution
from alternative sources may be of great value in minimizing energy price increases and eocnomic dislocations.

In the search for alternative energy sources it is useful to bear in mind
the three major kinds of activities involved in creating and diffusing new
technologies:

1.) The pursuit of knowledge about the physical world.

2.) The creation of new, practical devices and processes which lead to a more
     productive economy.

3.) The cost effective introduction of such new devices and processes into the
     economy.

Most of us at this conference are involved in one or more of these activities
and these activities comprise the business of science and technology. Each is
important in its own way. The realtionship among science, invention, and application is complex. Scientific knowledge does not automatically lead to invention and invention does not automatically flow into the economy. Each of us
should strive to establish and maintain both the atmosphere and the institutions
in which each activity can flourish and support the others. This requires, at
the very least, that we keep the lines of communication open and flowing and
that we have some empathy for what the other person is trying to accomplish.
Conferences such as this provide one means by which we attempt just such a task.

The human race now faces one of the greatest challenges it has confronted
in modern times: The challenge of creating and utilizing a new source of energy which is economic, non-polluting, and essentially infinite. We must now
demonstrate that we do, in fact, have the wherewithal to rise to this challenge.
It is the only way we can get from here to there. Thank you.

INSOLATION

An Algorithm to Estimate Long Term Solar Flux Incident on a Plane Anyhow Oriented Surface

GAETANO ALFANO, RITA MASTRULLO, and PIETRO MAZZEI
University of Naples
80125 Naples, Italy

ABSTRACT

An algorithm to estimate long-term monthly averages of the daily solar irradiation incident upon a plane anyhow oriented surface is presented. The experimental processed data are the monthly averages of daily global horizontal solar irradiation. The calculations of the possible periods of direct irradiation for the slope are expounded almost in detail since it seems that this problem has not been suitably exposed in literature. We present results for some significant cases: surfaces facing due south with a tilt angle equal to latitude or ten degrees grater, vertical surface with principal orientations. A comparison with experimental data available in literature is shown.

NOMENCLATURE

A   parameter eqn.(18)

B   parameter eqn.(19)

c   total atmospheric attenuation factor

G   solar flux density ($kJm^{-2}h^{-1}$)

H   daily global irradiation ($kWhm^{-2}$)

m   air mass

n   day of the year (n=1 on January 1)

r   daily eccentricity factor

R   horizontal to anyhow oriented global conversion factor

α solar altitude

γ surface azimuth

δ solar declination

θ surface incidence angle

θ_z zenith angle

ρ ground reflectivity

σ slope angle (relative to the horizontal)

ϕ latitude angle

ω hour angle

Subscripts

b   beam

bh   beam, horizontal

bd   beam, daily

d   diffuse

dh   diffuse, horizontal

f   last day of the month

h   horizontal

i   first day of the month

oh   outside atmosphere, horizontal

r   reflected

sc   solar constant

sr   sunrise

ss   sunset

sr1   first sunrise

sr2   second sunrise

ss1   first sunset

ss2   second sunset

Superscripts

*   relative to a non-horizontal anyhow oriented surface

-    monthly average of daily values

m   meridian (at solar noon)

1. INTRODUCTION

   Monthly averages of daily global solar irradiation incident upon
a horizontal surface are available for many locations and often for
many years. Such data are the base for long-term thermal performance
evaluation of solar systems. We present an algorithm which, from
this data, lead to an estimate of monthly average daily solar energy
incident upon plane anyhow oriented surface. The procedure was first

laid down by Liu and Jordan [1] and than extended by Klein [2]. Because the method scems particularly simple and reliable it was revised to allow the estimate for surfaces anyhow oriented and anywhere located. This aspect should be relevant for such application like heating and cooling loads evaluation were the receiving surfaces are variously oriented.

2. ANALYSIS

The method is based on the following steps: i) estimate of monthly averages of daily diffuse from daily global radiation experimental data for an horizontal surface, ii) calculation of the ratio of direct flux on the tilted surface to that on the horizontal surface by integration, iii) evaluation of diffuse and reflected fluxes incident on tilted surface, iiii) summation on the three components of the global flux.

2.1. Step One

Liu and Jordan [3] have presented a non-dimensional relation to estimate the diffuse and beam components of long-term monthly averages of the daily values of the global radiation incident on a horizontal surface. This correlation can be expressed [2] as

$$\overline{H}_{dh}/\overline{H}_h = 1.390 - 4.027\ \overline{K} + 5.531\ \overline{K}^2 - 3.108\ \overline{K}^3 \qquad (1)$$

where $\overline{H}_{dh}$ is the diffuse component of $\overline{H}_h$ and $\overline{K}$

is defined as

$$\overline{K} = \overline{H}_h/\overline{H}_{oh} \qquad (2)$$

where $\overline{H}_{oh}$ is the monthly mean of the daily values of the extraterrestrial irradiation. It holds

$$\overline{H}_{oh} = (n_f - n_i + 1)^{-1} \sum_{n=n_i}^{n_f} (H_{oh})_n$$

where n_i and n_f are respectively the days of the year at the start and end of each month, Table 1, and H_{oh} is the daily value of the extraterrestrial irradiation. This can be calculated as

$$H_{oh} = (24/\pi)G_{sc}\ r(\cos\phi\cos\delta\sin\omega_{sr} + \omega_{sr}\sin\phi\sin\delta) \qquad (3)$$

where G_{sc} is the solar constant - 4871 $kJm^{-2}h^{-1}$ - , r is the eccentricity factor, $\phi°$ is the latitude (north positive), δ is the solar declination (north positive), ω_{sr} (radians) is the sunrise hour angle (zero at noon, mornings positive) on which there will be later a detailed exposition -see equation (30)-. Following Kasten [4] and

Table 1. Days of the year at the start and end of each month.
Monthly mean values of the daily solar declination and the daily
eccentricity factor. Selected day for each month to evaluate $\bar{H}_{oh}$ as H_{ol}

	J	F	M	A	M	J	J	A	S	O	N	D
n_i	1	32	60	91	121	152	182	213	244	274	305	335
n_f	31	59	90	120	151	181	212	243	273	304	334	365
$\delta°$	-20.8	-13.3	-2.39	9.49	18.8	23.1	21.1	13.3	1.99	-9.85	-19.0	-23.1
r	1.03	1.03	1.02	1.00	.984	.971	.967	.971	.983	.999	1.02	1.03
n'	17	46	75	105	135	161	198	229	259	289	319	344

Cooper [5] δ (degrees) may be evaluated as

$$\delta = 23,45 \sin[(360/365)(284+n)] \qquad (4)$$

where n is the day of the year (n=1 on January 1, n=365 on December
31). The eccentricity factor can be calculated as

$$r = 1+0,033\cos(360\ n/365) \qquad (5)$$

The monthly mean value $\bar{H}_{oh}$ can be approximated as H_{oh} calculated
with the monthly mean value of the daily declination $\bar{\delta}$ and the dai
ly eccentricity factor $\bar{r}$, Table 1, or as H_{oh} calculated for a selected day , n' , for each month, Table 1. The correlation proposed
by Liu and Jordan seems to be still the most used. A more recent
correlation [6] in terms of the ratio of monthly average hours of
bright sunshine to the average hours of day-lenght seems to lead
to more conservative results.

2.2. Step Two

The ratio of beam solar flux density incident on the tilted sur
face to that incident on the horizontal surface

$$R_b = G_b^*/G_{bh} \qquad (6)$$

is a geometric function of the surface orientation and of the appa
rent position of the Sun

$$R_b = \cos\theta/\cos\theta_z \qquad (7)$$

where θ and θ_z are respectively the angles of incidence relative
to the slope and to the horizontal plane. These angles may be calculated as

$$\cos\theta = \sin\delta\sin\phi\cos\sigma - \sin\delta\cos\phi\sin\sigma\cos\gamma + \cos\delta\cos\phi\cos\sigma\cos\omega +$$
$$+ \cos\delta\sin\phi\sin\sigma\cos\gamma\cos\omega + \cos\delta\sin\sigma\sin\gamma\sin\omega \qquad (8)$$

$$\cos\theta_z = \cos\delta\cos\phi\cos\omega + \sin\delta\sin\phi \tag{9}$$

where σ is the slope angle of the surface ($0 \leqslant \sigma \leqslant \pi/2$), γ is the surface azimuth (zero due south, east positive, π due north).

Denoting by R_{bd} the ratio of the daily beam irradiation incident on the tilted surface to that incident on the horizontal surface

$$R_{bd} = H_b^*/H_{bh} \tag{10}$$

we have

$$R_{bd} = \frac{\int_{\omega_{ss}^*}^{\omega_{sr}^*} G_{sc} \, r \, \exp(-cm)\cos\theta \, d\omega}{\int_{\omega_{ss}}^{\omega_{sr}} G_{sc} \, r \, \exp(-cm)\cos\theta_z \, d\omega} \tag{11}$$

where c is the total atmospheric attenuation factor, m is the air mass, (ω_{sr}, ω_{ss}) and (ω_{sr}^*, ω_{ss}^*) are the daily possible direct irradiation periods, in terms of hour angle, respectively for an horizontal and a tilted anyhow oriented surface.

Following the first approximation hypothesis suggested by Liu and Jordan [1] of neglecting the atmospheric attenuation when (11) is to be evaluated on a long term statistical average basis, equation (11) becomes

$$\overline{R}_{bd} = \int_{\overline{\omega}_{ss}^*}^{\overline{\omega}_{sr}^*} \cos\overline{\theta} \, d\omega \, / \, \int_{\overline{\omega}_{ss}}^{\overline{\omega}_{sr}} \cos\overline{\theta}_z \, d\omega \tag{12}$$

where all the angles are calculated using the monthly mean value of the daily solar declination

$$\overline{\delta} = (n_f - n_i + 1)^{-1} \sum_{n=n_i}^{n_f} (\delta)_n \tag{13}$$

see Table 1.

From equations (8),(9) and (12) it follows

$$\overline{R}_{bd} = [(\overline{\omega}_{sr}^* - \overline{\omega}_{ss}^*)(\sin\overline{\delta}\sin\phi\cos\sigma - \sin\overline{\delta}\cos\phi\sin\sigma\cos\gamma) + \tag{14}$$

$$+(\sin\overline{\omega}_{sr}^* - \sin\overline{\omega}_{ss}^*)(\cos\overline{\delta}\cos\phi\cos\sigma + \cos\overline{\delta}\sin\phi\sin\sigma\cos\gamma) +$$

$$-(\cos\overline{\omega}_{sr}^* - \cos\overline{\omega}_{ss}^*) \cos\overline{\delta}\sin\sigma\sin\gamma]/[2(\overline{\omega}_{sr}\sin\phi\sin\overline{\delta} +$$

$$\sin\overline{\omega}_{sr}\cos\phi\cos\overline{\delta})]$$

where ω is in radians.
It may happen that in time interval ($\overline{\omega}_{sr}$, $\overline{\omega}_{ss}$) we have two different direct irradiation periods: ($\overline{\omega}_{sr1}^*$, $\overline{\omega}_{ss1}^*$) and ($\overline{\omega}_{sr2}^*$, $\overline{\omega}_{ss2}^*$).
This happens, for example, for a vertical surface facing north at a

latitude $\phi=40°N$ in the months between the spring and fall equinoxes. The generalization of the (14) is

$$\overline{R}_{bd} = [(\overline{\omega}^{-*}_{sr1}-\overline{\omega}^{-*}_{ss1}+\overline{\omega}^{+*}_{sr2}-\overline{\omega}^{-*}_{ss2})(sin\overline{\delta}sin\phi cos\sigma-sin\overline{\delta}cos\phi sin\sigma cos\gamma)+$$
$$+(sin\overline{\omega}^{*}_{sr1}-sin\overline{\omega}^{*}_{ss1}+sin\overline{\omega}^{*}_{sr2}-sin\overline{\omega}^{*}_{ss2})(cos\overline{\delta}cos\phi cos\sigma +$$
$$+cos\overline{\delta}sin\phi sin\sigma cos\gamma)-(cos\overline{\omega}^{*}_{sr1}-cos\overline{\omega}^{*}_{ss1}+cos\overline{\omega}^{*}_{sr2}-cos\overline{\omega}^{*}_{ss2})$$
$$cos\overline{\delta}sin\sigma sin\gamma|/|2(\overline{\omega}_{sr}sin\phi sin\overline{\delta}+sin\overline{\omega}_{sr}cos\phi cos\overline{\delta})] \qquad (15)$$

Once $\overline{R}_{bd}$ has been computed, the monthly mean value of the daily beam irradiation incident on a plane anyhow oriented surface is determined as

$$\overline{H}^{*}_{b} = \overline{R}_{bd} \; \overline{H}_{bh} \qquad (16)$$

To determine the conversion factor $\overline{R}_{bd}$ it is necessary to evaluate the possible direct irradiation period. This can be defined as the time interval during which the Sun is contemporary above the horizontal plane and above the plane which contains the surface.
From equation (8), when $\theta=90°$, we have

$$sin\omega +A cos\omega = B \qquad (17)$$

with

$$A = sin\phi/tan\gamma+cos\phi/tan\sigma/sin\gamma \qquad (18)$$

$$B = tan\delta(cos\phi/tan\gamma-sin\phi/tan\sigma/sin\gamma) \qquad (19)$$

Equation (17) may not be solved for $\sigma = 0°, \gamma = 0°, \gamma = 180°$, cases that will be considered later. In the other cases the solutions are

$$\omega'= 2 \ tan^{-1}[1+(A^2-B^2+1)^{\frac{1}{2}}]/(A+B) \qquad (20)$$

$$\omega''= 2 \ tan^{-1}[1-(A^2-B^2+1)^{\frac{1}{2}}]/(A+B) \qquad (21)$$

From equations (20) and (21) it follows
-When $(A^2-B^2+1)^{\frac{1}{2}} \leqslant 0$ and $cos\theta(\omega=0)>0$

$$\omega^{*}_{sr} = \omega_{sr} \ , \quad \omega^{*}_{ss} = \omega_{ss} = -\omega_{sr} \qquad (22)$$

-When $(A^2-B^2+1)^{\frac{1}{2}} \leqslant 0$ and $cos\theta(\omega=0)<0$, there is no direct irradiation

$$\omega^{*}_{sr} = \omega^{*}_{ss} = 0 \qquad (23)$$

-When $(A^2-B^2+1)^{\frac{1}{2}}>0$, denoting by ω^{+} and ω^{-} the positive and negative roots of (17), the following sub-cases arise

  a) $0<|\gamma|\leqslant 90°$

$$\omega^{*}_{sr} = MIN(\omega_{sr},\omega^{+}) \ , \quad \omega^{*}_{ss} = -MIN(\omega_{sr},|\omega^{-}|) \qquad (24)$$

b) $90° < |\gamma| < 180°$

b1) $(\omega^{\dagger}, \omega^{-})$ within $(\omega_{sr}, \omega_{ss})$ and $\cos\theta(\omega=0) < 0$

$$\omega^{*}_{sr1} = \omega_{sr}, \quad \omega^{*}_{ss1} = \omega^{+} ; \quad \omega^{*}_{sr2} = \omega^{-}, \quad \omega^{*}_{ss2} = \omega_{ss} \quad (25)$$

b2) (ω^{+}, ω^{-}) within $(\omega_{sr}, \omega_{ss})$ and $\cos\theta(\omega=0) > 0$

$$\omega^{*}_{sr} = \omega^{+}, \quad \omega^{*}_{ss} = \omega^{-} \quad (26)$$

b3) $\omega^{+} > \omega_{sr}, \quad |\omega^{-}| < \omega_{sr}$

$$\omega^{*}_{sr} = \omega^{-}, \quad \omega^{*}_{ss} = \omega_{ss} \quad (27)$$

b4) $\omega^{+} < \omega_{sr}, \quad |\omega^{-}| > \omega_{sr}$

$$\omega^{*}_{sr} = \omega_{sr}, \quad \omega^{*}_{ss} = \omega^{+} \quad (28)$$

b5) $\omega^{+} > \omega_{sr}, \quad |\omega^{-}| > \omega_{sr}$

$$\omega^{*}_{sr} = \omega^{*}_{ss} = 0 \quad (23)$$

When $\sigma = 0$, from equation (9), the solution is

$$\omega = \cos^{-1}(-\tan\phi\tan\delta) \quad (29)$$

and the following sub-cases arise

a) $-1 < -\tan\phi\tan\delta < 1$

$$\omega_{sr} = [\cos^{-1}(-\tan\phi\tan\delta)], \quad \omega_{ss} = -\omega_{sr} \quad (30)$$

b) $-\tan\phi\tan\delta < -1$ or $-\tan\phi\tan\delta > 1$, equation (29) has no solution. In the first case there are 24 hours of possible direct irradiation

$$\omega_{sr} = 180° = \omega_{ss} \quad (31)$$

in the second case there is no direct irradiation.

c) $-\tan\phi\tan\delta = 1$, there is no direct irradiation.

d) $-\tan\phi\tan\delta = -1$, 24 hours of possible direct irradiation.

When $\gamma = 0$, from equation (8), the solution is

$$\omega = \cos^{-1}[-\tan\delta\tan(\phi-\sigma)] \quad (32)$$

and the following sub-cases arise

a) $-1 < -\tan(\phi-\sigma)\tan\delta < 1$, denoting by ω_1 the positive solution of equation (32), we obtain

$$\omega_{sr}^* = \text{MIN}(\omega_{sr}, \omega_1) \quad , \quad \omega_{ss}^* = -\omega_{sr} \qquad (33)$$

b) $-\tan(\phi-\sigma)\tan\delta > 1$ or $-\tan(\phi-\sigma)\tan\delta < -1$, if $\cos\theta(\omega=0) > 0$

$$\omega_{sr}^* = \omega_{sr} \quad , \quad \omega_{sr}^* = \omega_{ss} \qquad (34)$$

if $\cos\theta(\omega=0) < 0$ there is no direct irradiation.

c) $-\tan(\phi-\sigma)\tan\delta = 1$, there is no direct irradiation.

d) $-\tan(\phi-\sigma)\tan\delta = -1$

$$\omega_{sr}^* = \omega_{sr} \quad , \quad \omega_{ss}^* = \omega_{ss} \qquad (35)$$

Where $\gamma = 180°$, from equation (8), the solution is

$$\omega = \cos^{-1} |-\tan\delta\tan(\phi+\sigma)| \qquad (36)$$

and the following sub-cases arise

a) $-1 < \tan(\phi+\sigma)\tan\delta < 1$, denoting by ω_2 and $-\omega_2$ the positive and negative solutions of (36) and by α^m the meridian solar altitude, we obtain

a1) the interval $(\omega_2, -\omega_2)$ is within the interval $(\omega_{sr}, \omega_{ss})$ and it is $\sigma < \alpha^m$

$$\omega_{sr}^* = \omega_2 \quad , \qquad \omega_{ss}^* = -\omega_2 \qquad (37)$$

a2) the interval $(\omega_2, -\omega_2)$ is within the interval $(\omega_{sr}, \omega_{ss})$ and it is $\sigma > \alpha^m$

$$\omega_{sr1}^* = \omega_{sr} \quad , \quad \omega_{ss1}^* = \omega_2, \quad \omega_{sr2}^* = -\omega_2, \quad \omega_{ss2}^* = \omega_{ss} \qquad (38)$$

a3) the interval $(\omega_{sr}, \omega_{ss})$ is within the interval $(\omega_2, -\omega_2)$ and it is $\sigma < \alpha^m$

$$\omega_{sr}^* = \omega_{sr} \quad , \quad \omega_{ss}^* = \omega_{ss} \qquad (39)$$

a4) the interval $(\omega_{sr}, \omega_{ss})$ is within the interval $(\omega_2, -\omega_2)$ and it is $\sigma > \alpha^m$, there is no direct irradiation.

b) $-\tan(\phi+\sigma)\tan\delta < -1$ or $-\tan(\phi+\sigma)\tan\delta > 1$

b1) $\sigma < \alpha^m$; $\omega_{sr}^* = \omega_{sr}$, $\omega_{ss}^* = \omega_{ss}$ $\qquad (40)$

Table 2. Summary for the evaluation of the possible direct irradiation periods.

eqn. no.	ω_{sr1}	ω_{ss1}	ω_{sr2}	ω_{ss2}
22	ω_{sr}	$-\omega_{sr}$	-	-
23	-	-	-	-
24	$MIN(\omega_{sr},\omega^+)$	$-MIN(\omega_{sr},\|\omega^-\|)$	-	-
25	ω_{sr}	ω^+	ω^-	$-\omega_{sr}$
26	ω^+	ω^-	-	-
27	-	-	ω^-	$-\omega_{sr}$
28	ω_{sr}	ω^+	-	-
30	ω_{sr}	$-\omega_{sr}$	-	-
31	$180°$	$-180°$	-	-
33	$MIN(\omega_{sr},\omega_1)$	$-MIN(\omega_{sr},\omega_1)$	-	-
34	ω_{sr}	$-\omega_{sr}$	-	-
35	ω_{sr}	$-\omega_{sr}$	-	-
37	ω_2	$-\omega_2$	-	-
38	ω_{sr}	ω_2	$-\omega_2$	$-\omega_{sr}$
39	ω_{sr}	$-\omega_{sr}$	-	-
40	ω_{sr}	$-\omega_{sr}$	-	-
41	ω_{sr}	$-\omega_{sr}$	-	-

b2) $\sigma > \alpha^m$; there is no direct irradiation.

c) $-\tan(\phi+\sigma)\tan\delta = 1$ or $-\tan(\phi+\sigma)\tan\delta = -1$

c1) $\delta > 0$; $\omega_{sr}^* = \omega_{sr}$, $\omega_{ss}^* = \omega_{ss}$ (41)

c2) $\delta \leqslant 0$; there is no direct irradiation.

These results are summarized in Table 2.

2.3. Step Three

Following the hypothesis of isotropic diffuse and reflected radiation fields [7], the monthly means of the diffuse and reflected irradiations may be evaluated as

$$\overline{H}_d^* = \overline{H}_{dh} \cos^2(\sigma/2) \tag{42}$$

$$\overline{H}_r^* = \rho\overline{H}_h \sin^2(\sigma/2) \tag{43}$$

where ρ is the ground reflectivity.

2.4. Step Four

The monthly mean value of the daily global irradiation incident upon a plane anyhow oriented surface is evaluated as

$$\overline{H}^* = \overline{H}_b^* + \overline{H}_d^* + \overline{H}_r^* \tag{44}$$

or, by substitution of equations (16), (42) and (43)

$$\overline{R} = \frac{\overline{H}^*}{\overline{H}_h} = \overline{R}_{bd}\left(1 - \frac{\overline{H}_{dh}}{\overline{H}_h}\right) + \cos^2\left(\frac{\sigma}{2}\right)\frac{\overline{H}_{dh}}{\overline{H}_h} + \rho\sin^2\left(\frac{\sigma}{2}\right) \tag{45}$$

where $\overline{R}$ is the global irradiation conversion factor. In Table 3 the four steps and the relative inputs and outputs are summarized.

Table 3. Algorithm I/O Summary

	INPUT	OUTPUT
Step one	$\overline{H}_h$ G_{sc} ϕ n	$\overline{H}_{bh}$ $\overline{H}_{dh}$
Step two	ϕ σ γ δ $\overline{H}_{bh}$	$\overline{H}_b^*$
Step three	$\overline{H}_{dh}$ $\overline{H}_h$ σ ρ	$\overline{H}_d^*$ $\overline{H}_r^*$
Step four	$\overline{H}_b^*$ $\overline{H}_d^*$ $\overline{H}_r^*$	$\overline{H}^*$

3. RESULTS

As an application of the previous analysis, numerical calculations have been carried out and some results are presented (a reflectivity of 20% has been used).

Measurements of the global irradiation incident upon a south facing vertical and horizontal surface were made for five years (1952-1956) at Blue Hill Mass. USA ($\phi = 42°13'N$), and the relative

averaged values are reported in [1]. Using the $\overline{H}_h$ values as an input
- see Tab.4 - the global irradiation conversion factor $\overline{R}$ was computed. In Fig.1 the experimental and theoretical factors are compared. A very good agreement exists. As expected the trend shows a
zero deviation during equinoxes and the maximum deviation during
solstices.

A second comparison between experimental and theoretical values
is reported in Fig.2. The measured global irradiation values were
for a north facing 60° slope and for an horizontal surface - see
Tab.4 - at Highett Victoria Australia [8]. Since the reported measurements were only for one year, the comparison displaies a greater
deviation especially during solstices. However the yearly trend is
similar to the previous one.

Long term (1958-1972) monthly averages of daily values of global
horizontal irradiation for Naples Italy $\phi=40°51'N$ - see Tab.4 - were
processed to compute $\overline{H}^*$ for the following cases: $\sigma=\phi$ and $\sigma=\phi+10°, \gamma=0$
- Fig.3 -; $\sigma=90°$ $\gamma=0$, 45°, 90°, 135°, 180° - Fig.4 -.

Table 4. Experimental monthly averages of daily values of global
irradiation (kWh/m²) incident on a horizontal surface for: 1. Naples
Italy $\phi=40°51'N$, 2. Blue Hill Mass. USA $\phi=42°13'N$, 3. Highett Victoria
Australia $\phi=37°5'S$.

$\overline{H}_h$	J	F	M	A	M	J	J	A	S	O	N	D
1	1.37	1.94	2.71	3.73	4.69	5.23	5.28	4.68	3.52	2.58	1.60	1.20
2	1.58	2.43	3.36	4.28	5.37	6.21	6.13	5.01	4.09	2.93	1.88	1.45
3	6.69	6.09	4.75	3.50	2.09	1.86	1.94	2.34	3.48	4.61	5.90	5.92

4. CONCLUSIONS

The agreement between the experimental and theoretical results,
although limited to surfaces tilted toward the equator because of
the lack of experimental data, seems satisfactory. Even if the presented algorithm is quite generally applicable to the design methods
of solar heating systems with receiving surfaces off-south/north
orientation, it seems almost straight forward and therefore suitable
or desktop calculators.

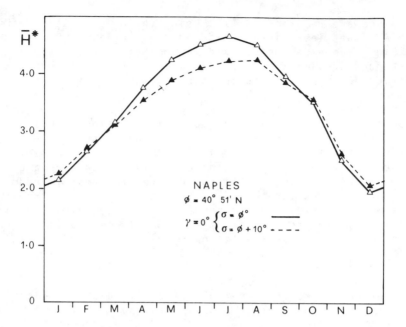

Fig.1. Experimental and Theoretical Global Irradiation Conversion
Factors Comparison

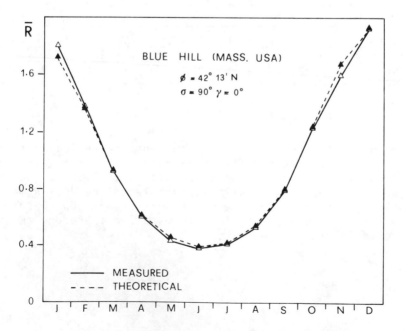

Fig.2. Experimental and Theoretical Global Irradiation Conversion
Factors Comparison.

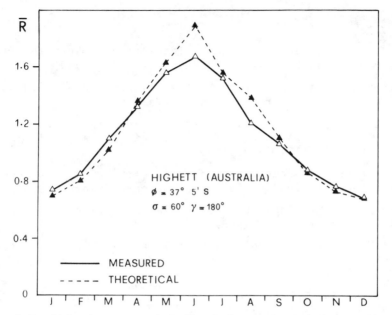

Fig.3.Monthly Average of Daily Values of Global Irradiance
      (kWhm^{-2}).

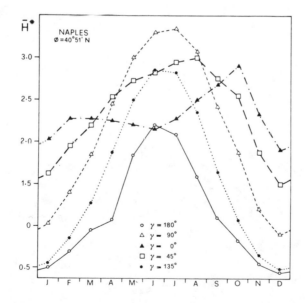

Fig.4. Monthly Average of Daily Values of Global Irradiance
       (kWhm^{-2}).

REFERENCES

1. Liu B.Y.H., Jordan R.C. - Daily Insolation on Surfaces Tilted
   Toward the Equator - Trans. ASHRAE, 526-541 (1962).

2. Klein S.A. - Calculation of Monthly Average Insolation on Tilted
   Surfaces - Solar Energy, 19, n°4, 325-329 (1977).

3. Liu B.Y., Jordan R.C. - The Interrelationship and Characteristic
   Distribution of Direct, Diffuse and Total Solar Radiation - Solar Energy, 4, n°3, 1-19 (1960).

4. Kasten F. - Measurement and Analysis of Solar Data - Commission
   of the European Communities Joint Research Centre, Ispra (1977).

5. Cooper P.I. - The Absorption of Solar Radiation in Solar Stills-
   Solar Energy, 12, n°3 (1969).

6. Iqbal M. - Estimation of the Monthly Average of the Diffuse Com
   ponent of Total Insolation on a Horizontal Surface - Solar Ener
   gy, 20, 101-105 (1978).

7. Kondratyev K.J., Manolova M.P. - The Radiation Balance of Slopes
   Solar Energy, 4, n°1 14-19 (1960).   .

8. Norris D.J. - Solar Radiation on Inclined Surfaces - Solar Energy,
   10,72-77 (1966).

Application of Simple Arima Models, with Respectively Constant and Seasonal Parameters, to Solar Meteorology

E. BOILEAU and B. GUERRIER
Laboratoire des Signaux et Systèmes, CNRS–ESE
Gif-sur-Yvette, France

ABSTRACT

Standard methods giving statistical models of time series have been tested for use in solar meteorology. Daily irradiation measurements from three different places were used with these methods for simulation and forecasting. We were led to modify somewhat these standard procedures and improvements of the methods were obtained.

1. INTRODUCTION

We test several statistical models of the daily irradiation on three typical sites: a temperate one (Trappes), a equatorial one (Huallao 12° latitude Sud. 3300 m altitude) and a mediterranean one (Carpentras). These models may be used for simulation (in order to optimize passive solar energy storage devices) and forecasting (useful for more elaborate adaptative devices).

A first test of the fitness of the simulations obtained consists in a qualitative inspection. On measured series of the daily irradiation I_i one observes that the values are not symmetrically distributed around the local mean value: the differences between small values and mean value are much more important than the ones between large values and mean value. With the models that we have examined, such a disymmetry cannot be directly obtained, and we have been led to use a preliminary transformation of irradiation to get a more symmetrical process. An exponential transformation :

$$z_i = A \, \text{Exp} \, (I_i) \tag{1}$$

gives satisfaction for the first two sites but not for the mediterranean one. Indeed, for this site, the high probability of clear days gives a particular distribution (see fig. 5a). Introducing a mean daily value K_i defined on clear days only, we use the transformation :

$$\left\{ \begin{array}{ll} z_i = I_i - K_i & \text{if } I_i < K_i \\ \\ z_i = h(I_i - K_i) & \text{if } I_i > K_i \end{array} \right. \tag{2}$$

and we adjust h in order that the distribution of z_i be roughly symmetrical.

In this communication we examine the possibility of obtaining a model for the irradiations I_i from a model of transformed series z_i built with the standard time-series modeling methods described by Box and Jenkins [1].

2. SEASONAL MODEL (2)

Our times series being obviously seasonal, we began by looking for a model working on the increment $\nabla_{365} z_i = z_i - z_{i-365}$. The second increment

$$w_i = \nabla\nabla z_i = z_i - z_{i-365} - (z_{i-1} - z_{i-366}) \qquad (3)$$

has approximately a suitable form [1]. Indeed, the estimated correlation coefficients are not signifiant, except $\gamma_0, \gamma_1, \gamma_{364}, \gamma_{365}$ and γ_{366} which are given in table 1.

Table 1: Correlation coefficients of $w_i = \nabla\nabla_{365} z_i$

	Huallao		Trappes		Carpentras	
	$z_i = I_i$	$z_i = Exp(I_i)$	$z_i = I_i$	$z_i = Exp(I_i)$	$z_i = I_i$	$z_i = Exp(I_i)$
0	1	1	1	1	1	1
γ_1	-0,42	-0,44	-0,37	-0,40	-0,37	-0,43
γ_2	-0,05	-0,03	-0,09	-0,07	-0,09	-0,01
γ_3	-0,04	-0,03	-0,04	-0,03	-0,04	-0,07
--	--	--	--	--	--	--
γ_{363}	0,04	0,04	0,01	0,00	0,10	0,04
γ_{364}	0,20	0,21	0,19	0,05	0,12	0,17
γ_{365}	-0,50	-0,51	-0,50	-0,51	-0,49	-0,48
γ_{366}	0,23	0,24	0,21	0,22	0,21	0,23
γ_{367}	0,01	0,00	0,04	0,02	0,01	0,00

We see that the exponential transformation does not much affect these coefficients. We are led [1] to use the model :

$$w_i = (1-\lambda B)(1-\mu B^{365})a_i$$
$$= a_i - \lambda a_{i-1} - \mu a_{i-365} + \lambda\mu a_{i-366} \qquad (4)$$

where a_i is a white noise, and the optimal values of the coefficients λ and μ must be calculated from a set $\{w_i, i\epsilon(1,n)\}$ (calculated with measured I_i) so as to minimize $S(\lambda,\mu) = \Sigma(a_i)^2$. The $\hat{a}_i$ are the best estimations of the a_i for given λ,μ and $w_i(i\epsilon(1,n))$.

Box and Jenkins suggest that these (a_i) should be calculated after some back-forecasts w_j for $j<o$. The values obtained for λ and μ respectively with back-forecasting [line (a)] and without back-forecasting [line (s)] are given in table 2.

With these coefficients and using N_a years of past measurements we calculate (I_{i+1}), the one-day-forecast for days i where we besides know the value I_{i+1} really obtained. We then get the forecast error $\varepsilon_i = (I_{i+1}) - I_{i+1}$ and deduce the r.m.s. $\sigma_p = \sqrt{E(\varepsilon_i^2)}$ for every month. The results are given on table 3 for Huallao.

Table 2: λ and μ calculated with (a) and
without (s) back-forecasting.

		λ	μ
site E	(a)	0,94	0,87
	(s)	0,96	0,72
site T	(a)	0,84	0,87
	(s)	0,84	0,72
site M	(a)	0,97	0,87
	(s)	0,97	0,69

Table 3: For Huallao, standard deviation of the forecast error (month M using Na=2, 4 or 6 years
of observation) with (a) without (s) back-
forecasting.

N_A M	2		4		6	
	(a)	(s)	(a)	(s)	(a)	(s)
1	190	155	158	144	148	136
2	182	163	157	145	150	146
3	152	133	129	119	119	116
4	137	129	120	114	104	106
5	94	89	78	77	73	75
6	91	88	83	86	78	79
7	90	90	79	81	68	69
8	109	99	97	96	97	96
9	124	111	113	110	111	113
10	150	149	126	119	124	122
11	165	149	142	132	122	120
12	183	165	163	158	154	153

We see that back-forecasting spoils the forecast instead of improving it
(except if we use at least 6 years of measurements; back-forecasting then does
not change significantly the error) and we decided to give up using back-forecasting.

The simulations given by this model, with a_i being a standard white noise, are good on periods of a few days. On longer periods, different problems appear and we must introduce the following modifications :
- The integrations performed on the simulated w_i to obtain the simulated irradiations I_i give rise to drifts that rapidly give aberrant values. To cancel these drifts we subtract their mean value from the successive sets of 30 consecutive I_i obtained (periodic recentring).
- To reproduce the seasonal variations of the mean and of the variance, an estimation of the daily mean $<I_i>$ (i=1,365) is used to initialize the simulation, and a modulation (month by month) of the entered white noise is done.
- Finally a gaussian white noise gives from time to time unreasonable peaks which we cut off.

3. SHORT RANGE MODEL (3).

To give forecasts of I_i, previous model needs one year of measurements at least, and the obtained forecasts are not very good. We test then models that no longer use ∇_{365}. We use now :

$$v_i = I_i - I_{i-1} \tag{5}$$

and estimate, for every 15-day-period of the year, the correlation coefficients of v_i. Results are given on figure 1. The precision is not very high because we take into account seasonal variations. After a few tests, we concluded that the first coefficient is the only significant one, and that the seasonal variation must be neglected here. Previsions calculated with AR, MA and ARMA models were examined. Finally the one coefficient MA model :

$$v_i = a_i - \theta\, a_{i-1} \tag{6}$$

gives the best results. The prevision error is not critically dependant on the θ value, and $\theta= .75$ can be used for the three sites.
With a given value of θ, the one-day-prevision (I_{i+1}) is calculated (from the (a_j)) using the measured irradiations I_i for days $j=i,i-1,...,i-n_a$. We observe that the forecast error σ_p decreases up to $n_a \simeq 15$ only. This improvement as n_a increases is evidently not observed with an AR1 model, but it must be noticed that with $n_a=2$, the previsions are better with an AR2 model (and for $n_a=1$ with an AR1 model) whereas for $n_a>3$ the best results are obtained with the above MA1 model (see figure 2).

4. THIRD MODEL.

A similar study (4) using reduced variables shows that models built with variables that have been centered by tranformations like

$$z_i = I_i - <I_i> \tag{7}$$

give analogous results. Besides, assuming that the correlation coefficients of z_i vanish after the second order, we explain (3) the order of magnitude of the correlation coefficients obtained with the previous variables v_i and w_i. Indeed the daily mean $<I_i>$ has small variations from one day to the next compared with the corresponding variations of I_i, and we may write :

$$v_i = I_i - I_{i-1} \simeq z_i - z_{i-1} \tag{8}$$

With the approximation (8), the correlation coefficients of v_i are easily calculated from the ones of z_i.

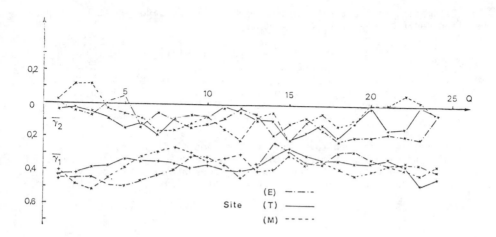

Figure 1. Correlation coefficients of $v_i = I_i - I_{i-1}$ estimated for every 15
days periods of the year.

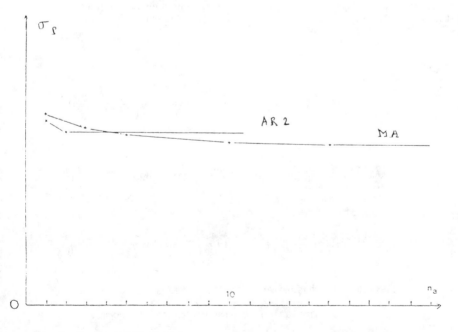

Figure 2. Variation of the r.m.s. of the forecast error with the number
of past measurements used $(n_a + 1)$

Estimating the latter (ρ_1 and ρ_2) for our three sites, we then deduced correlation coefficients of $v_i(\gamma_1^1, \gamma_2$ and $\gamma_3)$:

	ρ_1	ρ_2	γ_1	γ_2	γ_3
Huallao	0,22	0,09	-0,42	-0,03	-0,06
Trappes	0,32	0,14	-0,38	-0,03	-0,01
Carpentras	0,28	0,08	-0,36	-0,08	-0,06

The results are seen to agree with the ones of figure 1.

With the same approximation (8), we also calculate the correlation coefficients of w_i and the results agree with the ones obtained previously.

Thus it is worth while to study models of z_i (7). We analyse the simple model

$$z_i = a_i + P\, a_{i-1} \tag{9}$$

Calculations of forecasts have not been terminated yet, but simulations using this model give as good results as the first model (see fig. 3, 4 and 5) with the advantage that the simulated irradiation obtained now does not drift (no integration necessary). Thus the "recentring" described above is of no use.

5. CONCLUSION

The first model which we examined using the standard methods for the study of seasonal series is not the best. We described a model that gives better forecasts with only 15 days measurements. Besides, another model gives as good simulations as the first one with a much lighter procedure.

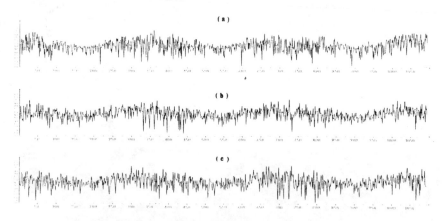

Figure 3. Huallao : a) measurements on three years
                    b) simulation obtained with the first model
                    c) simulation obtained with the third model

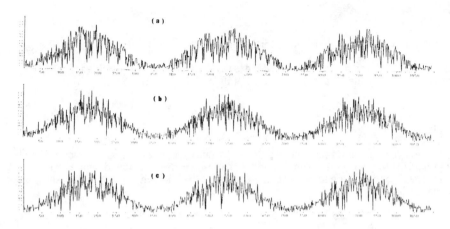

Figure 4. Trappes : a) measurements on three years
                    b) simulation obtained with the first model
                    c) simulation obtained with the third model

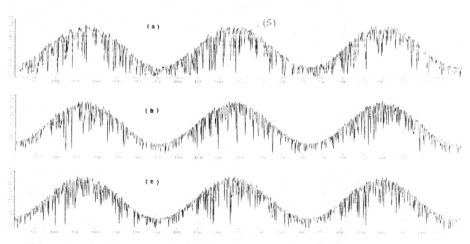

Figure 5. Carpentras : a) measurements on three years
                      b) simulation obtained with the first model
                      c) simulation obtained with the third model

REFERENCES

1   G.E.P. Box and G.M. Jenkins. Time Series Analysis. Holden Day 1970.

2   E. Boileau. Discussion d'un modèle statistique en météorologie solaire.
     Revue de Physique Appliquée 14, p. 145 à 152 (1979)

3   E. Boileau and B. Guerrier. Comparaison de modèles statistiques saison-
     niers et non saisonniers en météorologie solaire. La Météorologie. A pa-
     raître.

4   B. Guerrier, E. Boileau and C. Bénard. Analyse statistique temporelle de
     l'irradiation solaire global quotidienne : modélisation d'une variable ré-
     duite à l'aide de modèles stochastiques ARMA. Revue de Physique Appliquée
     A paraître.

Calculation of Hourly and Daily Available Solar Energy to a Flat Plate Collector Inclined by the Angle of Optimum Tilt in Iraq

H.R. HAMDAN and Z.M. MAJEED
Al-Mustansyriah University
Baghdad, Iraq

ABSTRACT

The maximum solar energy available to an inclined flat-plate collector is estimated by predicting the values of a conversion factor, R, which is the ratio of total solar radiation on a tilted plane to that on the horizontal plane. This factor is calculated for each hour and day and presented graphically as a function of collector tilts, hours in the day and months of the year.

Illustrating case is explained, taking the optimum tilt for the collector during months of the year, secured from the above mentioned graphs of R, to predict the maximum solar energy available to the optimum tilted flat-plate collector.

An exponential graph has been introduced to predict the value of the conversion factor, R, at any time for the optimum tilt angle at that time.

Verification of the results obtained for R, has been carried on for several months of the year using the Kipp and Zonen solarimeter, the empirical and the estimated data are found in good agreement.

INTRODUCTION

Daily and hourly average of solar radiation incident upon a horizontal plane are available for many locations . However radiation data on tilted planes are generally not available.

Estimates of the daily and hourly values of solar energy incident on a flat-plate collector having various inclination are required for solar energy design procedures and other applications .

A simple method of estimating the average daily solar radiation for each calender month on surfaces facing directly towards the equator has been developed by Liu and Jordan [1] , their method is applied here to estimate the daily and hourly amounts of total solar radiation which is incident on a collector of any inclination using a conversion factor , R .

Since the flat-plate collector responds to diffuse and beam radiation , the conversion factor , R , is defined as the ratio of total solar radiation on a tilted plane to that on a horizontal plane . So it can be estimated by individually considering the radiation on the tilted plane to be made up of three components ; the beam , diffuse and reflected solar radiation.

A plane tilted at slope , β , from the horizontal " sees " a portion of the sky dome given by ($1+\cos\beta$)/2 , if the diffuse solar radiation is uniformly distributed over the sky dome , this is also the conversion factor for diffuse radiation [2]. The tilted plane also "sees" a portion of the ground given by ($1-\cos\beta$)/2 and this is also the conversion factor for reflected radiation [3] . Considering these three components , R is given by ;

$$R = (1 - H_d/H)R_b + H_d/H(1+\cos\beta)/2 + \rho(1-\cos\beta)/2 \qquad (1)$$

Where;
    H ; Total solar radiation on the horizontal plane.
    H_d; Diffuse solar radiation on the horizontal plane.
    ρ ; Ground reflectance , it is taken to be 0.2 [4] .
    R_b; Ratio of beam radiation on a tilted plane to that on
        a horizontal plne .

The instantaneous value of R_b for a plane facing the equator is given by [5]

$$R_b = \frac{\cos(\phi - \beta)\cos\delta\,\cos\omega + \sin(\phi - \beta)\sin\delta}{\cos\phi\cos\delta\cos\omega + \sin\phi\sin\delta} \qquad (2)$$

Where;
    ϕ ; Latitude of the place .
    δ; Declination of the sun.
    ω; Hour angle .

An expression to estimate R_b for a certain period of time between two hour angles , ω_1 and ω_2 during the day for a plane facing the equator can be derived by integrating the beam solar radiation on the tilted and horizontal plane separately during that time interval , then ;

$$R_b = \frac{(\omega_2 - \omega_1)\sin(\phi - \beta)\sin\delta + \cos(\phi - \beta)\cos\delta(\sin\omega_2 - \sin\omega_1)}{(\omega_2 - \omega_1)\sin\phi \sin\delta + \cos\phi \cos\delta (\sin\omega_2 - \sin\omega_1)} \qquad (3)$$

In a course of the day ω_1 and ω_2 are the sunrise and sunset hour angles respectively , measured in radians ;

$$\omega_1 = \omega_s, \quad \omega_2 = -\omega_s, \text{ for horizontal plane },$$
$$\omega_1 = \omega'_s, \quad \omega_2 = -\omega'_s, \text{ for inclined plane }.$$

So equation (3) implies to ;

$$R_b = \frac{\omega'_s \sin(\phi - \beta)\sin\delta + \cos(\phi - \beta)\cos\delta \sin\omega'_s}{\omega_s \sin\phi \sin\delta + \cos\phi \cos\delta \sin\omega'_s} \qquad (4)$$

For a horizontal plane ω_s is given by ;

$$\omega_s = \cos^{-1}(-\tan\phi \tan\delta) \qquad (5)$$

For a tilted plane facing the equator , ω'_s can be found by considering the fact that $\omega'_s = \omega_s$ always , then ;

$$\omega'_s = \min\left[\omega_s, \ \cos^{-1}(\tan(\phi - \beta) \tan\delta)\right] \qquad (6)$$

Since measurements of diffuse solar radiation H_d , are rarely available , it must be estimated from measurements of the data of total radiation , a number of investigator [6,7] have founde that the diffuse fraction H_d/H , is a function of the fraction of total radiation H to the extraterrestrial radiation H_o ;

$$K_t = \frac{H}{H_o} \tag{7}$$

H_o is given by $[6]$;

$$H_o = 3600 \frac{12 S_c E}{\pi} \int_{\omega_1}^{\omega_2} (\cos\phi\cos\delta\cos\omega + \sin\phi\sin\delta\,)d\omega \tag{8}$$

Where $w = \frac{\pi}{12}h$, h ; number of hours from noon

S_c(Solar constant) $= 135.3$ mw/cm^2

E (Eccentricity) , is taken to be 1.0

For daily values of H_d/H , the Liu and Jordan expression$[4]$ is used ;

$$H_d/H = 1.390 - 4.027K_t + 5.531K_t^2 - 3.108K_t^3 \tag{9}$$

For hourly values of H_d/H Holland expression$[6]$is used;

$$H_d/H = 1.557 - 1.840K_t \quad \text{for } 0.35 < K_t < 0.75 \tag{10a}$$

$$H_d/H = 0.177 \qquad\qquad \text{for} \qquad K_t > 0.75 \tag{10b}$$

Before utilizing the flat-plate collector which is facing the equator it must be tilted with the horizontal plane by the angle of optimum tilt to receive the largest amounts of solar radiation . If the tilted collector is directed to the south or north the angle of incidence i is given by$[8]$;

$$\cos i = \cos\beta(\sin\phi\sin\delta + \cos\phi\cos\delta\cos\omega) + \\ \sin\beta(\sin\phi\cos\delta\cos\omega - \cos\phi\sin\delta) \tag{11}$$

For a plane tilted by an optimum tilt angle (β_o) , the angle of of incidence (i=0) ; and according to equation (11) ;

$$\cos \beta_o = \sin\phi\sin\delta + \cos\phi\cos\delta\cos\omega \qquad (12a)$$
$$\sin \beta_o = \sin\phi\cos\delta\cos\omega - \cos\phi\sin\delta \qquad (12b)$$

Since the largest amount of solar radiation reaching the earth is at noon , when $\omega=0$, then the optimum tilt angle for the collector at noon-time β_{on} is given by ;

$$\beta_{on} = \phi - \delta \qquad (13)$$

The purpose of this article is to introduce a relation between R and β_o, to find the value of R,for the optimum tilted flat-plate collector at any time , thereby the maximum solar energy available to the collector can be calculated.In addition the value of R related to any angle of tilt , β, can be found at any time of the day. Experimental verification of the result obtained for the hourly values of R was carried on during several mounths of the year. The study based on hourly and daily solar radiation data for Baghdad.

RESULTS AND DISCUSSION

Fig.1 plots the hourly values of the conversion factor R , as calculated by equs.(1) and (2) vs tilt angles of collector for different weather periods of the year represented by January (winter) , June (summer) , March and September (equinoxes).

It is noted from Fig.1 that in winter , the values of R decrease as the time of the day approaches noon , whereas in summer , the values of R increase as the time of the day approaches noon . During equinoxes the values of R   independent of day hours. It is noted that the hourly values of R in summer are generally small compared to that in winter , while they have nearly the same amounts during vernal and autumnal equinoxes .

The daily values of R are also calculated at the day 15th

of each month from eq.(1) , using the daily values of R_b (which
have been calculated from eq.(4) , they are presented graphically
vs tilt angles of the collector in Fig.2

Using Fig.2 , the average values of R , for any angle of
tilt , can be deduced . On the other hand the monthly average daily
values of β_o are predicted and listed in table 1 , together with its
values calculated at noon-time from equation (13). The predicted and
and the calculated values are in good agreement during the equinoxes
while in summer and winter the predicted values of β_o show great
deviations from values calculated at noon-time . These deviations
can be explained due to great flactuations of hourly values of R
during summer and winter months as seen from Fig.1.

Table 1

Predicted and calculated values of the optimum tilt angle
at noon-time β_{on} , β'_{on} respectively

Month	J	F	M	A	M	J	J	A	S	O	N	D
β_{on}	57	48	35	20	11	8	8	20	30	43	53	58
β'_{on}	54	46	35	23	14	10	12	19	31	42	52	57

Fig.1 and 2 are used to secure the values of R related to
any value of β_o at different hours and days during each month of the
year. When plotting R versus β_o an exponential relation is obtained as
as shown in Fig.3 , which can be used to find the value of R for
the optimum tilted collector at any time.

EXPERIMENTAL VERIFICATION

An experimental verification of the results obtained for R
by calculation , using the Kipp and Zonen Solarimeter for measuring total solar radiation . The whole period of test is 7 months
starting in June 1979 and ending in December of the same year .
A weekly measurements of the total solar radiation on the horizon-

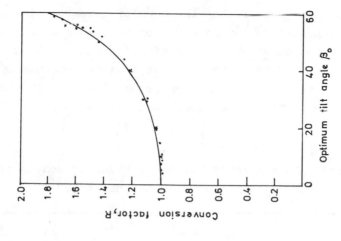

FIG.(3) TOTAL RADIATION CONVERSION FACTOR,R AS A FUNCTION OF COLLECTOR OPTIMUM TILT β_o.

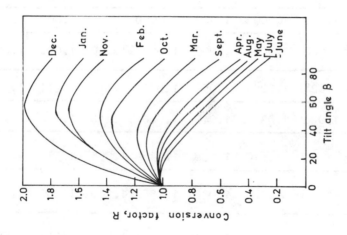

FIG.(2) DAILY VALUES OF THE TOTAL RADIATION CONVERSION FACTOR, R AS A FUNCTION OF TILT β.

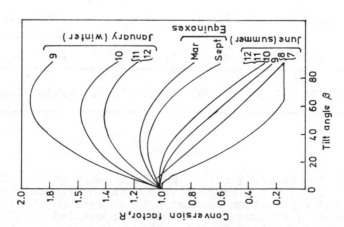

FIG.(1) VARIATION OF HOURLY VALUES OF THE TOTAL RADIATION CONVERSION FACTOR, R WITH TILT β & MONTHS.

43

tal and on tilted plane facing the equator have been taken simultaneously at noon-time during clear or nearly clear sky conditions
as possible . The emperical values of R designating by R´ are
tabulated in table 2 together with the estimated values . The
difference between R and R´ in November and December can be explained
due to unstable weather conditions during this period of the year.

Table 2

Emperical and Estimated values of R during months of
Experimental work as a function of β

Month β (degree)		J	J	A	S	O	N	D
Q	R	1.000	1.000	1.000	1.0000	1.000	1.000	1.000
	R´	1.000	1.000	1.000	1.000	1.000	1.000	1.000
10	R	1.028	1.025	1.058	1.092	1.113	1.211	1.310
	R´	1.020	1.030	1.040	1.080	1.110	1.160	1.160
20	R	1.019	1.010	1.058	1.146	1.255	1.378	1.522
	R´	1.020	1.000	1.030	1.150	1.200	1.280	1.290
30	R	0.986	0.975	1.027	1.157	1.314	1.521	1.690
	R´	0.980	0.980	1.030	1.160	1.250	1.360	1.400
40	R	0.890	0.909	0.995	1.135	1.314	1.682	1.850
	R´	0.900	0.920	0.980	1.150	1.270	1.430	1.440

APPLICATION

        Using the hourly (noon-time) and daily values of R ,
secured from Fig.1 and 2 respectively , the hourly and daily
amounts of solar radiation available to an optimum tilted flat
plate collector (RH) are estimated and presented graphically
in Fig.4a,b as a function of the months.

Study of Fig.4 showing that the use of a fixed collector which has an optimum tilt in a certain period of the year is the proper case during all months excluding summer solstice where the horizontal collector receives the same available energy as that received by the optimum tilted collector at that time.

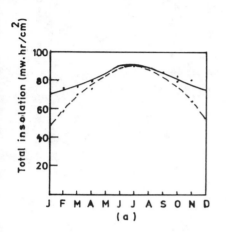 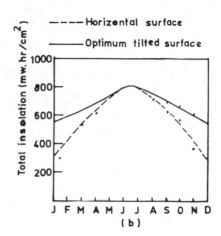

FIG (4) MONTHLY VARIATION OF TOTAL INSOLATION HORIZONTAL AND OPTIMUM TILTED COLLECTOR (a) HOURLY (non -time) INSOLATION. (b) DAILY INSOLATION

REFRENCES

1. B.Y.H.Liu and R.C.Jordan , " Daily insolation on surfaces tilted toward the equator ". Trans. ASHRAE (1962).

2. A.M.Zarem " Introduction to the utilization of solar energy " Mc Graw - Hill Book Company , Inc . New York (1963).

3. J.A.Duffie and W.A.Beckman , " Solar energy thermal processes", John Wiley and sons , (1974).

4. B.Liu and R.C.Jordan , " The interrelationship and characteristic distribution of direct , diffuse and total solar radiation ". Solar Energy vol.4 (1960).

5. S.A.Klein , " Calculation of monthly average insolation on tilted surfaces " . Solar Energy vol. 19 (1977).

6. J.F.Orgill and K.G.T.Hollands , " Correlation equation for Hourly diffuse radiation on a horizontal surface " Solar Energy vol.19

7. B.Anderson , " Solar Energy Fundamental of Building Design " , Mc Graw - Hill Book Comp. New York (1977) .

8. K.YA.Kondratyev , " Radiation in the atmosphere " , International Geophysics series vol. 12 (1969)

A Simulated Comparison of the Useful Energy Gain in Fixed and Tracking Flat Plate and Evacuated Tube Collectors

PETER DRAGO

United States Merchant Marine Academy
Kings Point, New York 11024, USA
Total Energy Corporation
Bohemia, New York 11716, USA

ABSTRACT

A study is presented to compare the useful energy gain in fixed, full-tracking and single-tracking (azimuthal angle) flat plate, and evacuated tube collectors. The study includes the consideration of using a single or double cover as well as an "ordinary" or a selective plate surface. A computer simulation was performed using actual hourly solar and weather data for the entire year of 1977, at Kings Point, N. Y. The hourly useful gain was found for each collection mode using a particular form of the Hottel-Whillier equation which fits that mode. The hourly values of, R, the ratio of solar radiation incident on a tilted surface to that incident on a horizontal surface was found for each collection mode using a set of computer programs for each mode. The results of this study give a detailed picture of a monthly, seasonally, and yearly comparison of various collection modes at a northern latitude. The results also show that a single-tracking (azimuthal angle) flat plate collector can provide a significant increase in useful energy gain over a fixed flat plate collector. A comparison of possible collector cost versus energy output shows that a single tracking collector can perform better than an evacuated tube collectors as regards K joules/m per initial dollar invested. While single-tracking collectors can only be used on flat roof or ground installations, the tracking collectors can only be used on flat roof or ground installations, the study seems to indicate that this mode of collection should be taken advantage of whenever possible.

NOMENCLATURE

H_o daily total extraterrestrial radiation (kj/m^2)

H total radiation on a horizontal surface (Kj/m^2)

ρ ground reflectance

H_D total diffuse radiation on a horizontal surface (kj/m^2)

S collector tilt angle (degrees)

R_b ratio of total beam radiation on a tilted surface to total beam radiation on a horizontal surface

δ declination (degrees)

ϕ latitude (degrees)

ω hour angle (degrees)

Γ_d diffuse enhancement factor

F_{pt} view factor

ρ_d reflectance of back panel

d   tube separation (m)

D_A tube absorber diameter (m)

D_6 Tube cover diameter (m)

θ_z zenith angle (degrees)

θ_T incident angle on tilted surface (degrees)

$g(\omega)$ shading factor

Q_u useful energy gain (kg)

A   collector area (m^2)

F_R heat removal factor

$(\tau\alpha)$ transmittance absorptance product

U_L total loss factor $(kj/m^2 \cdot C°)$

T   collector operating temperature $(C°)$

T_a ambient temperature $(C°)$

A_c tube array area (m^2)

L   tube length (m)

INTRODUCTION

       A study is presented here of the comparison in the useful energy gain
for eight different modes of solar collection. Three of these are fixed plate
of single cover, double cover and double cover with selective surface, facing
due south. Two are full-tracking flat plate of double cover and double cover
with selective surface. Two are single axis tracking (azimuth angle) flat
plate of double cover with selective surface, and one is an array of evacuated
tube collectors facing due south. The study was done at Kings Point, N. Y.
using actual hourly meterological data for the entire year of 1977, applied
to "theoretical" flat plate and evacuated tube collectors.

       This paper is divided into three parts. In the first part, the ratio
of hourly beam radiation on a tilted surface to that on a horizontal surface,

R_b, is calculated for a fixed and tracking flat plate and for evacuated tube collectors. The ratio of total hourly radiation on a tilted surface to that on a horizontal surface, R, is then found for various collection modes using R and a knowledge of the ratio of daily diffuse radiation to daily total radiation on a horizontal surface.

In the second part of this paper a digital computer is used to calculate the daily useful energy gain for the various collection modes, using hourly values of solar radiation on a tilted surface, ambient temperature and wind velocity. A study of this type was done by Drago [1] for a fixed and fully tracking collector for the month of January 1977, showing that a fully tracking flat plate collector could give a significant increase in useful energy gain over a similar fixed flat plate collector. This paper extends the time of study to one year and the number of collection models from four to eight.

In the third part of this paper an analysis is made of the collector cost in relation to the useful energy gain for the various collection modes. In this way one can find the collection mode that will yield the highest monthly energy gain per initial dollar invested. The results of this report give a detailed picture of the advantageous and disadvantages of various collection modes at a northern latitude (40.8 degrees).

DETERMINATION OF R FOR VARIOUS COLLECTION MODES

In order to find the ratio, R, of total radiation on a tilted surface to that on a horizontal surface, a method is used similar to that of Liu and Jordan [2] and Klein, Duffie and Beckman [3].

The ratio of total radiation on a horizontal surface, H, to daily total extraterestrial radiation H_o, is first found from the equation,

$$K_T = \frac{H}{H_o} \tag{1}$$

Figure 7 in Liu and Jordan is then used to estimate the daily diffuse radiation, H_D, as a function of K_T. The hourly diffuse and total radiation on a horizontal surface can then be estimated from the daily diffuse and total radiation using Figure 15 in Liu and Jordan. It was found by Drago, however, that the ratio of daily diffuse radiation to daily total radiation, H_D/H, is very close to the ratio of hourly diffuse radiation to hourly total radiation H_{Dr}/H. The difference between the two values was well within the estimated error of the graph of Figure 7 in Liu and Jordan. The value of H_D/H was therefore used for each hour of the day instead of H_{Dr}/H.

The total radiation for each hour can then be found from the equation

$$H_{(TOT)} = HR \tag{2}$$

where for a flat plate collector, R, is given as in Liu and Jordan by

$$R = (1 - \frac{H_D}{H})R_b + \frac{H_D}{2H}(1 + \cos S) + \frac{\rho}{2}(1 - \cos S) \tag{3}$$

where ρ is the ground reflectance, taken to be 0.2 in this report.

Therefore hourly values of R for fixed and tracking flat plate collectors can be found provided that the appropriate hourly values of R_b are known for each collection mode.

In order to find R_b for a fixed tilted surface facing due south one may use the well known method of Duffie and Beckman [4].

If the collector is not fixed but instead tracking on two axes, it was shown by Drago [1] that hourly values of R_b can be found by finding the hourly values of tilting angle, S_m, and the surface azimuthal angle, y_m, which will give a maximum value of R_b for that hour. These values are given by

$$S_m = \cos^{-1}(\sin\delta \sin\phi + \cos\delta \cos\phi \cos\omega) \tag{4}$$

$$\gamma_m = \sin^{-1}(\frac{\cos\delta \sin\omega}{\sin S_m}) \tag{5}$$

If the collector tracks only on the azimuthal axis then the tilting angle S is held constant and the hourly values of y_m from eq. (5) are used to find hourly values of R_b.

In order to find R for an array of evacuated tube collectors, a method is used similar to the one used by Mather and Beekley [5]. In this method R is given by

$$R = (1 - \frac{H_D}{H})R_b + \Gamma_d((\frac{H_D}{2H})(1 + \cos S) + \frac{\rho}{2}(1 - \cos S)) \tag{6}$$

where Γ_d, the diffuse enhancement factor is given by

$$\Gamma_d = \frac{d}{D_4}(1 - (1 - \rho_d F_{pt})(1 - F_{pt})) \tag{7}$$

where the view factor F_{pt}, is given by

$$F_{pt} = 1 - \frac{D_4}{d}[((\frac{d}{D_4})^2 - 1)^{1/2} - \cos^{-1}(\frac{D_4}{d})] \tag{8}$$

If the array of evacuated tubes is facing south the hourly values of R_b can be found from the equation

$$R_b = R_T + 0.6R_p(2 - \frac{1}{\cos\omega}) \tag{9}$$

where R_p and R_T are given by

$$R_p = \frac{\cos \theta_T}{\cos \theta_z} \tag{10}$$

$$R_T = g(\omega)\frac{[1 - (\sin(S-\theta)\cos\delta\cos\omega + \cos(S-\theta)\sin\delta)^2]^{1/2}}{\cos \theta_z} \tag{11}$$

The term $g(\omega)$ is a shading factor given in terms of tube spacing, d, and the tube absorber diameter, D_6, as

$$g(\omega) = 1 , \qquad \omega < \omega_0 \tag{12}$$
$$g(\omega) = \frac{d}{D_4} \cos\omega + \frac{1}{2} (1 - \frac{D_6}{D_4}) , \qquad \omega > \omega_0$$

where

$$\omega_0 = \cos^{-1}(\frac{D_4 + D_6}{2d}) \tag{13}$$

the values of R_b for the various collection modes were calculated using a digital computer for the various values of hour angle and declination at a latitude of 40.8 degrees (Kings Point, New York).

DETERMINATION OF USEFUL ENERGY GAIN

In order to determine the useful energy gain for a flat plate collector the Hottel and Whillier (6) collector equation is used. This expressed the rate of energy collection per m^2 as

$$\frac{Q_u}{A} = F_R[HR(\tau\alpha) - U_L(T - T_a)] \tag{14}$$

where the product and the total loss coefficient, U_L can be found by the method of Duffie and Beckman [4].

For an array of evacuated tube collectors it was shown by Mather and Beekley [5] that the rate of energy collection per m^2 can be given by

$$\frac{Q_u}{A} = F_R[HR(\tau\alpha) - \pi U_L(T - T_a)] \tag{15}$$

where R is found from eq. (6) and A_c is the effective tube collector area given by

$$A_c = nD_4L \tag{16}$$

where n is the number of tubes in the array, D_4 is the diameter of the absorber and L the length of each tube. In this paper an array will be considered similar to the Owens-Illinois "Sunpak" array. The center to center tube spacing will be assumed to be given by $d = 2D_6$. Therefore, the area of the array A_m, will be given by

$$A_m = A_c + nD_6L \tag{17}$$

But using the parameters in TABLE III it can be seen that $D_6 = 1.23 \ D$. Therefore, the rate of energy collection per m^2 of array can be written in terms of
the rate of energy collection per m^2 of effective tube collector area as

$$\frac{Q_u}{A_m} = \frac{1}{2.23} \frac{Q_u}{A_c} \tag{18}$$

Thus the useful energy gain per m^2 of array area will be given by

$$\frac{Q_u}{A_m} = 0.448F_R[\ HR(\tau\alpha) - \pi U_L(T - Ta)] \tag{19}$$

It should be noted that the term HR in eq. (19) represents the total incident
insolation per m^2 of collector area, Ac. If we let (HR) be the total incident
insolation per m^2 of array area, A_m, then

$$(HR)_m = 0.448(HR) \tag{20}$$

The method used in computing the total useful energy gain for the various
collection modes is similar to the method used in Duffie and Beckman [4]. The
hourly radiation on a horizontal surface at Kings Point, New York, was determined using an Eppley pyronometer and a chart recorder. The ambient temperature and wind velocity were found for each hour. It was shown by Duffie and
Beckman that U is not a strong function of plate, cover or ambient temperature for either a single cover collector of high plate emittance, or for a
double cover collector of high or low plate emittance. Therefore, in finding
U for flat plate collectors for each hour, the plate and cover temperature
were considered fixed and only variations in wind velocity were taken into
account. The fluid inlet temperature was also kept constant throughout the
analysis.

For evacuated tube collectors, it was shown by Mather and Beekley [5]
that for low operating temperatures UL is not a strong function of plate,
cover or ambient temperature or of wind velocity. Thus, U_L was considered
constant during this analysis.

The ratio of total radiation on a horizontal surface to that on a tilted
surface, R, is found for each hour using the appropriate equation for each
collection mode along with the appropriate Rb. A computer program was set up
to measure U_L as a function of wind velocity for each hour flat (flat plate
collectors), R as a function of Rb and HD/H for each hour, and finally these,
alsong with hourly data for H and Ta to find Q_u/A (Q_u/A_m for an array of evacuated tube collectors) for each hour of the day. The daily useful energy gain
was found by adding the hourly gains, where only positive hourly gains are
considered. An example of this method is shown by Drago [1] for a fixed and
a fully tracking flat plate collector.

The daily energy gains were added for each collection mode to give monthly energy gains and finally yearly energy gains.

USEFUL ENERGY GAIN VS. COLLECTOR COST

The mode of collection which yields the highest useful energy gain is not necessarily the most advantageous. In any practical applications of these collection modes, cost must also be considered. A highly energy efficient collector which also carries with it a very high initial cost may not be as economical to use as a less energy efficient collector. Although the cost of individual collectors is something that changes yearly and also varies from supplier to supplier, one can still make an approximate comparison of energy yield to cost ratios for the various collection modes tested. One way of ana-lyzing the useful energy gain in relation to collector cost is to first define an initial cost factor given by

$$\text{initial cost factor} = \frac{\text{initial collector cost per } m^2}{\text{collector lifetime in months}} \tag{21}$$

We can define a monthly energy/cost factor given by

$$\text{MONTHLY ENERGY/COST FACTOR} = \frac{\text{monthly energy gain } (kJ \times 10^{-3})}{\text{initial cost factor}} \tag{22}$$

We can also define a year average monthly energy/cost factor by

$$\text{M.E./C.F.} = \frac{\text{yearly energy gain } (kJ \times 10^{-3})}{12 \times (\text{initial cost factor})} \tag{23}$$

Example A single cover flat plate collector costs $140/$m^2$ and is assumed to have a lifetime of 20 years (240 months), the monthly energy gain for this collector in February 1977, is 32.75 x 10^{3}kJ and the yearly gain for 1977, is 869.75 x 10^3 kJ. Find the monthly energy/cost factor for February and the year average M.E.C.F. for 1977.
Solution Using eq. (21) we obtain,

$$\text{initial cost factor} = \frac{140}{240} = 0.583 \frac{\text{dollars/}m^2}{\text{month of lifetime}}$$

The monthly energy/cost factor is then given by eq. (22) as

$$\text{MONTHLY ENERGY/COST FACTOR} = \frac{32.75}{0.583} = 56.18 \frac{kJ \times 10^{-3}}{\$/\text{month of lifetime}}$$

The year average M.E.C.F. is given by eq. (23) as

$$(\text{M.E.C.F.})_{Av.} = \frac{869.75}{12(0.583)} = 124.32 \frac{kJ \times 10^{-3}}{\$/\text{month of lifetime}}.$$

The significance of the energy/cost factors defined by eqs. (22) and (23) is that one can then compare collection modes and find the most advantageous mode to use for any particular month or for the entire year. The lifetime of the collector is also considered since a poorly made collector may be inexpensive, but it may also have a relatively short lifetime. Of course the M.E.C.F. for any particular month and the (M.E.C.F.)$_{Av.}$ for any particular year for any particular collection mode will vary from year to year, but the object of this study is not so much to find the most accurate value of the M.E.C.F. for any particular collection mode, but instead to compare various collection modes in order to find the one that is most advantageous to use at the site tested. For example if it was desired to use solar collectors to heat domestic hot water then the collection mode with the highest $(M.E.C.F.)_{Av.}$ would be the one that would yield the most kilojoules per initial dollar invested and therefore the shortest payback time. Of course the physical limitations of any particular site might rule out using the collection mode with the highest (M.E.C.F.). But then all one has to do is go down the list and find the collection mode with the highest $(M.E.C.F.)_{Av.}$ that will fit that particular site.

RESULTS

TABLES I, II and III give the parameters for the flat plate and evacuated tube collectors used.

TABLE IV gives the monthly useful energy gain for each mode of collection as well as the year and heating season totals for 1977, at Kings Point. New York. The heating season at Kings Point is assumed to be from October through May.

TABLE V shows the energy gain for each collection mode as compared to a fixed single cover flat plate collector with $\varepsilon_p = 0.95$.

TABLE VI shows the energy gain for each collection mode as compared to a fixed double cover collector with $\varepsilon_p = 0.95$.

TABLE VII shows the energy gain of each collection mode as compared to a fixed double cover collector with $\varepsilon_{\bar{p}} = 0.07$.

TABLE VIII shows a comparison of monthly energy cost/factors for a fixed double cover flat plate collector with $\varepsilon_p = 0.95$ and a single axis (azimuthal) tracking double cover flat plate collector with $\varepsilon_p = 0.95$. The value of monthly energy cost factor for the other collection modes can be found by simply using the monthly energy gains from TABLE IV and dividing by the appropriate cost factor from TABLE IX.

TABLE IX shows a comparison of year average energy/cost factors for each collection mode. An average monthly energy/cost factor is also found for the heating season by adding the values of M.E.C.F. for the months of October through May dividing by eight. The cost of a single cover flat plate collector is found by comparing various collectors now on the market to be about $140/m^2$. It is then assumed that adding a second glass cover will add about $20/m^2$ to the cost, and adding a selective coating to the absorber plate will add another $20/m^2$ to the cost.

The cost of allowing a flat plate collector to track on a single axis (azimuthal) can be approximated from the cost of such a tracking collector now on the market and produced by American Solar Systems [7]. They provide a tracking system at a cost of $500 which can track from one to fifteen collectors, each of which are about $5.6m^2$ in collector area. Thus the more collectors the less is the tracking cost area. Thus the more collectors used the less is the tracking cost per m. For example if eight collectors are used then the cost is about $11/m^2$. If four collectors are used then the cost is $22/m^2$. In this report the year average M.E.C.F. is found for assumed tracking costs of $10, $25, and $40 per m^2. The cost of allowing a flat plate collector to

Table I. Single Cover Collector Parameters.

F_R	0.85
T	60° C
$(\tau\alpha)$	0.87
ε_p	0.95
ε_c	0.88
$U_b + U_e$	1 w/m^2 ^{o}C
T_{plate}	65^oC
T_{cover}	23^oC
L	2.5 cm

Table II. Double Cover Collector Parameters

F_R	0.85
T	60^oC
$(\tau\alpha)$	0.75
S (fixed and single tracking)	50.8^o
ε_p(selective)	0.07
ε_p	0.95
ε_{c1}	0.88
ε_{c2}	0.88
$U_b + U_e$	1 w/m^2 ^{o}C
T_{plate}	65^oC
$T_{cover\ 1}$	45^oC
$T_{cover\ 2}$	15^oC

Table III. Evacuated Tube Collector Array Parameters

Absorber tube diameter D_4	4.3 cm
Cover tube diameter D_6	5.3 cm
U_L	0.4 w/m^2 °C
Tube spacing d	10.6 cm
F_R	0.98
$(\tau\alpha)$	0.79
Γ_d	1.88
T	60°C

Table IV. Useful Energy Gain for Eight Collection Modes (kJ/m^2 x 10^{-3})

Mode	Year	Heating Season
HR - Fixed	4440	2928
QU/A Single cover ε_p = .95	870	488
QU/A Double cover ε_p = .95	1237	767
QU/A Double cover ε_p = .07	1822	1178
HR - Single Tracking	5385	3508
QU/A Double Cover ε_p = .95	1832	1111
QU/A Double Cover ε_p = .07	2464	1559
HR - Full Tracking	5915	3885
QU/A Double Cover ε_p = .95	1994	1197
QU/A Double Cover ε_p = .07	2631	1636
$(HR)_m$ Evacuated Tube	3497	2275
QU/A_m d = $2D_6$	2452	1583

Table V. Energy Gain Normalized to Single Cover Fixed Collector

Collection mode	HR (year)	Q_u/A(year)	HR (Heating Season)	Q_u/A (Heating Season)
Fixed double cover ε_p = 0.95	1.00	1.00	1.00	1.00
Fixed double cover ε_p = 0.95	1.00	1.42	1.00	1.57
Fixed double cover ε_p = 0.07	1.00	2.09	1.00	2.41
Single tracking double cover ε_p = 0.07	1.21	2.84	1.20	3.19
Full tracking double cover ε_p = 0.95	1.33	2.29	1.33	2.45
Full tracking double cover ε_p = 0.07	1.33	3.03	1.33	3.35
Evacuated Tube Array $d - 2D_6$	0.79	2.82	0.78	3.24

Table VI. Energy Gain Normalized to Double Cover Fixed Collector

Collection Mode	HR (year)	Q_u/A(year)	HR (Heating Season)	Q_u/A (Heating Season)
Fixed double cover $\varepsilon_p = 0.95$	1.00	1.00	1.00	1.00
Fixed double cover $\varepsilon_p = 0.07$	1.00	1.47	1.00	1.54
Single tracking double cover $\varepsilon_p = 0.95$	1.21	1.48	1.20	1.45
Single tracking double cover $\varepsilon_p = 0.07$	1.21	1.99	1.20	2.03
Full tracking double cover $\varepsilon_p = 0.95$	1.33	1.61	1.33	1.56
Full tracking double cover $\varepsilon_p = 0.07$	1.33	2.13	1.33	2.13
Evacuated Tube Array $d = 2D_6$	0.79	1.98	0.78	2.06

Table VII. Energy Gain Normalized to Double Cover Fixed Collector with Selective Surface

Collection Mode	HR (year)	Q_u/A (year)	HR (Heating Season)	Q_u/A (Heating Season)
Fixed double cover $\varepsilon_p = 0.07$	1.00	1.00	1.00	1.00
Single tracking double cover $\varepsilon_p = 0.07$	1.21	1.35	1.20	1.32
Full tracking double cover $\varepsilon_p = 0.07$	1.33	1.44	1.33	1.39
Evacuated Tube Array $d = 2D_6$	0.79	1.35	0.78	1.34

Table VIII. Comparison of Monthly Energy/cost Factors for Two Modes of Collection $\left(\dfrac{(kJ/m^2 \times 10^{-3}}{\text{dollar/month of lifetime}} \right)$

	Cost $/m^2$	Cost Factor	Jan.	Feb.	March	April	May
Fixed cover $\varepsilon_p = 0.95$	160	0.667	83.49	116.08	208.26	204.25	231.62
Single tracking $\varepsilon_p = 0.95$	170	0.798	108.16	147.0	294.44	306.5	351.19

	June	July	Aug.	Sept.	Oct
Fixed cover	200.13	209.0	169.24	125.73	118.74
Single tracking	303.83	314.25	232.0	168.25	138.26

	Nov.	Dec.	Year Av.	Heating Season Average
Fixed cover	102.91	85.41	154.58	143.85
Single tracking	117.45	106.31	215.62	196.15

Table IX. Comparison of Average Monthly Energy/Cost Factors for Eight
Collection Modes (kj x 10^{-3}/dollar/month of lifetime)

Mode	Cost ($/m^2)	Cost Factor	Year Average	Heating Season Average
Fixed single cover ϵ_p = .95	140	0.583	124.32	104.7
Fixed double cover ϵ_p = .95	160	0.667	154.58	143.85
Fixed double cover ϵ_p = .07	180	0.75	202.45	196.41
Single tracking double cover ϵ_p = .07	170 185 200	0.708 0.771 0.833	215.62 198.0 183.26	196.15 180.12 166.71
Single tracking double cover ϵ_p = .07	190 205 225	0.792 0.854 0.938	259.33 240.9 177.17	246.1 228.24 207.80
Full tracking double cover ϵ_p = .95	190 205 225	0.792 0.854 0.938	209.82 194.59 177.17	188.89 175.17 159.48
Full tracking double cover ϵ_p = .07	210 225 245	0.875 0.938 1.021	250.57 233.34 214.74	233.66 217.97 200.25
Evacuated Tube Array	220 270 320	0.917 1.125 1.333	222.82 181.62 153.28	215.86 175.95 148.5

track on two axes is an unknown quantity since none are now commercially available. But as a point of comparison the year average M.E.C.F. is found for two
axis tracking costs of $30, $45, and $65 per m^2.

The cost of an array of evacuated tube collectors can be approximated
from the current market price (Owens-Illinois "Sunpak") as about $270/m^2 of
array. As a point of comparison the year average is also given for the evacuated tube array for costs both above and below this figure.

CONCLUSIONS

An examination of TABLES IV through IX will show that a single cover
fixed flat plate collector with a non-selective coating is outperformed by a
wide margin both in useful energy gain and yearly average M.E.C.F. by the
other collection modes. The addition of a second cover at this northern latitude provides significant increases in useful energy gain. One reason for
this is that the loss factor U_L for a single cover collector is much more sensitive to changes in wind velocity than for a similar double cover collector.
The analysis shows that a second cover for a flat plate collector would be a
worthwhile investment at a northern latitude. At an assumed cost of $20/m^2
a selective absorber coating again would seem to be a worthwhile investment
at a northern latitude. It should be noted that it is assumed here that the
lifetime of the coating will be 20 years. If the cost is higher or the lifetime shorter, then the year average M.E.C.F. will decrease for all modes of
collection with a selective surface in comparison to other modes of collection.

Although an array of evacuated tube collectors gives significantly more
useful energy gain than any of the fixed collector modes, the high initial
cost of the array give it a year average M.E.C.F. which is lower than a fixed
double cover collector with a selective surface and any of the tracking modes.
Of course it should be realized that the main benefit of evacuated tube collectors is that they may be used at relatively high operating temperatures without much decrease in efficiency. This is not true of the other collection
modes unless the cost of evacuated tube arrays can be decreased. However it
would not be advantageous to use this type of collection mode for low operating temperatures.

Allowing a flat plate collector to track on a single azimuthal axis or
double axes can yield a significant increase in useful energy gain. Ultimately the cost of allowing a flat plate collector to track on one or two axes
will determine just how advantageous these modes of collection will be. From
TABLE IX it can be seen that if the cost of a single axis, two cover tracking
collector is about $190/m then this mode of collection will yield the highest
year average and heating season average M.E.C.F. and will thus be the best
collection mode to use at the site tested. It would therefore seem that this
mode of collection should be tested. It would therefore seem that this mode
of collection should be taken advantage of whenever the physical properties
of a site allow tracking to be done without too much difficulty.

It should be noted that this analysis was done over a period of one year
and that the yearly energy gains, and thus the year average M.E.C.F., for
each collection mode will vary from year to year. But over the year period
there were enough days of widely varied conditions tested so that the values
of energy gain and year average M.E.C.F. of one collection mode relative to
another would remain just about the same.

It should also be noted that the monthly energy/cost factors as defined here do not include the installation costs of any of the collection modes tested. Installation costs depend on many factors and can vary widely from site to site. However, if at any particular site one had to choose which collection mode to use, the choice would be the mode which yields the highest monthly energy/cost factor for the fraction of the year that it is to be used. Once that mode is chosen the system can then be sized to give the desired percentage of heating load to be taken over by the system.

REFERENCES

1. P. Drago. A simulated comparison of the useful energy gain in a fixed and a full tracking flat plate collector. Solar Energy 20(6), 419(1978)

2. B. Y. H. Liu and R. C. Jordan, The interrelationship and characteristic distribution of direct, diffuse and total solar radiation. Solar Energy 4(3) (1960)

3. S. A. Klein, J. A. Duffie and W. A. Beckman, A design procedure for solar heating systems, Solar Energy 18 (21), 113 (1976).

4. J. A. Duffie and W. A. Beckman, "Solar Energy Thermal Processes" Wiley, New York (1974).

5. G. R. Mather and P. C. Beekley, Analysis and experimental tests of a high-performance evacuated tube collector, Owens-Illinois Inc., Toledo, Ohio. (1976)

6. H. C. Hottel and W. Whillier, Evaluation of flat plate solar collector performance, Trans Conf. Use of Solar Energy - II - Thermal Processes, pp 74-104. Univ. of Arizona, Tempe, Arizona (1976).

7. American Solar Systems Inc. Arroyo Grande, California

FLAT PLATE COLLECTORS

Static Endo-Absorbent Flat Solar Collector

LUCIANO N. BLANCO
University of Miami
Coral Gables, Florida 33124, USA

ABSTRACT

The possibility of absorbing solar radiation by means of discrete micro-absorbers completely immersed in a heat transfer fluid is analyzed. This is termed the endo-absorbent concept and does not include designs such as the overlapped-glass plates or the blackened metal screen.

The closest realization of this concept is the black liquid collector in which the micro-absorbers are permanently suspended in the heat transfer fluid and travel in it via a closed circuit.

As opposed to the above, the static endo-absorbent concept is presented in which the micro-absorbers are not permanently dispersed or suspended in the heat transfer fluid but are instead kept within the limits of the solar radiation absorbing volume while the heat transfer fluid circulates through it. The performance of a collector of this type is studied analytically as well as experimentally.

It is shown that designs based on this concept allow not only a higher efficiency but also several other advantages when compared to a traditional tube-and-fin collector and to the black liquid concept.

INTRODUCTION

At the present stage of solar technology it is worth examining new alternatives in solar radiation collection, especially if these alternatives could bring substantial improvements in efficiency, simplicity in construction, and important reduction in cost.

The purpose of the present work is to examine, analytically and experimentally, the possibility of absorbing solar radiation by means of micro-absorbers completely immersed in a heat transfer fluid but kept within the limits of the solar radiation absorbing volume. This is termed the static endo-absorbent concept to emphasize the fact that the micro-absorbers do not circulate with the heat transfer fluid as in the black liquid concept (1,2,3,4), and also that absorption of radiation takes place within the heat transfer fluid in contrast to a conventional tube-and-fin collector.

The basic unit considered here, among other possibilities, consists of a transparent glass or plastic tube which contains a certain amount of micro-absorbers. The micro-absorbers may be particles of a wide variety of sizes, configurations and materials such as small spheres of a blackened plastic. These particles are dispersed, filling the entire radiation absorbing volume of the tube as the heat transfer fluid is passing through it. To prevent the particles from being removed from the tube by the circulating fluid, a special filter or suitable trap is provided at the ends of the tube.

Figure 1 illustrates the basic unit proposed. The complete collector consists of an array of such basic units attached to two manifolds through which the transfer fluid is carried.

This basic unit, although suitable as a component for a flat-type collector, has good characteristics for a receiver in a focusing-type collector.

THERMAL ANALYSIS

In the absence of micro-absorbers, the basic unit, Figure 1, if only clear water circulates through it, absorbs a minimum amount of radiation. This is due to absorption in the infrared region while the other components of the spectrum pass through, practically unabsorbed.

As soon as some micro-absorbers are introduced into the exposed volume, absorption takes place according to the expression:

$$\Delta I_x = I_{xo} (1 - e^{-kx}) \quad \text{or} \quad \propto = \frac{\Delta I_x}{I_{xo}} = (1 - e^{-kx})$$

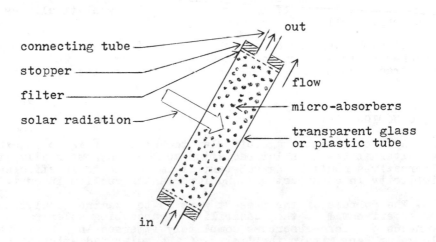

FIG. 1   BASIC UNIT PROPOSED FOR THE STATIC ENDO-ABSORBENT CONCEPT

The extinction coefficient k has the form (5):

$$k = c\,\rho^{-1}\,T^n\,d^{-2n}\mu \quad ;$$

and it can be seen that if k and x are properly maximized as allowed by the design contraints, their product kx can be made great enough as to render α practically equal to 1.

On the other hand, if the thermal network represented in Figure 2 is applied to the basic unit, the following energy balance equations can be established:

$$h_r(T_p-T_c)+h_1(T_f-T_c) = U_t(T_c-T_a) \qquad \text{tube wall}$$

$$I_a = h_r(T_p-T_c)+h_v(T_p-T_f) \qquad \text{micro-element}$$

$$h_v(T_p-T_f) = I_u + h_1(T_f-T_c) \qquad \text{fluid}$$

solving these equations, the following expression is obtained:

$$I_u = F'\left[I_a - U_1(T_f-T_a)\right]$$

where

$$F' = \frac{h_r h_1 + h_r h_v + h_1 h_v + h_v U_t}{h_r h_1 + h_r h_v + h_1 h_v + h_v U_t + h_r U_t}$$

and

$$U_1 = \frac{U_t(h_r h_1 + h_r h_v + h_1 h_v)}{h_r h_1 + h_r h_v + h_1 h_v + h_v U_t}$$

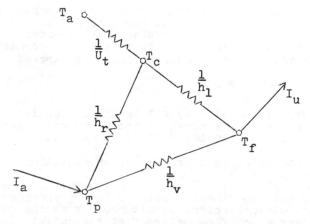

FIG. 2 THERMAL NETWORK FOR BASIC UNIT PROPOSED

In a first approximation (6)

$$h_v \simeq C \ \dot{m}^n \ A^{-n} \ d^{-n}$$

and therefore it can take very high values. On the other hand, h_r is a very small coefficient and U_t is also small taking into consideration the relatively high thermal resistance coupling the surface of the tube to the ambient. As a consequence $h_r U_t$ in the denominator of the expression defining F' is negligible and F'=1.

As a result of the higher values for α and F' when compared to those of a tube-and-fin design, a significant efficiency enhancement should be expected. Furthermore, the tubular design should also improve the efficiency for solar positions away from the plane normal to the collecting surface.

EXPERIMENTAL PROCEDURE

In order to test experimentally the feasibility of the concept, a prototype was constructed and tested simultaneously with a tube-and-fin design of high efficiency and the same geometrical parameters.

Two identical wooden frames 1.20 x 0.70 x 0.15 m provided the housing for the collectors. Each of them were equally insulated in the interior (sides and back) with 7 cm thick polyurethane insulation. An aluminum sheet of 0.5 mm thickness covered the back of each frame and supported the insulating material. Both were single-glazed with 3 mm tempered glass.

One frame contained the static endo-absorbent version which consisted of 15 paralell Pyrex glass tubes of 30 mm outside diameter, 1.8 mm wall thickness and 100 cm long attached to two common headers by means of 9.5 mm internal diameter glass tube connectors. Each tube had a filter at both ends in the form of a plastic disc with holes of 0.5 mm in diameter. Neoprene rubber stoppers suported the connections.

Each tube was filled with black Neoprene rubber particles of about 0.6 to 1 mm in diameter in amount equal to 87% of the exposed volume. In order to provide for the passing fluid a cross sectional area in the absorbing tubes equivalent to that of the connecting tubes, the particle volume was calculated from the expression:

$$v = \frac{A-a}{A} \times 100$$

The other frame was occupied by a tube-and-fin collector of traditional design comprised of paralell copper tubes soldered to a copper sheet providing a calculated F' of about 0.95.

Both versions had an exposed area to solar radiation equal to 0.45 m^2.

Figure 3 shows the experimental set up. Pump p provided a continuous flow for both collectors simultaneously at the same temperature T_i. Valves v and flow meters f at the outlet of each collector allow the flow through each collector to be regulated

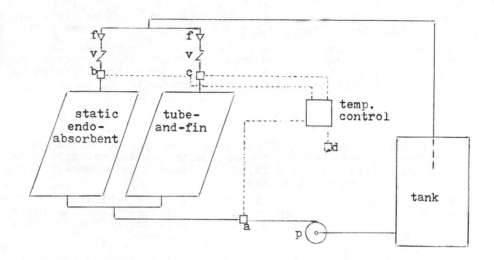

FIG. 3 EXPERIMENTAL SET UP. STATIC ENDO-ABSORBENT COLLECTOR
TESTED SIMULTANEOUSLY WITH A TUBE-AND-FIN DESIGN

to the desired value. In all cases the flow was regulated to the
same value for both collectors. Sensors a,b,c and d made possible
the determination of temperatures T_i, T_{so}, T_{to} and T_a respectively.

Both collectors were installed facing due south and tilted
25° from the horizontal during the tests.

Tests were performed at different times of the day for different solar positions either with a clear sky or with a stable
sky cover whenever possible. Data were also taken for various
values of T_i and T_a.

RESULTS AND DISCUSSION

The outlined experimental procedure allows the calculation
of the ratio η_s/η_t based on the following equations (6):

$$\eta_s = \frac{\dot{m}\, C_p (T_{so}-T_i)}{I_i} = \frac{\dot{m}\, C_p\, \Delta T_s}{I_i} \quad \text{and}\quad \eta_t = \frac{\dot{m}\, C_p (T_{to}-T_i)}{I_i} = \frac{\dot{m}\, C_p\, \Delta T_t}{I_i}$$

and therefore, $\eta_s/\eta_t = \Delta T_s/\Delta T_t$.

Figure 4 shows the value of η_s/η_t and its variation for different values of T_i-T_a. As expected, this ratio is always greater
than one and increases when T_i-T_a increases. This suggest that
the static endo-absorbent concept becomes a better choise for
higher T_i-T_a differences. The efficiency increases approximately
from 10 to 15 % in the T_i-T_a range from 10 to 40 $^\circ$C.

Figure 5 shows the value and variation of η_s/η_t at different solar positions during the day but same T_i-T_a. It is shown again that this ratio is always greater than one and increases somewhat for solar positions away from normal incidence. The asymmetry of the curve with respect to 12 noon is probably due to different sky conditions during the morning and afternoon tests.

CONCLUSION

This work shows that a collector of the type developed and studied here performs well as compared with traditional tube-and-fin type collectors and presents, as the black liquid concept does, the following important advantages:

a.- Greater efficiency. An extra 10 to 15 % is achieved.
b.- Since no metals are needed in the active sections, an impor-
     tant reduction in cost is obtained while corrosion problems
     are eliminated.
c.- Construction is simplified.
d.- Due to the above mentioned reasons, final cost is reduced
     not only in terms of cost per unit collector area but also
     due to the smaller total collector area needed as a result
     of its better efficiency.

Furthermore, the static endo-absorbent concept presents the following additional advantages when compared to the black liquid concept:

a.- Designs are possible for both air or liquid heat transfer
     fluid.
b.- No heat exchange is always necessary. This could result in
     an extra reduction in cost for a given installation.
c.- No sedimentation problems are to be taken into account with
     improved maintenance and system simplification.
d.- Much greater flexivility when selecting material, size and
     configuration of the micro-absorbers.

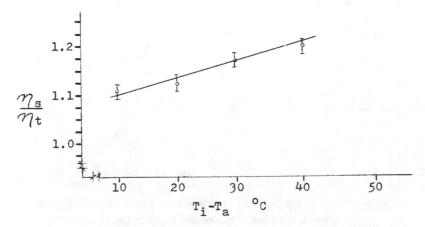

FIG. 4 EFFICIENCY RATIO vs T_i-T_a FOR THE STATIC
ENDO-ABSORBENT AND THE TUBE-AND-FIN COLLECTOR

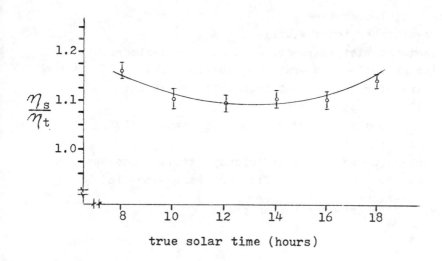

FIG. 5 EFFICIENCY RATIO vs TRUE SOLAR TIME FOR THE STATIC
ENDO-ABSORBENT AND THE TUBE-AND-FIN COLLECTOR

      Further research is necessary and is being carried on for the
proper optimization of this type of collector.

NOMENCLATURE

A   Cross sectional area in absorbing tube
a   Cross sectional area in connecting tube
C   Constant
C_p Specific heat
d   Micro-absorber average diameter
F' Collector efficiency factor
h_1 Heat transfer coefficient fluid-to-tube wall
h_r Heat transfer coefficient absorber-to-tube wall
h_v Heat transfer coefficient absorber-to-fluid
I_a Solar radiation absorbed
I_i Incident solar radiation
I_u Useful heat from collector
I_x Solar radiation along coordinate x
I_{xo} Solar radiation at $x = x_o$
k   Extinction coefficient
ṁ   Mass flow rate
n   Constant
T   Absolute temperature
T_a Ambient temperature
T_f Fluid temperature
T_i Collector inlet temperature

T_c Tube wall temperature
T_p Micro-absorber temperature
T_{so} Collector outlet temperature (static endo-absorbent)
T_{to} Collector outlet temperature (tube-and-fin)
U_t Heat transfer coefficient tube wall-to-ambient
U_1 Overall heat loss factor
V Total micro-absorber volume in % of total radiation absorbing
          volume
α Absorptivity
η_s Instantaneous collector efficiency (static endo-absorbent)
η_t Instantaneous collector efficiency (tube-and-fin)
μ Concentration of micro-absorbers
ρ Density of micro-absorber material

REFERENCES

1.- Minardi, J.E. and Ghuang, H.N., Solar Energy, 17, 179 (1975)

2.- Trentleman, J. and Wojciechowski, P.H., Proceedings of the
    International Solar Energy Society, June 1977

3.- Anderson, J.H., Solar Heating Cell, U.S. Pat. 4,047,518 (1977)

4.- Blanco, L.N., Unpublished work, (1978)

5.- Kutateladze, S.S., Fundamentals of Heat Transfer, Academic
    Press Inc., New York, N.Y. (1963)

6.- Kreith, F. and Kreider, J.F., Principles of Solar Engineering,
    Hemisphere Publishing Corp., Washington, D.C. (1978)

A Figure of Merit for Solar Collectors with Several Separate Absorber Segments

R.P. PATERA and H.S. ROBERTSON
University of Miami
Coral Gables, Florida 33124, USA

ABSTRACT

A figure of merit for solar collectors is proposed and examined. A figure of merit for solar collectors is a parameter that is indicative of thermal performance. Typically, the efficiency itself is used as a figure of merit. In comparison of collectors having several thermally separated absorber segments, the computation of efficiency becomes unwieldy. Therefore, there is a need for a figure of merit that is easily computed, even for collectors with several absorber segments.

The proposed general figure of merit, Q, has information about the input distribution, the concentration and the acceptance function built into its definition. In addition, we propose the use of a channel matrix to characterize a collector. The channel matrix enables the figure of merit to be readily calculated for various input distributions.

MODERATELY CONCENTRATING COLLECTORS

Robertson has shown that it is possible to construct a collector which concentrates the direct solar radiation and collects all of the diffuse radiation.[1] Theoretical analysis of this radiation collection problem is aided by Liouville's phase space theorem which states that under a transformation phase space areas are preserved. Figure 1 illustrates the phase space of radiation present at the collector aperture where the phase space area associated with the direct solar radiation is shown explicitly. The concentrator produces a transformation of the phase space distribution. The optimum concentrator would produce a transformation which yields a phase space distribution at the absorber as shown in Fig. 2. Optically, this transformation corresponds to a perfect imaging system such as the Luneburg Lens.[2] Perfect imaging systems are not practical for solar energy collection purposes. The concept of concentrating the beam radiation on one absorber segment while collecting all of the diffuse radiation on all of the segments is important. Even a very poor imaging system, which moderately concentrates the beam radiation, can achieve a significant improvement in efficiency over a typical flat plate collector. We anticipate the use of moderately concentrating collectors for solar absorption air conditioning systems. Thus, there is a need for a simple method of estimating the collector performance of collectors having several thermally separate absorber segments.

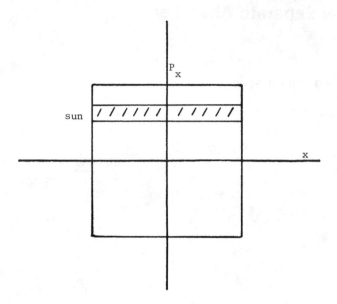

Figure 1. Phase space at the collector aperture for a trough-like collector.
The direct solar component is shown explicitly.

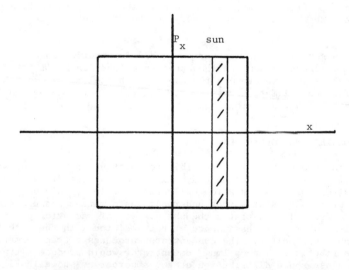

Figure 2. Phase space at the absorber for a perfect trough-like collector
channel.

DISCUSSION

There are three factors that influence a collector's performance, in addition to the factors that depend on what the concentrator is being used for; i.e., high or low operating temperature. These three factors are the input distribution of radiation, the concentration and the acceptance function, $I(\theta)$.

The input distribution is given and is the same for concentrators under comparison. The input distribution, however, determines how the other factors are weighted in the total collector performance. For example, a concentrating collector offers no advantage over a flat plate collector if the input radiation is isotropic, i.e., uniformly diffuse with no beam component.

The concentration is an indicator of collector performance. There are at least two definitions of concentration. The geometric concentration is the ratio of the aperture area to the absorber area; it can have any positive value. The optical concentration is equal to the geometric concentration times the fraction of rays within the collecting angle that get transmitted to the absorber.

The acceptance function, $I(\theta)$, is defined as the probability that a photon having an incident angle of θ at the aperture strikes the absorber.

The usual theoretical method of comparing concentrators is to specify the geometric concentration and the acceptance function. We know of no procedure for combining these factors into one figure of merit. As a result, it is impossible to determine which concentrator will perform better if these two factors contradict one another. Furthermore, the effect of the input distribution is ignored.

We propose Q as a general figure of merit, where

$$Q = \sum_r b_r \, Q_r \qquad\qquad (1)$$

The index r labels each separate absorber segment. b_r is defined as the probability that a photon, after having entered the aperture, strikes the $\underline{r}$ th absorber segment. Q_r is defined as

$$Q_r = \frac{A\, b_r}{\sigma_r} \; , \qquad\qquad (2)$$

where A is the aperture area and σ_r is the area of the $\underline{r}$ th absorber segment.

This figure of merit was deduced from information theory considerations which will not be presented in this paper.[3]

Q_r is equal to the geometrical concentration only if b_r equals unity. Q_r can be considered another definition of concentration. It is the geometrical concentration times the fraction of the total radiation entering the aperture that strikes the $\underline{r}$ th absorber. If Q_r has a value of four it means that the radiation striking the $\underline{r}$ th absorber is four times as intense as the radiation entering the aperture. A value of four for Q_r represents greater intensity than an optical concentration of four because Q_r includes diffuse radiation in its definition. If the input radiation is isotropic, the maximum value of Q_r is one, whereas the optical concentration can be very large.

Q is the mean of the set of Q_rs, where each Q_r is weighted by b_r. Therefore, Q can be considered an average concentration that is representative of the collector as a whole.

The lowest value of Q subject to the constraint that b is equal to a constant is

$$Q = \frac{b^2 A}{\sigma} \quad , \tag{3}$$

where

$$b = \sum_r b_r \tag{4}$$

and

$$\sigma = \sum_r \sigma_r \quad . \tag{5}$$

This represents a uniform distribution of radiation over the absorber which can be represented by a single segment absorber. For a single absorber segment, Eq.1 yields Eq. 3.

There is a maximum possible value of Q which depends on the input distribution of radiation. We divide the sky into a number of solid angular regions labeled by the index i. These regions may not all be the same size. Each solid angular region corresponds to a region in four dimensional phase space which is labeled with the index i. The maximum value of Q is given by

$$Q_{max.} = \sum_i \frac{a_i^2 \, \pi \, p^2}{p_i} \quad , \tag{6}$$

where p is the momentum of the photons, p_i is the area of the $\underline{i}$ th momentum region in the four dimensional phase space of radiation at the collector aperture and a_i is the probability that a photon is in the $\underline{i}$ th momentum region. Therefore the range of values that Q can have is given by

$$\frac{b^2 A}{\sigma} \leq Q \leq \sum_i \frac{\pi a_i^2 \, p^2}{p_i} \quad . \tag{7}$$

If $\sum b_r < 1$ the maximum value of Q is reduced by choosing a_i so that the right side of Eq. 7 is maximum with the constraint

$$\sum a_i = \sum b_r \quad . \tag{8}$$

The computation of the upper limit on Q is usually not important. It is important however, to note that there is an upper limit which is determined by the distribution of radiation at the aperture.

It is instructive to evaluate Q for two simple cases. In the case of a flat plate collector:

$$b_r = 1 \tag{9}$$

$$\sigma_r = 1 \tag{10}$$

$$r = 1 \quad . \tag{11}$$

Using these values in Eq. 1 and Eq. 2 we find that

$$Q = 1 \quad . \tag{12}$$

Q is always equal to one for a flat plate collector regardless of the input distribution.

Values of Q for other collector designs are dependent on the input distribution. For example, the value of Q for the ideal compound parabolic collector

(CPC) proposed by R. Winston[4] depends strongly on the input distribution. Let s be the relative amount of direct solar radiation and let d be the relative amount of diffuse solar radiation defined as

$$s = \frac{\text{direct solar radiation}}{\text{total solar radiation}} \qquad (13)$$

$$d = \frac{\text{diffuse solar radiation}}{\text{total solar radiation}} \qquad (14)$$

By definition, $s + d = 1$.   (15)
For an ideal CPC the relative amount of diffuse radiation striking the absorber is

$$d/c \quad , \qquad (16)$$

where c is the geometrical concentration defined as

$$c = A/\sigma_r \quad . \qquad (17)$$

If the sun is within the collector acceptance angle, we find after simplication that

$$Q = \frac{(1 - s + sc)^2}{c} \qquad (18)$$

Q is greater than one only if

$$s > \frac{1}{\sqrt{c} + 1} \quad . \qquad (19)$$

Thus if cloudiness reduces s such that

$$s < \frac{1}{\sqrt{c} + 1} \quad , \qquad (20)$$

then the figure of merit falls below that of a flat plate collector. We suggest that, wherever possible, collectors should be designed such that the value of Q never falls significantly below unity even in the event of cloudiness.

Since Q is proportional to the mean concentration on the absorber, it is thought to be a good measure of collector performance. A collector's performance, however, depends on its operating temperature. Generally speaking, the higher the operating temperature, the lower the efficiency. An exact measure of performance should include the collector operating conditions. Therefore, Q represents, at best, the average collector performance over a range of operating conditions.

One can easily find a measure valid for low operating temperatures. At low temperatures, collector losses are minimal. The most important factor is not the average concentration but the total amount of radiant energy transmitted to the absorber. Therefore, a useful measure is just b.

At high temperatures, the collector losses dominate. The most important factor is the term in Q that represents the absorber segment with the greatest concentration. A useful measure is $b_r^2 A/\sigma_r$, where r indicates the segment with the greatest concentration.

At intermediate temperatures, one needs the exact measure of performance. We will consider collector operation where the increase in the working fluid temperature across the collector, $T_{out} - T_{in}$, is relatively small. Therefore the temperature dependence of the overall thermal loss coefficient, U, can be neglected and we can use the standard flat plate collector equations.

The standard expression for the flat plate collector output temperature in degrees above ambient is

$$T(\sigma) = (T(0) - h Q) g^{\sigma} + Q h \qquad (21)$$

where $g = \exp(-U/\dot{m} C_p)$, $h = I F' \alpha\tau/U$.

σ is the area of the absorber, $T(0)$ is the input fluid temperature in degrees above ambient, U is the heat transfer coefficient from the fluid to the ambient air, $\dot{m}$ is the transfer fluid flow rate, C_p is the specific heat of the transfer fluid at constant pressure, F' is the ratio of the heat transfer coefficient from fluid and from absorber to ambient air, I is the solar insolation (power/unit area), α is the absorptance of the absorber, τ is the transmittance of the collector cover, and Q is the concentration of the radiation as defined earlier. It can be noted that the values of g can lie between zero and one.

For a multisegmented absorber, the fluid is first passed through the segment with the lowest concentration and then the segment with the next highest concentration and so on until it is finally passed through the segment with the highest concentration. The output temperature of the fluid, after being passed through all n segments, is

$$T_n = T_o g^{\sigma} + \sum_{r=1}^{n} h Q_r (1 - g^{\sigma r}) g^{\psi r} \quad , \qquad (22)$$

where $\psi_r = \sum_{r+1}^{n+1} \sigma_k$, $\sigma_{n+1} = 0$, T_o is the input fluid temperature and Q_r and σ_r are as defined earlier. From Eq. 21, h can be identified as the stagnation temperature of a flat plate collector, that is, the equilibrium temperature when $\dot{m}=0$. This is the maximum temperature that the collector will reach and depends on the insolation. Equation 22 can be rewritten as

$$d = e g^{\sigma} + W \quad , \qquad (23)$$

where $d = T_n/h, \qquad e = T_o/h, \qquad (24), (25)$

and

$$W = \sum_{r=1}^{n} Q_r (1 - g^{\sigma r}) g^{\psi r} \quad . \qquad (26)$$

The parameters d and e are very useful since they are independent of the actual input and output fluid temperatures. Notice that d and e are always less than one for useful collector operation. W is a parameter that depends on the distribution of radiation over the set of absorber segments. In some cases it is advantageous to eliminate flow through segments with lower concentration since the first term in Eq. 23 increases if a segment is eliminated and $T_o > 0$, since $g < 1$.

The efficiency is defined as

$$\eta = \frac{\dot{m} C_p (T_n - T_o)}{I A} \qquad (27)$$

where A is the area of the aperture. For simplicity we will assume that $A = 1$ for all collectors under consideration. Equation 27 reduces to

$$\eta = F'\alpha\tau(d - e)/\ln(1/g) \quad , \qquad (28)$$

where use has been made of Eqs. 24, 25 and 21. For a typical collector W is computed from Eq. 26. Equation 23 is used to solve for g since d, e, and σ are known. Equation 28 is then solved to determine the efficiency. To illustrate this procedure we will consider the case of a flat plate collector. From Eq. 26 one finds

$$W = 1 - g \quad . \tag{29}$$

Solving for g in Eq. 23, one obtains

$$g = (1 - d)/(1 - e) \quad . \tag{30}$$

The efficiency becomes

$$\eta = F'\alpha\tau(d - e)/\ln((1-e)/(1-d)) \quad . \tag{31}$$

If we expand the logarithmic term in powers of (d-e)/(1-d) and keep the first two terms in the expansion, the efficiency becomes

$$\eta \simeq F'\alpha\tau(1 - (d+e)/2) \quad . \tag{32}$$

Thus with a knowledge of d and e one can find the efficiency of a typical flat plate collector.

We would like to compare the performances of several different collectors under identical operating conditions so that d and e are constant for all collectors under consideration. In addition, we will assume the same quality of materials and construction for each collector so that F', α, and τ are the same for each collector. A useful parameter to consider is the ratio of efficiency of a collector to the efficiency of a flat plate collector. Using Eq. 28, one finds

$$\eta/\eta_f = \ln(g_f)/\ln g \quad , \tag{33}$$

where the subscript f refers to a flat plate collector. The parameter, η/η_f, represents an exact figure of merit for collector comparison. Thus one can check to see how well Q predicts collector performance by comparing the values of η/η_f and Q for a group of collectors. The advantage in using Q instead of η/η_f is that Q can be computed much easier than η/η_f.

In comparing various collectors we will define A, the aperture area, to be one for each collector. For simplicity we will consider concentrators with only up to three absorber segments. We will merely assume that the sets $\{b_r\}$ and $\{\sigma_r\}$ are known. The format for presenting examples is: $\{b_r\}/\{\sigma_r\}$. The cases are shown in Table I. The collectors are compared under three ranges of collector operation defined as:

        low temperature -   e = .4,   d = .55
     medium temperature -   e = .5,   d = .65
       high temperature -   e = .6,   d = .75   .

An absorber segment can add heat to the transfer fluid only if $Q_r > e$. Therefore the criterion for keeping an absorber segment in the heat transfer fluid flow path is

$$Q_r > e \quad . \tag{34}$$

An absorber segment that does not satisfy Eq. 34 will not be used in determining Q or η/η_f.

Table 1. Cases under comparison.

Case Number $\{b_r\}/\{\sigma_r\}$
_____

case 1 - $\{.4, .5\}$
                               $\{.383, .383\}$
case 2 - $\{.1, .4, .5\}$
                               $\{.333, .333, .333\}$
case 3 - $\{.2, .2, .6\}$
                               $\{.333, .333, .333\}$
case 4 - $\{.1, .4, .5\}$
                               $\{.234, .383, .383\}$
case 5 - $\{.2, .8\}$
                               $\{.5, .5\}$
case 6 - $\{.2, .3, .5\}$
                               $\{.333, .333, .333\}$
case 7 - $\{.49, .5\}$
                               $\{.333, .5\}$
case 8 - $\{.4, .5\}$
                               $\{.333, .333\}$
case 9 - $\{.2, .2, .6\}$
                               $\{.5, .5, .5\}$
case 10- $\{1\}$
(flat plate) $\{1\}$

    Table II contains W as a function of g for each case where all the absorber
segments have been retained for completeness. Eliminating absorber segments
merely means eliminating corresponding terms in the W function. For each case
the correct W function is inserted in Eq. 23, which is then solved for g. g_f is
determined from Eq. 30. These two values of g are used in Eq. 33 to obtain η/η_f.
Values of Q are simply determined from

$$Q = \sum_r b_r^2/\sigma_r. \tag{35}$$

    Figure 3 illustrates η/η_f vs. d for given values of e for case 1. At low
temperatures, e and d are small and η/η_f falls below unity. The reason why η/η_f
can be less than one is that b<1. For medium and higher temperatures, η/η_f increases sharply due to the increasing importance of concentration at the higher
temperatures.

Table II. W as an explicit function of g for the ten cases listed in Table I.

CASE 1   W = $(1.044\ g^{.383} + 1.305)\ (1 - g^{.383})$
CASE 2   W = $(.3\ g^{.667} + 1.2\ g^{.333} + 1.5)\ (1 - g^{.333})$
CASE 3   W = $(.6\ g^{.667} + .6\ g^{.333} + 1.8)\ (1 - g^{.333})$
CASE 4   W = $.427\ (1 - g^{.234})\ g^{.766} + W$ from case 1
CASE 5   W = $(.4\ g^{.5} + 1.6)\ (1 - g^{.5})$
CASE 6   W = $(.6\ g^{.667} + .9\ g^{.333} + 1.5)\ (1 - g^{.333})$
CASE 7   W = $(1.47\ g^{.333} + 1.5)\ (1 - g^{.333})$
CASE 8   W = $(1.2\ g^{.333} + 1.5)\ (1 - g^{.333})$
CASE 9   W = $(.4\ g + .4\ g^{.5} + 1.2)\ (1 - g^{.5})$
CASE 10 W = $(1 - g)$

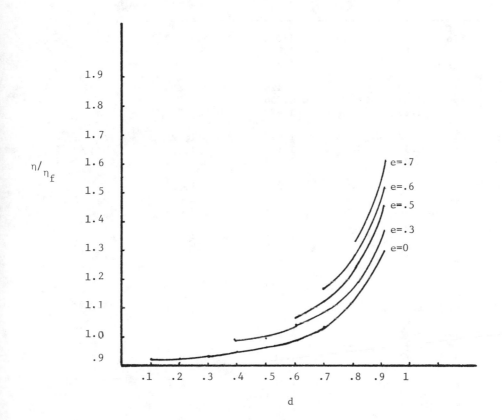

Figure 3. η/η_f versus d for given values of e for case 1, where d is the output temperature to stagnation temperature ratio and e is the input temperature to stagnation temperature ratio.

Table III contains values of η/η_f, Q and the absorber segments that have been retained for each case at the indicated temperature of operation. Q' is the value of Q computed with the more stringent criterion for retaining segments of $Q_r > d$. Q' appears in Table III only when it differs from Q. Notice that in two of the three cases where Q and Q' differ, η/η_f lies between Q and Q'.

Table IV contains the cases ordered according to their relative merit of η/η_f for the medium temperature range. Clearly there is a very strong correlation between Q and η/η_f. In fact, in most cases the value of Q is nearly equal to η/η_f. The ten cases were thought to provide a tough test for Q. Q seems to provide an excellent figure of merit even for the low and high temperature ranges, as shown in Table V. In the high temperature region, where it is clear which absorber segments should be eliminated, the agreement between Q and η/η_f is nearly perfect. In addition, where Q and η/η_f disagree, the differences are not

Table III. η/η_f, Q, Q', and the absorber segments retained for each case at three operating temperatures.

case		Q	Q'	η/η_f	segments retained
1	low	1.07		1.034	1 & 2
	med.	1.07		1.092	"
	high	1.07		1.213	"
2	low	1.23		1.124	2 & 3
	med.	1.23		1.236	"
	high	1.23		1.422	"
3	low	1.32		1.071	1, 2 & 3
	med.	1.32	1.08	1.105	1, 2 & 3
	high	1.08		1.173	2 & 3
4	low	1.113	1.07	1.044	1, 2 & 3
	med.	1.07		1.092	2 & 3
	high	1.07		1.213	2 & 3
5	low	1.28		1.077	2
	med.	1.28		1.216	"
	high	1.28		1.446	"
6	low	1.14		1.053	1, 2 & 3
	med.	1.14	1.02	1.08	1, 2 & 3
	high	1.02		1.133	2 & 3
7	low	1.47		1.29	1 & 2
	med.	1.47		1.44	"
	high	1.47		1.689	"
8	low	1.23		1.124	1 & 2
	med.	1.23		1.236	"
	high	1.23		1.422	"
9	low	.72		.693	3
	med.	.72		.739	3
	high	.72		.817	3
10	low	1		1	1
	med.	1		1	1
	high	1		1	1

significant. In no case is the Q-ordering in major disagreement with ordering in terms of η/η_f .

The advantage of using Q instead of η/η_f as a figure of merit is the ease of computation. Given the sets $\{b_r\}$ and $\{\sigma_r\}$, Q can be computed quickly. The evaluation of η/η_f was lengthy even for only three absorber segments.

A PRACTICAL EXAMPLE

The usefulness of a channel matrix can be seen by examining a design patented

Table IV. Cases ordered according to their relative merit of η/η_f for the medium temperature range.

case	Q	Q'	η/η_f
7	1.47		1.44
2	1.23		1.236
8	1.23		1.236
5	1.28		1.216
3	1.32	1.08	1.105
1	1.07		1.092
4	1.07		1.092
6	1.14	1.02	1.08
10	1		1
9	.72		.739

Table V. Cases ordered according to their relative merit of η/η_f for the low and high temperature ranges.

low temperature

case	Q	Q'	η/η_f
7	1.47		1.29
2	1.23		1.124
8	1.23		1.124
5	1.28		1.077
3	1.32		1.071
6	1.14		1.053
4	1.113	1.07	1.044
1	1.07		1.034
10	1		1
9	.72		.693

high temperature

case	Q	Q'	η/η_f
7	1.47		1.689
5	1.28		1.446
2	1.23		1.422
8	1.23		1.422
1	1.07		1.213
4	1.07		1.213
3	1.08		1.173
6	1.02		1.133
10	1		1
9	.72		.817

by Dr. Robertson.[1] The cross section of the collector is shown in Fig. 4. The semi-circular structure has a reflective inner surface. The absorber segments bisect the semicircle. Each absorber segment (a, b, c, d) is insulated from the other absorber segments. For a sample analysis the input distribution is made discrete by dividing the sky into twenty angular regions. There are twenty

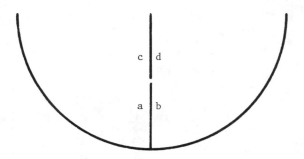

Figure 4. Robertson's moderately concentrating flat plate collector. a, b, c and d label the absorber segments.

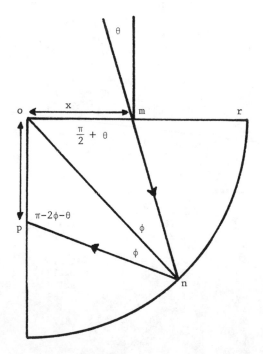

Figure 5. Geometry of a ray undergoing a single reflection before striking the absorber.

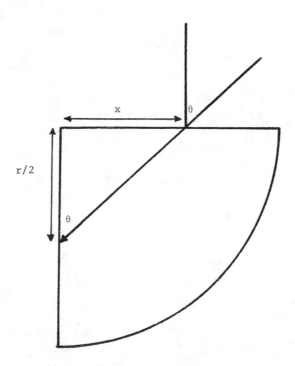

Figure 6.   Geometry of a ray striking the absorber directly.

equal momentum intervals which form the input channel structure. The four absorber segments form the output channel structure. Figure 5 illustrates the
geometry of an incident ray which undergoes a single reflection before striking
the absorber. Figure 6 illustrates the situation for a negative incidence angle.

Table VI contains the input alphabet ($\sin(\theta)$), θ, ϕ and the channel matrix,
M_{ir}, where θ and ϕ are defined in Fig. 5. At any given angle of incidence the
channel matrix contains the probability of a photon striking each absorber segment. The sum of the elements in each row of the channel matrix equals one as
it should. We have neglected all reflection losses. If reflection losses were
included they would not significantly affect the channel matrix since the average
number of reflections is on the order of one. The channel matrix characterizes
the geometry of the collector. Once the channel matrix is computed it can be
used with any input distribution to compute $\{b_r\}$. $\{\sigma_r\}$ is fixed by the geometry
of the collector. Table VII contains the input probability distribution and the
matrix C_{ir} defined by

$$C_{ir} = a_i \, M_{ir} \quad . \tag{36}$$

b_r is obtained from C_{ir} by summing the columns,

Table VI.   Input alphabet and channel matrix for Robertson's collector.

sin(θ)	θ	φ	M_{ir} d	b	c	a
.95	71.81		.5	0	.5	0
.85	58.21	27.4	.435	.065	.4035	.0965
.75	48.59	28.722	.363	.137	.2835	.2165
.65	40.54	29.537	.3245	.1755	.4275	.0725
.55	33.37	29.946	.299	.201	.1645	.3355
.45	26.74	29.943	.2795	.2205	.126	.374
.35	20.49	29.433	.2625	.2375	.0935	.0465
.25	14.48	28.187	.244	.256	.0645	.4355
.15	8.63	25.628	.2185	.2815	.038	.462
.05	2.87	19.47	.167	.333	.0125	.4875
-.05	-2.87	19.47	.0125	.4875	.167	.333
-.15	-8.63	25.628	.038	.462	.2185	.2815
-.25	-14.48	28.187	.0645	.4355	.244	.256
-.35	-20.49	29.433	.0935	.4065	.2625	.2375
-.45	-26.74	29.943	.126	.374	.2795	.2205
-.55	-33.37	29.946	.1645	.3355	.299	.201
-.65	-40.54	29.537	.4275	.0725	.3245	.1755
-.75	-48.59	28.72	.2835	.2165	.363	.137
-.85	-58.21	27.4	.4037	.0965	.435	.065
-.95	-71.81		.5	0	.5	0

Table VII.   Input alphabet and the joint probability matrix for Robertson's collector.

sin(θ)	a_i	$C_{ir} \times 10^{-2}$ d	b	c	a
.95	.01	.5	0	.5	0
.85	.01	.435	.065	.4035	.0965
.75	.01	.363	.137	.2835	.2165
.65	.01	.3245	.1755	.4275	.0725
.55	.01	.299	.201	.1645	.3355
.45	.01	.2795	.2205	.126	.374
.35	.01	.2625	.2375	.0935	.4065
.25	.01	.244	.256	.0645	.4355
.15	.01	.2185	.2815	.038	.462
.05	.01	.167	.333	.0125	.4875
-.05	.81	1.0125	39.4875	13.527	26.973
-.15	.01	.038	.462	.2185	.2815
-.25	.01	.0645	.4355	.244	.256
-.35	.01	.0935	.0465	.2625	.2375
-.45	.01	.126	.374	.2795	.2205
-.55	.01	.1645	.3355	.299	.201
-.65	.01	.4275	.0725	.3245	.1755
-.75	.01	.2835	.2165	.363	.137
-.85	.01	.4035	.0965	.435	.065
-.95	.01	.5	0	.5	0

$$b_r = \sum_i C_{ir} \quad .$$

The input distribution, $\{a_i\}$, in Table VII consists of two parts, the diffuse radiation and the direct radiation. The diffuse radiation is isotropic and contributes .01 to each a_i and the direct component contributes an additional .8 to the input element represented by $-2.87°$. We can alter $\{a_i\}$ to be consistent with any condition of the sky that is desired.

From the assumed input distribution given in Table VII, and Eq. 37 we find

$$\{b_r\} = \{d = .06, \; b = .44, \; c = .1836, \; a = .3164\}. \quad (38)$$

Since the aperture is equal to one,

$$\{\sigma_r\} = \{.25, \; .25, \; .25, \; .25\} \quad . \quad\quad\quad (39)$$

With these values of $\{b_r\}$ and $\{\sigma_r\}$, Eq. 35 yields

$$Q = 1.324$$

or $$Q = 1.31 \quad \text{without the first segment.} \quad\quad (40)$$

CONCLUSION

We believe the figure of merit, $Q = \sum_r \dfrac{b_r^2}{\sigma_r} A$, is a quick and easy method for comparing collectors when a general indication of performance is needed. It is a meaningful definition of the average concentration of radiation on the absorber segments. In addition, it is in excellent agreement with the exact results of collector efficiency. The use of a channel matrix to characterize a collector enables the figure of merit to be readily calculated for any desired input distribution.

ACKNOWLEDGEMENT

We would like to thank Laurence Poteat of the University of Miami for making several useful suggestions.

REFERENCES

1. H.S. Robertson, Proceedings of the Solar Cooling and Heating forum, Dec.13-15, 1976, Miami Beach, Florida (Hemisphere Publishing Corp., Washington, D.C. 1978), Vol. 2, p. 377.
2. W.T. Welford, R. Winston, The Optics of Nonimaging Concentrators (Academic Press, New York, 1978), p. 18.
3. R.P. Patera, Information Theory Applied to Solar Radiation Concentration, (Doctoral Dissertation, University of Miami, 1979), P. 122.
4. Roland Winston, Solar Energy 16, 89 (1974).

Performance Comparison of Flat Plate Collector Absorber Coatings Utilizing NBS Standard 74-635 and the ASHRAE Collector Performance Method

H.A. INGLEY, E.A. FARBER, and R. REINHARDT
University of Florida
Gainesville, Florida 32611, USA

ABSTRACT

In determining the effect of selective surfaces upon flat plate collector efficiency, six surfaces were evaluated. The surfaces investigated were two highly selective surfaces (black chrome and copper oxide), two thought to be mildly selective (Chemglaze Z306 paint and Solarsorb C-1077 paint), along with an industrial flat black paint and an untreated copper panel.

The mechanisms that produce selective surfaces were studied, and the theoretical effect of each surface upon collector efficiency was determined, using both the National Bureau of Standards (NBS) and the American Society of Heating, Refrigeration and Air-Conditioning Engineers (ASHRAE) evaluation procedures.

The results reported in this paper indicate that changes in surface selectivity are detectable by the NBS and ASHRAE evaluation procedures. Two of the major parameters determining the differences in performance are the surface's solar absorptivity, which affects the y-intercept of the performance curve, and infra-red emissivity, which affects the slope of the performance curve. Highly selective surfaces with α_s/ε_i ratios greater than 9.0 were not evaluated. The moderately selective surfaces evaluated (with α_s/ε_i less than 2.0) resulted in efficiencies lower than that of the flat black paint for normal operating temperatures less than 180°F.

INTRODUCTION

The advantages of solar energy utilization are becoming increasingly important as our needs for energy increase while our supplies of conventional fuels dwindle. Although experts disagree as to how long our fossil fuel reserves will last, there is no disagreement with the fact that these energy sources are being rapidly depleted. In the future, therefore, as fossil fuels become more and more difficult to obtain, it will be necessary to convert to alternate energy sources. With solar energy, we have the advantages of an inexhaustible source of supply and minimal adverse environmental effects.

One way to utilize solar energy is with flat plate solar collectors. These are basically insulated boxes having a cover plate that is transparent to solar radiation, and containing an absorber plate through which fluid passes. As the fluid travels from inlet to outlet, heat is transferred to it from the absorber plate. This heat can then be used for a wide variety of purposes, including space heating, domestic hot water, and air conditioning.

To facilitate widespread use of solar energy, efforts are continually being made to increase the efficiency of flat plate collectors without appreciably increasing the cost. This can be done by decreasing any of the three modes of heat loss from the collector to its surroundings: convection, conduction and radiation. As the operating temperature of the collector is increased, the percentage of losses due to radiation is increased. Therefore, any method that would enable us to limit radiation losses from the collector merits investigation.

There are two ways that radiation losses can be appreciably lessened: 1) by decreasing radiation from the cover plate, and 2) by decreasing radiation from the underlying absorber plate. Research is underway in both areas, but in this paper, it is the latter method that is of interest.

The radiation to and from a collector absorber plate each have a characteristic wavelength. The incoming solar (high temperature) radiation is in the 0.0 to 3.0 μ region, while most of the lower temperature radiation from the absorber plate is in the 3.0 to 30 μ, infrared region. Since the function of a collector is to absorb and retain as much as possible of the radiant energy that falls upon it, a good surface for reducing radiation losses in a solar collector would be one which had a high absorptivity in the solar spectrum (α_s) and a low emissivity in the infra-red range (ϵ_i). The surface would then absorb most of the energy falling upon it, but would emit only a very small amount. Surfaces of this type are called selective surfaces. (As a measure of the selectivity of a surface, the quantity α_s/ϵ_i is used.) Surfaces with high α_s/ϵ_i values are the most selective, ratios of 20 and higher being most desirable.

It is well known that glass has selective transmitting properties. The 'greenhouse effect' describes the property that the glass has of transmitting solar wavelengths but not infra-red wavelengths. Therefore, with solar collectors, we are actually working with a composite of selective parameters: The glass cover plate, the absorber coating, and the underlying metallic substrate.

It should be noted that actually all surfaces are selective. The appearance of different colored objects is due to the selective absorption of these surfaces, and even objects that appear black to the human eye can have varying spectral absorptance in wavelength ranges that are beyond the visible.

In order to enhance the absorption of the absorber plate, as well as to protect it from corrosion, the plate has been traditionally coated with black paint. However, black paint normally has both a high value for solar absorptivity α_s and a high infra-red emissivity ϵ_i. Thus, regular black paint with α_s/ϵ_i at approximately 1.0, cannot take advantage of the spectral variations in solar and infra-red radiation.

To be a useful selective surface for flat-plate collectors, the coatings must have a high α_s value as well as a high α_s/ϵ_i ratio. In addition, it must have longterm durability, so that its absorptive properties will not change over a period of time. The factors that tend to rapidly degrade the surfaces (not the collector) are prolonged exposure to high temperature (250-400·F), moisture and atmospheric pollutants, and ultraviolet radiation.

In this study, six surfaces (some more selective than others) were examined. The six surfaces studied were: flat black paint, two paints thought to be mildly selective (Chemglaze Z306 polyurethane and Solarsorb C-1077), and two highly selective surfaces (black chrome and copper oxide), as well as an untreated copper plate that had been exposed in a collector for several months. All of these surfaces are commercially available. The different types of surfaces and the factors involved in determining their spectral properties were investigated and the findings reported. [1] An experimental evaluation of the surfaces was performed, using both the National Bureau of Standards (NBS) Method of Testing for Rating Solar Collectors Based on Thermal Performance, and ASHRAE Standard 93-77 Method of Testing to Determine the Thermal Performance of Solar Collectors.

METHODOLOGY

1. <u>NBS-ASHRAE Performance Evaluation</u>

Until recently, there was no generally accepted procedure for evaluating and comparing the performance of flat plate solar collectors. In the past few years, however, two testing procedures have emerged: the National Bureau of Standards Method of Testing for Rating Solar Collectors Based on Thermal Performance, and the American Society of Heating, Refrigeration and Air Conditioning Engineers Methods of Testing to Determine the Thermal Performance of Solar Collectors. These tests are now being widely used and accepted as an appropriate way to determine and report results on solar collector efficiency.

The theoretical basis of these performance tests is the same, and is based on work done by several investigators [2, 3, 4]. The tests utilize values of instantaneous efficiency over a 15 minute interval. (Although actual steady state does not occur in solar collectors, it has been found adequate [5] to assume steady-state conditions for periods of 15 to 60 minutes.)

The basic parameters involved in collector performance equation are the solar energy absorbed, (q_a), the energy lost to the surroundings, (q_L), and the useful energy collected (q_u), where:

$$q_u = q_a - q_L \tag{1}$$

Therefore, the test requirements involve measuring the rate of incident solar radiation, and the rate of energy transfer to the fluid as it passes through the collector, all under steady state conditions. Let

$$q_a = I_t \, (\tau\alpha)_e A_a \tag{2}$$

where, I_t = total incident solar radiation.
$(\tau\alpha)_e$ = effective solar transmissivity of cover times solar absorptivity of collector plate.
A_a = transparent frontal area for a flat plate collector.

And,

$$q_L = U_L A_a \, (t_p - t_a) \tag{3}$$

where, U_L = solar collector heat transfer loss coefficient
t_p = average temperature of the absorber plate
t_a = ambient air temperature.

Then, equation (1) becomes:

$$\frac{q_u}{A_a} = I_t (\tau\alpha)_e - U_L(t_p - t_a) = \frac{\dot{m}}{A_a} C_p (t_{f,e} - t_{f,i}) \tag{4}$$

where, $\dot{m}$ = mass flow rate of the heat transfer fluid.

C_p = average specific heat of the transfer fluid.

$T_{f,e}$ = temperature of the heat transfer fluid leaving the collector.

$T_{f,i}$ = temperature of the heat transfer fluid entering the collector.

Unfortunately, the average temperature of the absorber plate is not easily determined. The NBS and the ASHRAE procedures resolve this in different ways. The NBS procedure uses the average temperature of the transfer fluid, t_f, as determined by the equation:

$$\bar{t}_f = t_{f,i} + \Delta t_f / 2 \tag{5}$$

where, Δt_f = the change in temperature of the transfer fluid from the inlet to outlet

and inserts a collector efficiency factor, F', into equation (4) such that

$$\frac{q_u}{A_a} = F'\{ I_t(\tau\alpha)_e - U_L(\bar{t}_f - t_a)\} = mC_p(t_{f,e} - t_{f,i}) \tag{6a}$$

The collector efficiency factor, F', accounts for the heat transfer resistance between the absorber plate and the transfer fluid, and is a function of plate geometry and the overall heat transfer coefficient U_L.

The F' for three collector configurations is given in Figure (1). The F' factor for the Roll-Bond absorber plates used in this study being that of case 'c'. [4] As can be seen in the equations, the variables determining F' are the heat transfer coefficient, h, between the tube wall and the transfer fluid, the bond conductance, c, and the fin efficiency. The fin efficiency accounts for the fact that the entire plate is not at the temperature of the fluid-carrying tubes.

The ASHRAE procedure uses the inlet fluid temperature, $t_{f,i}$ and heat removal factor, F_R in equation (1) such that

$$\frac{q_u}{Ag} = F_R\{ I_t(\tau\alpha)_e - U_L(t_{f,i} - t_a)\} = \dot{m}C_p(t_{f,e} - t_{f,i}) \tag{6b}$$

The heat removal factor, F_R, is determined by the equation:

$$F_R = F' \times F'' \tag{7}$$

where, F' = the collector efficiency factor, and

F'' = the flow rate factor.

The flow rate factor, F'', is a common occurrence in heat exchanger calculations, and is determined by the flow rate of the transfer fluid within the collector plate.

The following equation can be used to determine F_R:

$$F_R = \frac{G}{U_L}\{ 1 - \exp(-F'U_L/G)\} \tag{8}$$

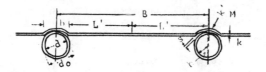

Fig. 1a – Tubes below plate

$$F' = \frac{1}{\dfrac{BU_L}{\pi dh} + \dfrac{BU_L}{c} + \dfrac{B}{b+2L'F}}$$

where

c kb/s, bond conductance

F' $\dfrac{\tan haL'}{aL'}$, fin efficiency

a^2 U_L/kM

Fig. 1b – Tubes above plate

$$F' = \frac{BU_L}{\pi dh} + \frac{1}{\dfrac{do}{B} + \dfrac{1}{(BU_L/c) + (B/2L'F)}} - 1$$

Fig. 1c – Tubes integral with plate

$$F' = \frac{1}{\dfrac{BU_L}{\pi dh} + \dfrac{B}{b+2L'F}}$$

Fig. 1 – Collector efficiency factors for typical absorber plates

where, G = the flow rate per unit area of the collector.
       U_L = the overall heat transfer coefficient.
       F' = the collector efficiency factor.

Equation (6) can be used to determine the efficiency of a collector. The NBS and ASHRAE vary in their definitions of efficiency, the two definitions given as:

$$\eta_{NBS} = \frac{q_u}{A_a I_t} \tag{9a}$$

and

$$\eta_{ASHRAE} = \frac{q_u}{A_g I_t} \tag{9b}$$

where, A_a = the transparent frontal area.
       A_g = the gross collector area.

Using equation (6), the efficiency equation for the NBS procedure is thus:

$$\eta_{NBS} = \dot{m}C_p(t_{f,e} - t_{f,i}) = F'\{(\tau\alpha)_e - \frac{U_L(\bar{t}_f - t_a)}{I_t}\} \tag{10a}$$

and the efficiency for the ASHRAE procedure is determined by:

$$\eta_{ASHRAE} = \dot{m}C_p(t_{f,e} - t_{f,i}) = \frac{F_R A_a}{A_g}\{(\tau\alpha)_e - \frac{U_L(t_{f,i} - t_a)}{I_t}\} \tag{10b}$$

Equations (10 a) and (10 b) are the basis for the method of efficiency presentation in the NBS and ASHRAE procedures. It can be seen that equation (10) is the equation for a straight line of the form:

$$y = b + mx \tag{11}$$

Thus, the data taken during the performance evaluations are presented as on the vertical axis, versus the parameter 'A' on the horizontal axis, where:

$$\eta \text{ NBS} = (\bar{t}_f - t_a)/I_t \tag{12a}$$

$$\eta \text{ ASHRAE} = (t_{f,i} - t_a)/I_t \tag{12b}$$

The resulting points should scatter closely along a straight line.

As stated above, the accepted NBS and ASHRAE evaluation procedures were used in evaluating the effect of the selectivity of the absorber plate upon the efficiency of a flat plate collector. The same collector box was used in each test, and the same type of copper Roll-Bond absorber plate. Therefore, the only change in the collector configuration from test to test was the coating that was applied to the absorber plate. For this reason, it was felt that the results show the influence of coating selectivity upon collector efficiency.

The purpose of the evaluation procedure was to determine 15 minute steady-state efficiencies for varying conditions of solar radiation, I_t, and inlet fluid temperature, $t_{f,i}$.

In order to fulfill the steady state requirements, the tests were run under conditions of constant flow rate of the transfer fluid, constant inlet temperature, and a steady value of insolation (solar radiation). The experimental procedure involves the measurement and recording of:

t_a – ambient temperature
$t_{f,i}$ – transfer fluid inlet temperature
Δt_f – change in fluid temperature from inlet to outlet
I_t – total solar radiation
I_d – diffuse solar radiation
m – flow rate of transfer fluid
V_w – wind velocity

Equations (10 a) and (10 b) were used to plot the collector efficiency η, versus the parameter $t_f - t_a/I_t$. It can be seen then that the only other factors needed are C_p, the specific heat of the transfer fluid at its average temperature inside the collector, and A, the collector area, which can be determined separately after the test has been run.

To fulfill the ASHRAE evaluation requirements, the collector with the absorber plate to be evaluated was pre-conditioned prior to the test. This pre-conditioning involved exposing the collector in a dry, non-operating condition for three days. The cummulative mean solar radiation for the three days must be at least 1500 Btu/ft^2 day.

During the actual test, the collector was rotated to follow the sun in such a way that the incident angle (measured from the normal to the collector surface) was always less than 30·.

A total of 16 data points were taken during the test, including 4 different inlet temperatures, and for each temperature, 4 data points were taken (2 preceding solar noon and 2 after solar noon). The efficiency η vs. $t_f - t_a/I$ curve was determined from these data points, the efficiency being obtained by dividing the useful energy collected, $m C_p(t_{f,e} - t_{f,i})$, the integrated average value of incident solar energy.

2. Coating Specifications

2.1 Black Chrome Coating Specifications

The proprietary black chrome surface evaluated was applied by Olympic Plating Company, using Harshaw Chemical Co. "Chrome Onyx" black chrome electroplate. The process reported by several investigators [6, 7, 8, 9] obtained highly selective black chrome coatings follows this general procedure:
1. A thin layer (.005 inch) of nickel is electroplated first.
2. The black chrome is then plated on at a high current density (200 ampere per ft^2) in a chromic solution bath.

2.2 Copper Oxide

The copper oxide coating used in this investigation is a proprietary coating developed and applied by Enthone, Incorporated.

Copper oxide coatings have been produced [10] by baking absorber panels in Enthone's Ebonal-C solution at elevated temperatures (135·F - 150·F) for approximately 6 minutes.

2.3 Painted Coatings

The paints evaluated were: Benjamin Moore Plat Black, Chemglaze Polyure-thane Z306, and Caldwell Chemical Co. Solarsorb C-1077. The same procedure was used to prepare the three panels to be painted. The copper panels were scrubbed with steel wool to remove corrosion and dirt. The panels were then wiped free of dust and remaining steel wool particles, and then painted.

Literature [11, 12] on the paints believed to be mildly selective (the Chemglaze and Solarsorb paints) both specified coating thickness of approxi-mately 1 mill (0.001 inch). In order to obtain this critical thickness, the coatings were spray-painted, using 3 coats, with a mixture of 4 parts thinner to 1 part paint. Instructions for applying the Solarsorb C-1077 [11] paint noted that a good way to determine when the proper thickness has been reached is to stop applying the paint after the color of the coating changes from brown to black. The procedure was followed for both paints. However, subse-quent measurements showed that:
   1) the thickness of the coating varied considerably from point to point on the absorber plate (0.0003 in. to 0.0008 in. for the Z306 paint, and 0.0004 in. to 0.001 in. for the C-1077 paint), and
   2) at a majority of the points, the thickness was less than 1 mill.

The flat black paint was used as a control. It was meant to represent typical collector coatings presently used, and therefore no effort was made to control the thickness. One coat of the flat black paint was applied with a brush.

2.4. Bare Copper Panel

The untreated Roll-Bond panel was simply installed in the collector and exposed for three months prior to the evaluation test.

3. Spectral Analysis

To determine the solar absorptance, α_s, and infra-red emittance, ε_i, of the coatings, spectral analysis of a sample of each of the coatings was per-formed.

The instruments used to measure α_s and ε_i were the Willey Alpha Meter and the McDonald emissometer.

DISCUSSION

The experimental performance lines in Figures (2) and (3) were each determined by a least-squares method fit of the 16 data points that were

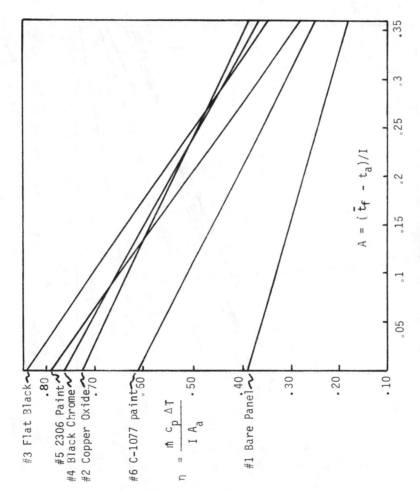

Fig. 2 – Comparison of Surfaces, NBS Procedure

97

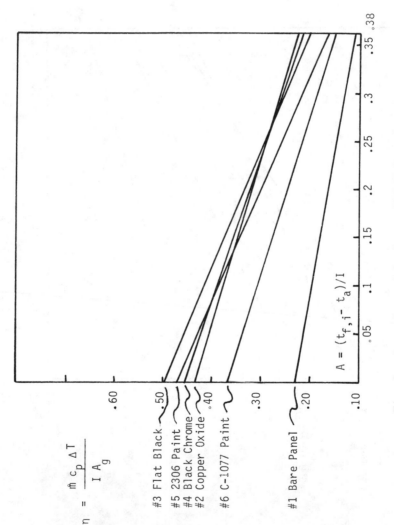

Fig. 3 – Comparison of Surfaces, ASHRAE Procedure

$$\eta = \frac{\dot{m} \, c_p \, \Delta T}{I \, A_g}$$

#3 Flat Black

#5 2306 Paint

#4 Black Chrome

#2 Copper Oxide

#6 C-1077 Paint

#1 Bare Panel

.60

.50

.40

.30

.20

.10

$$A = (t_{f,i} - t_a)/I$$

.05   .1   .15   .2   .25   .3   .35 .38

taken. Table 1 illustrates the linear coefficient of correlation determined
for each plot. The maximum difference between the actual experimental points
and the line drawn to fit through them was +0.05, or 5 absolute percentage
points.   Thus, one performance line is not significantly different from
another unless they are more than 5 percentage points apart. For example, the
three highest experimental curves at A = 0.35 for both the NBS and ASHRAE procedures are less than 5 percentage points apart. Therefore, their differences
at that point are not statistically significant. This should be remembered
when looking at Figures (2) and (3).

The ASHRAE procedure gives lower efficiencies and less inclined slopes
than the NBS procedure, due to their different definitions for efficiency
and the parameter 'A'. For low values of frontal transparent area over gross
collector area, the differences in slopes and efficiencies are quite
pronounced. The NBS procedure is more advantageous when the average fluid
temperature obtainable versus efficiency needs to be known.

TABLE 1
Linear Correlation Coefficient, "r"

Surface	NBS Procedure	ASHRAE Procedure
1	-0.973	-0.974
2	-0.971	-0.972
3	-0.977	-0.979
4	-0.978	-0.981
5	-0.989	-0.990
6	-0.944	-0.946

CONCLUSIONS

The graphs of Figures (2) and (3) illustrate that changes in surface
selectivity of the absorber plate can be detected by the NBS and ASHRAE
collector evaluation procedures. The magnitude of the differences in the
performance curves depends both upon the absorber plate solar absorptivity
(which affects mainly the y-intercept), and infra-red emissivity (which
affects mainly the slope).

It is mainly the infra-red emissivity ε_i of the absorber plate which
determines differences in the slopes of the performance curves. This is
illustrated in the figures, which show cross-over patterns for the performance
lines of the four surfaces with the highest y-intercepts. The two surfaces
with the lowest infra-red emissivity, copper oxide and black chrome, ranked
fourth and third in efficiency at the y-intercept, while at A = .35, they
ranked first and second. This is explained by the fact that the performance
lines of the surfaces with lower ε_i's have slopes that are less steep, causing
them to "cross over" performance lines with slopes of higher magnitude.

None of the surfaces experimentally evaluated showed a significant
increase in performance over the flat black paint in the temperature ranges
studied.   This indicated that moderately selective surfaces (such as the
altered black chrome or the copper oxide with relatively high ε_i) did not
significantly improve the efficiency of the flat plate solar collector
considered in this evaluation.

The untreated copper panel and the C-1077 paint (as applied in this
study) resulted in significantly lower performance than the flat black paint.

This was true over the complete temperature range studied, with inlet temperatures varying from 85·F to 160·F.

REFERENCES

1.   R. Reinhardt, "The Effect of Selective Surfaces Upon The Efficiency of Flat Plate Solar Collectors," unpublished master's thesis, University of Florida, 1978.

2.   H.C. Hottel and B.B. Woertz, "The Performance of Flat-Plate Solar-Heat Collectors," Transactions of ASME, 64, pp 1959.

3.   R. Bliss, "The Derivations of Several 'Plate-Efficiency Factors' Useful in the Design of Flat Plate Solar Heat Collectors," Solar Energy, 3:4, pp. 55-64, Dec. 1955.

4.   A. Whillier, "Design Factors Influencing Solar Collector Performance," Applications of Solar Energy for Heating and Cooling of Buildings - ASHRATE Grp. 170, pp. 8-1 to 8-13.

5.   S.A. Klein, J.A. Duffie, and W.A. Bechman, "Transient Considerations of Flat-Plate Solar Collectors," Transactions of ASME Serial A, 96:2, pp. 109-113, April 1974.

6.   D.M. Mattox, G.J. Kominiak, "Deposition of Semiconductor Films with High Solar Absorptivity," Journal of Vacuum Science Technology, 12:1, pp. 182-185, Jan./Feb. 1975.

7.   S.W. Moore, "Results Obtained from Black Chrome Production Run of Steel Collectors," Proceedings from American Electroplaters' Society, Inc.'s, Coatings for Solar Collectors Symposium, pp. 57-58, Nov. 1976.

8.   M.C. Keeling, R.K. Asher, and R.W. Gurtler, "High Selective Absorbers Utilizing Electrodeposited Black Chrome," Proceedings from American Electroplaters' Society, Inc.'s, Coatings for Solar Collectors Symposium, pp. 57-58, Nov. 1976.

9.   A.C. Benning, "Black Chromium - A Solar Selective Coating," Proceedings from American Electroplaters' Society Inc.'s, Coatings for Solar Collectors Symposium, pp. 57-58, Nov. 1976.

10.   H. Mar, "Research on Optical Coatings for Flat Plate Solar Collectors," Quarterly Summary Report, Contract No. NSF-C957 (AER-74-09104), TD-26840, Dec. 1974 to March 1975.

11.   "Chemglaze (Z-line) Polyurethane Coatings," Technical Bulletin 7037P, Hughson Chemicals, Lord Corporation, Erie, Pennsylvania.

12.   J.A. Ytterhus. "Performance of a Mildly-Selective Coating from the Caldwell Chemical Coatings Corporation," Proceedings from American Electroplaters' Society Coatings for Solar Collectors Symposium, p.43, Nov. 1976.

Performance of a Flat Type Solar Collector Composed of the Selective Transparent and Absorbing Plates

KIMIO KANAYAMA, HIROMU BABA, and HISASHI EBINA
Kitami Institute of Technology
Kitami 090, Japan

ABSTRACT

The analytical and experimental results for the performance of a flat-plate type solar collector composed of a selective transparent plate and a selective absorbing plate are reported. The analysis of thermal property of the solar collector is spectrally made by considering heat balance of thermal process caused by radiation, convection and conduction on both transparnt and absorbing plates. Comparing the analytical result with the experimental result on a glass-covered collector with selective absorbing plate, the good agreement between both was obtained through the climate of four seasons so that the reliability of analytical method was verified. The performance of various kinds of collectors in different combinations with transparent and absorbing plates was checked according to this analytical method using weather data in the winter.

1. INTRODUCTION

To increase the thermal efficiency of a flat-plate type solar collector, frequently the selective characteristic of absorbing plate is made of producing microrough surface or multilayered films on it by means of chemical treating or coating. Combining the absorbing plate with any transparent material like a glass plate of which characteristic is also selective, the performance of such a structure will become more effective due to improvement of the mechanism of radiation heat transfer including less convection heat loss.

In this paper, heat balance on the flat-plate type collector composed of a selective transparent plate and a selective absorbing plate is analyzed, the relation between the optical property and the collector efficiency on various combinations of selctive transparent and absorbing plates is clarified, and then making clear an influence of cut-off wavelength on the collector efficiency when both plates have idealized selectivities the optimum cut-off wavelength is discussed.

To confirm appropriateness of the present method of analysis, a comparison between the calculated and measured results by an experiment on some commercial collector installed on the roof of our laboratory was performed. Besides, the spectral transmittance of various transparent materials used in the analysis and experiment as specimens was measured by ourselves.

2. ANALYTICAL METHOD

According to the experimental results in our laboratory, it was clarified that the efficiency of a double covered collector with two transparent plates was lower than that of a single covered collector except for on unusual weather of which outside temperature is extremely low and wind velocity is excessively speedy, and therefore, only single covered collector is taken into account as the subject for analytical method and experimental specimen in this report. Figure 1 shows the mechnism of heat transfer and relation of heat balance on the collector model composed of a transparent plate and an absorbing plate.

Though temperature of the transparent and absorbing plates makes change with time, on the assumption of that the temperature on each part is nearly constant within a short time, the equations of heat balance on quasi-steady state of both transparent and absorbing plates which interfere in contact with water and air are as follows:

Heat balance equation on the absorbing plate is

$$(Q_{PA1}+Q_{PA2}+Q_{PA3}+Q_{PA4})-(Q_{PCV}+Q_{PR}+Q_{PCO}+Q_{PCD}+Q_{PW}+Q_{PL})=0 \qquad (1)$$

and heat balance equation on the transparent plate is

$$(Q_{GA1}+Q_{GA2}+Q_{GA3}+Q_{GA4}+Q_{PCV})-(Q_{GR}+Q_{GCVO}+Q_{GRO})=0 \qquad (2)$$

From the experimental condition that a volume of water within the storage tank is 55 liters and a flow rate of water is 2.3 liters /min, ten minutes were determined as a unity of time period. With respect to the propagation process of radiation incident within the collector, it is assumed that the radiation beam undergoes by multiple reflections between both infinite parallel plates. When water circulation was stoped, the absorbed energy into water, Q_{PW}, and the equivalent piping heat loss within collector, Q_{PL}, are omitted from Eq.(1).

Procedure of the calculation is as following ways: First, the temperature of a transparent plate, t_g, equals outside temperature, the temperature of an absorbing plate, t_p, equals water temperature at the entrance of collector, and both are assigned as the initial temperatures respectively. Next, considering heat balance between absorbing tube and water on a differential element dx as shown in Fig.2, Eq.(3) yields.

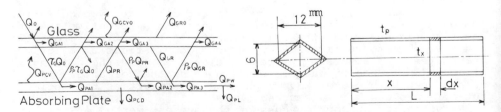

Fig.1 Heat transfer process on the col-
    lector model

Fig.2 Details of the absorbing tube

$$h_i A_{pc} dx (t_p - t_x) = c_w G \, dt_x \tag{3}$$

where, heat transfer coefficient of inside surface of an absorbing tube, h_i, is given by

$$h_i = 1.063 \frac{k}{d} (R_{ed} P_r \frac{d}{x})^{1/3}$$

Applying the boundary conditions that at $x=0$, $t_x=t_{wen}$ and at $x=L$, $t_x=t_{wex}$, under the condition of constant temperature on the inside surface of absorbing tube, t_p, integration of Eq.(3) by means of variable separation produces water temperature of collector exit, t_{wex}, as

$$t_{wex} = t_p - (t_p - t_{wen}) (e^{-1.595 A_{pc} (R_{ed} P_r d)^{1/3} k L^{2/3} / c_w G \, d})$$

Namely, absorbed thermal energy into water, Q_{PW}, is

$$Q_{PW} = c_w G (t_{wex} - t_{wen})$$
$$= c_w G (t_p - t_{wen}) (1 - e^{-1.595 A_{pc} (R_{ed} P_r d)^{1/3} k L^{2/3} / c_w G \, d}) \tag{4}$$

If heat balance is not satisfied when substituting t_{wex} and Q_{PW} into Eqs.(1) and (2), both temperatures of t_p and t_g are gradually increased, and when each of these equations converged smaller than ± 1.0 Kcal/h, the satisfaction of heat balance is attained. Ultimately, the values of t_p, t_g, t_{wex}, Q_{PW}, and Q_{PL} can be determined. In this case, water temperature within a tank, t_{end}, which is also the initial temperature for next step of unit period can be obtained by Eq.(5), in addition to the temperature increase for this period.

$$t_{end} = t_{int} + \frac{c_w G (t_{wex} - t_{wen}) - Q_{WL}}{c_w W} \Delta \tau \tag{5}$$

With determining the value of Q_{PW}, we can get the collector efficiency for one hour, η_{ct}, by Eq.(6) for the running period of a circulation pump.

$$\eta_{ct} = \frac{Q_{PW}}{Q_O} = c_w G (t_{wex} - t_{wen}) / A_O J \tag{6}$$

At the numerical calculation, an element of wavelength increment for the incident, reflected, and reradiated beams is 0.02 microns on the wavelength region of 0.2 ∿ 1.0 microns, and the element is 0.2 microns on the wavelength region of 1.0 ∿ 25 microns.
An incident beam of solar radiation on the inclined surface of collector is defined by distributing the total energy of insolation proportionally to Planck's function for surface temperature of 6000 K, and the absorbed, transparent, and reflected components of incident energy are determined by multiplying these values by optical properties α, τ, and ρ respectively. Reradiated energies from each plate are calculated by Planck's function corresponding to the surface temperature. Where, the directional dependence of optical properties is not accounted for the calculation, and the integration over wavelengths was carried out using Simpson's formula. Heat transfer coefficient around the collector, h_o, was

given by McAdams' equation[1], and thermal conductance of natural convection within a closed rectangular space between transparent and absorbing plates was given as a proportional value to its inclined angle referring the experimental equations[2] for horizontal and vertical cases. Hausen's equation[3] was used to evaluate the heat transfer coefficients for an equivalent piping heat loss in the collector and for a piping heat loss on the equipment.

3. EXPERIMENTAL EQUIPMENT AND PROCEDURE

A schematic diagram of the experimental equipment collecting solar energy is shown in Fig.3. The temperatures of water at entrance and exit of the collector, of water within the storage tank, inside and outside the room were measured. The insolation energies on the four items of three kinds, that is, horizontal total and inclined total incidents (at an elevation angle of 60 deg. and at the same inclined surface as collector), and direct incident have been measued, and these measurements also are continued now. Only wind velocity was obtained from the data at Kitami Weather Station neighboring with our Institute.

Two kinds of the flat-plate type solar collectors were used as specimens for the experiment. One, collector A, is composed of a high tempered glass plate and a slective absorbing plate, and its wavelength dependence is shown in Fig.4. The other, collector B, is composed of a compound structure of acryl and polycarbonate with honeycomb sandwich and a nonselective plate, and its wavelength dependence is shown in Fig.5.

An autorecording spectrometer, SR-3 (JASO) is used to measure the spectral transmittance, a tungsten lamp was provided as a light source for visible and near-infrared regions, and a siliconit lamp was provided for infrared region. The spectral properties of selective transparent materials have been measured at an air conditioning room in which temperature is kept at 20°C and humidity is kept under 50 per cent through the year. Some measurements are shown in Fig.6.

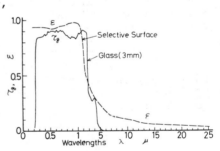

Fig.4 Wavelength dependence of collector A

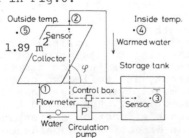

Fig.3 Schematic diagram of experimental equipment for collecting solar energy

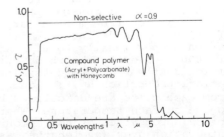

Fig.5 Wavelength dependence of collector B

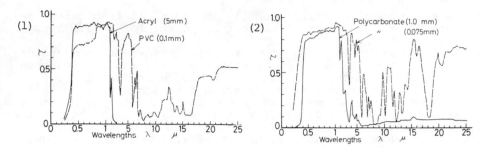

Fig.6 Wavelength dependence of various specimens of polymers

4. RESULTS AND DISCUSSION

According to the NBS display[4], the experimental results on collector efficiency per hour for collectors A and B are shown in Figs.7 and 8 comparing with the calculated results respectively. The experiment of collector was carried out through the four seasons from 1978 to 1979, and it was recognized that the collector performance did not depend on the season. Hence, in Fig.7 for collector A, only the experimental results on five days for the period from August 3rd to 11th are plotted, and the calculated result using weather data on Aug. 10th, '79, is shown together with a nominal performance presented by the maker. In Fig.8 for collector B, the experimental results on 17 days from Apr. to June, '79, are plotted and the calculated result on Jan. 29th is shown. As a comparison between the measured and the calculated values in both figures, one closely coinsides with the other over all temperature difference, Δt, and therefore, the reliability of this analysis was verified.

The calculation of collector performance was made under the following conditions:
1) Energy distribution of insolation is similar to that of Planck's law.

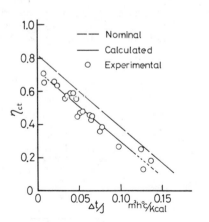

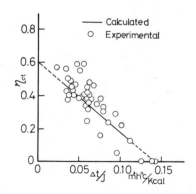

Fig.7 Comparison between the measured and the calculated for performance of collector A

Fig.8 Comparison between the measured and the calculated for performance of collector B

2) The measured value of horizontal hemispherical insolation consists of only the direct incident component for a collector
surface.
3) Sky temperature is equal to surrounding atmosphere temperature.
    With respect to the item 1) of above conditions, a spectrum
of solar radiation on the earth has an irregular distribution
different from that defined by Planck's function caused by that
water vapor, carbon dioxide, and ozone contained in the air absorb
a particular part of solar radiation over infrared region. On the
item 2), according to the measured result of insolation, the
direct component of horizontal hemisphercal insolation is about
90 per cent even if on a clear day, and the residual is scattered
component. Therefore, the insolation must be treated dividing
into the direct incidence and scattered component strictly.
On the item 3), since the transmittance of air in Kitami district
is very high, that is, 0.753 on average of maximum value each
month through one year, it is estimated that sky temperature is
fairly lower than surrounding temperature, and it needs to adjust
the temperature difference. However, it is too difficult to do
the complete analysis considering all these conditions, and thus
using the present analytical mehtod the performance of a collector
composed of the selective transparent plate and absorbing plate,
and the optimum cut-off wavelength are discussed as folowing
descriptions.
    First, supposing several kinds of collectors composed of
some polymer plate and a nonselective absorbing plate, these collector efficiencies are compared in Fig.9. The efficiency for a
glass or compound plate is highest, those for the acryl and polycarbonate covers are middle, thin polycarbonate of which efficiency
is lower than thick one has lower value, and PVC cover has the
lowest value of these materials. Since the thick acryl cover has
short cut-off wavelength of 1.5 ∿ 2.0 microns inspite of high
transparence in the visible and near-infrared regions, it is
estimated the efficiency of acryl cover can not be rised so much
due to absorption and extinction of incident energy at infrared
region. From this fact, the effectiveness of glass cover as a
transparent plate is certainly recognized.
    Next, idealizing the selective characteristics of the trans-
parent and absorbing plates likewise a figure of step, the optimum
cut-off wavelength, λ_c, which gives the maximum efficiency will be
obtained. The calculated results in
which both plates have the idealized
selective characteristics are shown
in Fig.10 indicating initial water
temperature as a parameter. The collector efficiency is so affected by
initial water temperature that the
initial temperature is lower, the
efficiency is higher.
    In all cases, it is sure that
the value of optimum cut-off wavelength,
λ_c, is nearly equal to 4 microns. In the
combination of two plates of which one
has a selectivity and the other has a
non-selectivity, the efficiency is
shown in Fig.11 with the result for
both selective plates. Comparing
the magnitude of the efficiencies

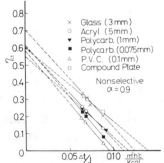

Fig.9 Comparison of collector efficiencies in case of various
transparent materials

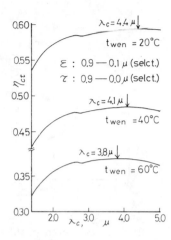

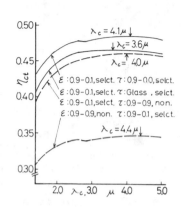

Fig.10 Optimum cut-off wavelength in case that both transparent and absorbing plates are selective

Fig.11 Optimum cut-off wavelength in case that either both plates are selective or one is selective

on the combination of selective and nonselective plates each other, the efficiency of a collector made of selective absorbing plate is much higher than that of a collector made of selective transparent plate. Also in both cases, the optimum cut-off wavelength is nearly same as 4 microns.

From the analytical results of above, it was clarified that the optimum wavelength is nearly equal to 4 microns having no relation to either that both transparent and absorbing plates have selectivities or that only one of these two plates has selectivity. Moreover, a peak of efficiency curve is widely flat around the optimum cut-off wavelength. This means that the curve of selective property may slowly slope centering on the 4 microns, and this also means it is not so difficult to make more effective selectivity for the absorbing structure.

In combination with a selective absorbing palte and a non-selective transparent plate, $\tau=0.9$, the relation between the incline of idealized selectivity around 4 microns and collector efficiency is shown in Fig.12. The wavelength dependence of the idealized selective absorbing plate should be changed like a step slope of No.5 across 3.6 to 4.2 microns.

As an example of the evaluation of performance on the real selective surface, six kinds of selective absorbing plates prepared by NASA[5], Black Chrome, are referred and the wavelength dependences are shown in Fig.13 with a parameter of plating time. The calculated efficiencies of the collector composed of this absorbing plate of Black Chrome and a transparent plate of glass are shown in Fig.14. In this result, the collector efficiency for Black Chrome of plating time of 2 minutes

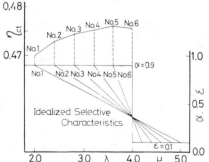

Fig.12 Relation between collector efficiency and inclined angle of the idealized selectivity

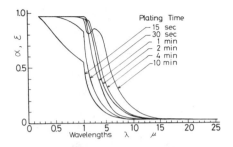

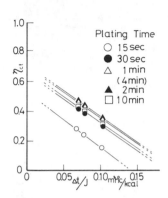

Fig.13 Wavelength dependence of Black Chrome
with the parameter of plating time

Fig.14 Comparison of performance of
Black Chrome with plating time

is highest and that of 15 seconds
is lowest.

Table 1 Comparison of the radiative
properties of Black Chrome
with plating time

On the other hand, the
absorptance α obtained by integrating over shorter wavelength
region, emittance ε obtained by
integrating over longer wavelength region, and its ratio
α/ε are described in Table 1 by
two calculation ways of NASA's
and ours. Judged the performance
of Black Chrome from the only
ratio α/ε as a usual method,
the value of α/ε at plating

Plating Time		15sec	30sec	1 min	2 min	4 min	10 min
NASA's	α	0.64	0.87	0.96	0.96	0.95	0.94
	ε	0.04	0.06	0.10	0.12	0.17	0.34
	α/ε	16.00	14.50	9.60	8.00	5.59	2.76
Authors'	α	0.62	0.81	0.89	0.91	0.91	0.90
	ε	0.04	0.05	0.07	0.09	0.12	0.22
	α/ε	15.50	16.20	12.71	10.11	7.58	4.09

time of 15 or 30 seconds is highest and the value at plating time
of 10 minutes is lowest. Comparing the present method of ours and
usual simple mehtod to judge the performance of selective surface,
the former is more reasonable because the process of heat transfer
is considered by energy balance.

On a view point using the collector for winter, weather conditions on January 29th, 1979, were adopted to calculate the effi-
ciecy after the Fig.9.

5. CONCLUSION

Thermal property of the flat-plate type solar collector
composed of the selective transparent and absorbing plates was
spectrally analyzed. As a result, the calculated values of collector efficiency for a region of usual use agreed exactly with
the measured ones.

Using this analytical method to check the influence of wavelength dependence of various transparent plates on the collector
efficiency, it was clarified that the effect of glass which is
most practical is highest and the effect of PVC is lowest.

Assumed the idealized selectivity on the absorbing plate,
the optimum cut-off wavelength was nearly equal to 4 microns,
so that the cut-off wavelength of 1.5 ∿ 2.5 microns is not suitable for a flat-plate type soalr collector, and the curve of selective property must be shifted to longer wavelengths.

As a subject of future study, the analysis on a cloudy day
will be necessary dividing the insolation into two parts of direct

and scattering components. Moreover, the analysis for the winter
season should be made by considering the effect of reflection from
the snow field[6].

NOMENCLATURE

J : insolation energy on the same inclined surface as collector, $Kcal/m^2h$

Q_O : incident energy upon the collector, $Kcal/h$

$Q_{PA1}, Q_{PA2}, Q_{PA3}$: absorbed energy within absorbing plate, $Kcal/h$

Q_{PCV} : convective heat flow from absorbing to transparent plates, $Kcal/h$

Q_{PR} : radiative heat flow from absorbing to transparent plates, $Kcal/h$

Q_{PCD} : conductive heat loss through bottom of collector, $Kcal/h$

$Q_{GA1}, Q_{GA2}, Q_{GA3}, Q_{GA4}$: absorbed energy within transparent plate, $Kcal/h$

Q_{GR} : radiative heat flow from transparent to absorbing plates, $Kcal/h$

Q_{GCVO}: convective heat loss from transparent plate to air, $Kcal/h$

Q_{GRO} : radiative heat loss from transparent plate to air, $Kcal/h$

Q_{PW} : absorbed energy into water, $Kcal/h$

Q_{PL} : equivalent piping heat loss within the collector, $Kcal/h$

Q_{WL} : heat loss from piping and tank, $Kcal/h$

d : equivalent diameter of absorbing tube, m

x : distance from entrance of absorbing tube, m

L : length of absorbing tube, m

t_p : temperature of absorbing plate, °C

t_g : temperature of transparent plate, °C

t_x : water temperature at point x of absorbing tube, °C

t_{wen} : water temperature at entrance of collector, °C

t_{wex} : water temperature at exit of collector, °C

t_{end} : final water temperature within tank at each unit time, °C

t_{int} : initial water temperature within tank at each unit time, °C

$\Delta\tau$: unit time, h

Δt : difference between average temperature of absorbing plate and outside temperature, deg_2

A_o : surface area of collector, m^2

A_{pc} : cross sectional area of absorbing tube, m^2

h_i : heat transfer coefficient inside absorbing tube, $Kcal/m^2h°C$

k : thermal conductivity of water, $Kcal/mh°C$

c_w : specific heat of water, $Kcal/kg°C$

G : flow rate of water in weight, Kg/h

W : water within tank in weight, Kg

λ : wavelength of radiation, μ

λ_c : cut-off wavelength, μ

η_{ct} : collector efficiency for one hour,

τ : transmittance of transparent plate,

ρ : reflectance of transparent and absorbing plates,

α : absorptance of absorbing plate,

ε : emittance of absorbing plate,

R_{ed} : Reynolds Number based on equivalent diameter of absorbing tube,

P_r : Plandtl Number of water,

REFERENCES

1. Duffie,J.A., and Beckman,W.A.,:" Solar Energy Thermal Process",
   p.83, John Wiley & Sons, (1974).
2. Tanaka, S.,:"Heating and Cooling System by Solar Energy", p.76,
   Ohm Publ. Co., Tokyo, (1977).
3. Edited and Issued by JSME,:"Data Book of Heat Transfer", 3rd
   Revised Edition, p.25, (1975).
4. Hill,J.E., and Streed,E.R.,:"A Method of Testing for Rating
   Solar Collector Based on Thermal Performance", Solar Energy,
   18-5, (1976), p.421.
5. McDonald,G.E., and Curtis,H.B.,:"Variation of Solar-Selective
   Properties of Black Chrome with Plating Time", NASA Technical
   Memorandom, TMX-71731, (1975).
6. Kanayama,K., et al.,:"Discussion on Energy and Shining Time of
   Insolation to Evaluate the Collector Performance", Proceeding
   of 3rd Meeting of JSES, (1977), p.69.

Solar Collector Performance without Flow Measurement

PIO CAETANO LOBO
Universidade Federal da Paraiba
58.000 Joao Pessoa
Paraiba, Brasil

ABSTRACT

A method is described for characterizing solar collector performance in four series of experiments with temperature and radiation measurements. The proposed method eliminates the requirement for mass flow rate meters and is therefore suited to small thermosyphon flow collection circuits. Experimental measurements on a specific system were not reliable because of the occurrence of internal mass transfers between collector and storage reservoir. Suppression of these transfers by insertion of an isolating valve at collector outlet should permit the acquisition of more reliable data.

NOMENCLATURE

A_c collector receiving area

   specific heat of collector fluid at constant pressure

   solar radiation intensity incident on collector receiving

$\dot{m}$ mass flow rate of collector fluid

m_c equivalent mass of abosrber with fluid

$\dot{q}$ heat transfer rate from or to collector

T temperature

U_l overall absorber plate to ambient air heat transfer coefficient

   time

   transmissivity absorbtivity product

SUBSCRIPTS

a   absorber, ambient air
e   fluid entering, effective
f   fluid mean
l   loss
p   absorber plate
s   fluid leaving, stored
u   useful

INTRODUCTION

Solar collector performance may be determined experimentally by measuring either useful thermal output or heat losses. The collector loss coefficient and transmisivity - absorbtivity product may be evaluated from measurements of incident solar radiation and plate temperature in the zero-flow condition. These coefficie can then be used to determine mass flow rates from temperature and radiation data with the collector in normal operation. The method is of particular interest in applications where mass flow rate measurements would not be feasible.

STANDARD TEST PROCEDURE

The standard test procedure involves measurement of incident solar radiation intensity, collecting fluid inlet and outlet tem peratures and pressures and mass flow rate. The incident solar radiation intensity must lie within a prescribed range, fluid inlet temperature controlled to ± 0.5°C and the flow rate maintained constant. The performance test may take weeks to produce enough valid data for an evaluation of the collector.

NO-FLOW TEST PROCEDURE OF MOREHOUSE & VACHON

Morehouse and Vachon [1] proposed the measurement of collector heat loss coefficients instead of thermal output. Collector output or performance is then determined by difference. However, part of their procedure involves measurement of mass flow rates, which may not always be feasible, especially in remote areas and installations not equipped with flowmeters.

The test procedure outlined in this paper proposes the determination of overall collector loss coefficients — U_l — and the effective transmissivity absorbtivity product — $(\tau\alpha)_e$ —in two sets of zero - flow tests, one without and the other with incidence of solar radiation, followed by performance tests in which these values of overall loss coefficient and transmissivity-absorbtivity product are used to determine the mass flow rate.

BASIC EQUATIONS

The energy equation for a solar collector may be written:

$$\dot{q}_u \quad = \quad \dot{q}_a \quad - \quad \dot{q}_s \quad - \quad \dot{q}_l$$

| (useful | = | (energy | − | (stored | − | (thermal |
| energy) | | absorbed) | | energy) | | losses) |

$$\dot{m}c_p(T_s - T_e) = A_c I(\tau\alpha)_e - {}_mc_p \, dT_f/d\Theta - A_c U_l (T_p - T_a)$$

where $\dot{m}$ = mass flow rate of collector fluid

T_s, T_e, T_f = fluid outlet, inlet and mean temperatures respectively

A_c = collector receiving area

I = solar energy intensity on collector receiving area

$(\tau\alpha)_e$ = effective transmissivity - absorbtivity product

${}_mc_p$ = thermal capacity of collector with fluid

Θ = time

U_l = overall plate to ambient air heat transfer coefficient

T_p, T_a = absorber plate and ambient air temperatures respectively.

Since mean and inlet fluid temperatures may be easier to measure than absorber plate temperature, Eq (1) is often rewritten:

$$\frac{\dot{m}c_p}{A_c} (T_s - T_e) = T_f \left[I \, (\tau\alpha)_e - \frac{{}_mc_p}{A_c} \frac{dT_f}{d\Theta} - U_l \, (T_f - T_a) \right] \ldots \quad (1a)$$

or

$$\frac{\dot{m}c_p}{A_c} (T_s - T_e) = F_R \left[I (\tau\alpha)_e - \frac{\dot{m}c_p}{A_c} \frac{dT_f}{d\theta} - U_1 (T_c - T_a) \right] \cdots \text{(1b)}$$

where F', the "efficiency factor" corrects for the difference
       between the plate and the fluid temperature

and F_R, the "heat removal factor" corrects for difference between
       fluid inlet and plate temperature.

In an exhaustive analysis of the Hottel, Whillier and Bliss method
of characterising flat plate collector performance, Duffie   and
Beckman | 2 | present equations and graphs for the determination of
F' and F_R as functions of absorber geometry, overall heat transfer
coefficient, absorber plate to fluid heat transfer coefficient and
fluid thermal capacity.

    In principle, equation (1) or variations (1a) and (1b) permit
the evaluation of U_1, $\dot{m}$, $(\tau\alpha)_e$ and F' or F_R in a series of static
(zero flow) and flow tests with and without solar radiation
incidence:

   A. If there is no flow or solar radiation input ($\dot{m} = 0 = I$),Eq (1)
       becomes

$$U_1 (T_p - T_a) = - \frac{\dot{m}c_p}{A_c} \frac{dT_f}{d\theta} \qquad\qquad (2)$$

So if $(T_f - T_a)$ and $dT_f/d\theta$ are measured ever a convenient   time
period, $\dot{m}c_p/A_c$ being known or measured for a given   collector,
U_1 can be estimated and plotted as a function of $(T_p - T_a)$

   B. With flow through the collector but no solar radiation input
       $\dot{m} \neq 0, I = 0$)

$$\frac{\dot{m}c_p}{A_c} (T_s - T_e) 1 - \frac{\dot{m}c_p}{A_c} \frac{dT_F}{d\theta} - U_1 (T_p - T_a)$$

substituting for U_1 from the plot obtained in the first series of
tests and measuring T_s, T_e, T_p and T_a, $\dot{m}$ can be calculated   and,

in the case of thermosyphon flow, plotted against $T_e - T_s$.

C. In a zero flow test with finite solar radiation input
    ($\dot{m} = 0$, $I \neq 0$)

$$I\,(\tau\alpha)_e = \frac{m_c c_p}{A_c} \frac{dT_f}{d\theta} - U_1\,(T_p - T_a)$$

Substituing U_1 from the plot of U_1 vs $(T_p - T_a)$ and measuring the temperatures and solar radiation intensity I, $(\tau\alpha)_e$ can be obtained, if necessary as a function of angle of incidence.

D. Finally, in a flow test with solar radiation incidence
    ($m \neq 0$, $I \neq 0$) U_1, $(\tau\alpha)_e$ and measured values of I and
    temperatures can be inserted into Eq (1) to obtain $\dot{m}$ in
    normal use. If the system components are symmetrical
    fluidically, ie. have the same resistance in either flow
    direction, the values of $\dot{m}$ should agree with corresponding
    values obtained in tests with no radiation incidence
    excepting a reversal in sign, for thermosyphon flow.

In case zero - flow tests are not feasible with input of solar radiation, then Eq (4) is not available and if we can assume flow symmetry, Eqs (1), (2) and (3) can be solved for each value of $(T_p - T_a)$ to obtain U_1, $\dot{m}$ and $(\tau\alpha)_e$. Since plate temperature T_p is not always convenient to measure, it may sometimes be preferred to use average fluid temperature T_f as a substitute. This approximation is reasonably accurate in flow tests with liquids, but may be inaccurate in non flow situations.

In case Eq 1(b) is preferred, the corresponding relations are:

$$\dot{m} = 0 = I : \quad U_1\,(T_e - T_a) = -\frac{m_c c_p}{A_c} \frac{dT_f}{d\theta} \quad \ldots\ldots\ldots\ldots \quad (2a)$$

$$\dot{m} \neq 0 \quad I = 0: \frac{\dot{m}c_p}{A_c}\,(T_s - T_e) = F_R\left[-\frac{m_c c_p}{A_c} \frac{dT_f}{d\theta} - U_1\,(T_e - T_a)\right] \quad (3a)$$

$$\dot{m} = 0 \quad I \neq 0: I\,(\tau\alpha)_e = \frac{m_c c_p}{A_c} \frac{dT_f}{d\theta} + U_1\,(T_e - T_a) \quad \ldots\ldots\ldots \quad (4a)$$

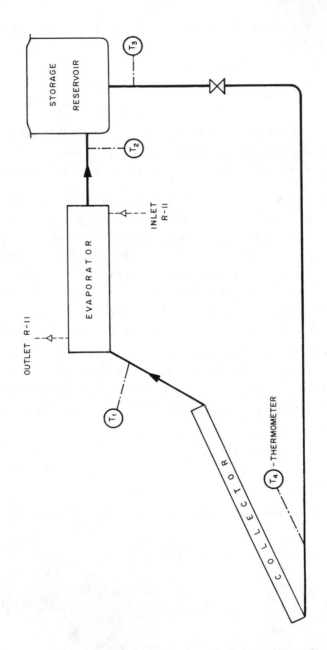

Fig. 1. SKETCH OF SOLAR ENGINE COLLECTOR CIRCUIT.

$$\dot{m} \neq 0 \quad I \neq 0: \quad \frac{\dot{m}c_p}{A_c}(T_s - T_e) = F_R \left| I(\tau\alpha)_e - \frac{m_c c_p}{A_c}\frac{dT_f}{d\Theta} - U_l(T_e - T_a) \right| \quad \ldots \ldots \quad (1b)$$

Equations (1b) to (4a) are solved simultaneously, at similar values of $(T_e - T_a)$ and $(T_s - T_e)$, for U_l, $\dot{m}$, $(\tau\alpha)_e$ and F_R.

SYSTEM ANALYZED

The system analysed is the collector of a research prototype solar engine being constructed at the Solar Energy Laboratory of the Universidade Federal da Paraíba. The water circuit described in Fig. 1 consists of the collector, R-11 vapour generator, a hot water reservoir to act as a thermal capacitor and connecting tubing. Water temperatures are measured at points indicated by T_1, T_2, T_3 and T_4. The collector consists of ten flat plate modules, each with a collector opening of $3m^2$. Each module contains four aluminium roll-bond absorbers painted matt black, 180cm long and 45cm wide with 5cm glass wool insulation on the back and double glazing with spacings of 4cm between cover to cover and cover to absorber. Side walls are insulated with 4cm expanded polystyrene. The collector elements are housed in asbestos cement casings, economical and weather resistant but permeable to moisture, so that a loss in collector performance is observable after a heavy shower until all the water has evaporated and left, through small holes drilled for this purpose in the end walls. A photograph of the collector is shown in Fig. 2.

TEST PROCEDURE

The water circuit is filled with cold water which the collector is allowed to heat during the day. At sunset the control valve between thermal storage reservoir and collector inlet is shut and temperatures are measured every half hour at the four points indicated in Fig. 1 --- collector inlet and outlet manifolds, reservoir inlet and outlet. This corresponds to zero flow and zero solar energy input.

It was not possible to obtain reliable plate temperature measurements, so these values were not recorded. The thermal

Fig. 2 VIEW OF COLLECTOR SYSTEM

capacitance of the collector is calculated from the fluid and
absorber plate masses and specific heats. The thermal capacitance
of other elements includes the containment wall but not the
insulation. Readings are continued until the temperature
difference in the interval falls below the measurement resolution.

In the next test run the collector is allowed to heat water
with the control valve still closed, the same temperature measurements
are repeated and solar radiation income is evaluated for the half
hour intervals. In case of rapid changes in radiation intensity,
shorter intervals may be used. This run corresponds to zero flow
with finite solar energy input. Outlet temperatures above $90^0 C$
should be avoided as vapour formation may introduce large errors
in the equations due to changes in U_l and to neglect of energy
transfer in evaporation.

The third series of measurements are performed with the
control valve open i.e. with reverse thermosyphon flow,   after
sunset and with the circuit in the cooling mode as in the first
test. The same temperature measurements are taken.

In the final series of tests the control valve is open and
the collector operares with incidence of solar radiation, i.e. it
performs in the normal thermosyphon mode. Readings are taken as
for the second test run.

RESULTS AND OBSERVATIONS

Conclusive evidence of the practical value of this test
procedure was not obtained, since it was not possible to secure
zero flow in the collector nor to measure absorber plate temperature.
Mean measured values of fluid temperatures would yield an overall
collector loss coefficient in the range 0.3 to 0.5 W/m^2, or about
10% of the expected figure for this type of collector. It is
therefore likely that the thermal capacitance involved in the heat
transfer is about ten times larger than that of the collector alone,
i.e. mass transfers occur between storge reservoir, evaporator and
collector. Corroborating evidence of these transfers is revealed
in the more or less random variations of temperatures T_1, T_2 and
T_4 observable in Fig. 3, which is an extract of recorder traces
for the case $m \neq 0$ $I \neq 0$. The variations are larger in the zero

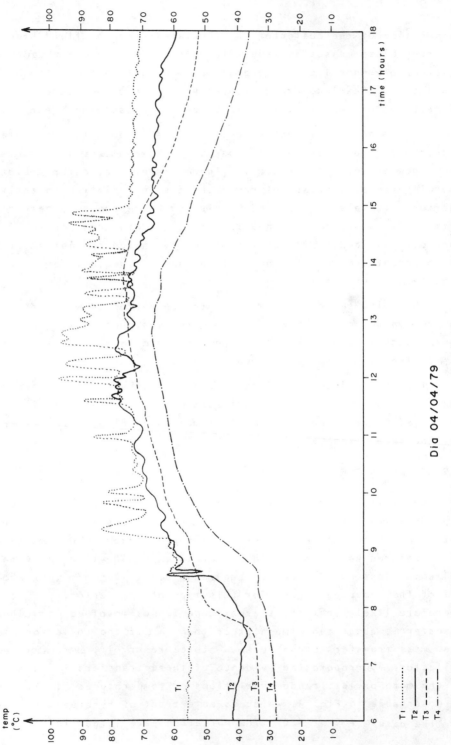

Fig. 3 TEMPERATURE – TIME CURVES FOR COLLECTOR CIRCUIT.

Dia 04/04/79

142

flow casa ($\dot{m}$ = 0).   Installation of a gate valve at collector
exit should suppress or greatly attenuate these mass transfers and
yield more reliable results Statistical analysis of a large number   of
readings taken at short intervals of time could substantially improve
the accuracy of measured values and speed validation of the method.

CONCLUSIONS

A method has been proposed to estimate loss coefficients,
transmissivity-absorbtivity product and mass flow rate from temperature and incident solar radiation intensity readings in four
series of dynamic tests, two without and two with incidence   of
solar radiation. The validity of the method could not be established due to mass transfer between storage reservoir, and
collector during the "zero-flow" tests. Nevertheless the   method
warrants further investigation since it is easily applicable   to
thermo syphon installations, especially in smaller units where
flow measurement is inconvenient during operation. If   Sunshine
data for each site can be interpolated from measured values in the
radiation measuring network, all installations equipped   with
thermometers at collector inlet and outlet and provided with suitably
trained operators to record temperatures at specified   intervals
could be included in a wider evaluation programme to determine
performance of several different collector designs under different
conditions. This aspect of possible applications of the method is
considered of greatest value.

Effects of Dust on the Performance of Thermal and Photovoltaic Flat Plate Collectors in Saudi Arabia: Preliminary Results

BRUCE NIMMO and SEID A.M. SAID
Research Institute
University of Petroleum and Minerals
Dhahran, Saudi Arabia

ABSTRACT

The effect of dust accumulation on the surfaces of flat plate thermal and photovoltaic collectors has been studied and preliminary results are presented. One photovoltaic and two thermal panels were tested and the degradation in performance due to surface dust was determined. The thermal panels showed a decrease in efficiency of about 26 percent as a result of several months of accumulated dust from outdoor exposure. The photovoltaic panel efficiency decreased by 40 percent over a period of about six months.

NOMENCLATURE

α	Absorptivity of absorber surface for thermal panels
Γ	Enhancement factor for evacuated tube collector
d/D	Ratio of absorber tube diameter to tube centerline spacing for evacuated tube collector
F_R	Heat removal factor for thermal panels
I	Total solar radiation in plane of collector (W/M^2)
η	Panel efficiency (output divided by solar radiation input)
η_{clean}	Panel efficiency determined for clean panel (Photovoltaic)
T_i, T_o	Inlet and outlet temperatures for thermal panels (^oC)
τ	Cover glass transmissivity for thermal panels

1. INTRODUCTION

The effect of dust accumulation on the surfaces of flat
plate thermal and photovoltaic collectors has been studied at
the Research Institute/UPM, Dhahran and preliminary results are
now available. The program was originated because there appear
to be substantial differences in the literature on this topic;
major solar projects requiring the information are underway in
the Kingdom of Saudi Arabia; and dust storms are a characteristic
climatological factor over much of the country. Recommendations
in the literature on collector performance dust effects range
from a correction factor of two percent [1,2] to a factor of
thirty percent [3] . The two percent recommendation appears to
be based on the work of Hottel and Woertz [4] in the Boston
area of the United States while the thirty percent degradation
in performance over a period of three days was described by
Sayigh [3] from results obtained in Riyadh, Saudi Arabia.
Garg [5] has measured the normal transmittance of direct radiation through glass and found that over a period of thirty days
the transmittance decreased from ninety percent to thirty
percent for a horizontal mounting. This data was obtained in
Roorkee, India. The performance on actual collectors would
probably not be affected to the same extent as the direct
transmittance because the collector absorber plate would be
expected to capture the forward scattered radiation which would
not be available to the normal incidence pyrheliometer. The
Sayigh and Garg results were obtained during the months of
April, May, and June when dust storms are frequent and rainfall
is minimal (or zero) in these areas. The results of reference
 [3] are somewhat difficult to interpret because the conclusions
are based on an undefined temperature measurement in the collector
while the degradation is described in terms of "heat collected",
also undefined. Results are also presented in [3] for effects
of dust on a photovoltaic array. Here again, however, inter-
pretation is difficult since the dirty and clean tests were
done on different days and it does not appear that isolation
measurements were made to reduce the data to a common base.
We should note that information on changes in power generation
for photovoltaic arrays is particularly important in view of
the planned 350 KW solar village project to be located outside
of Riyadh as part of the Saudi Arabian/U.S. Joint Solar Program
[6].

It appears from the information in the literature that, as
one would suspect, dust effects are site specific and that
additional work in assessing these effects is called for.

2. DESCRIPTION OF EXPERIMENT

Two commercially available thermal solar collector panels
and one photovoltaic panel were tested during the course of the
measurements. The first thermal panel was double glazed, had a
non-selective surface steel absorber and was housed in an insulated fiberglass shell. The absorber plate was formed with
integral headers and parallel risers. This unit had an aperture
area of about 1.1 m^2. The second thermal panel was an evacuated
tube, selective surface, reflector augmented unit with a gross
area of approximately 2.98 m^2. Both panels were tested outdoors

under natural sun conditions using our solar laboratory recirculat-
ing test loop with water as the heat transfer fluid. The test
procedure followed was based on the U.S. National Bureau of
Standards recommendations [7] . Efficiencies of the collectors
under both clean and dirty conditions were determined by dividing
the useful energy collected by the total insolation measured in
the plane of the collector. Temperature rise across the collectors was measured with either a calibrated differential thermo-
pile or thermistors, flow rate was determined using a calibrated
venturi flowmeter with a capacitance differential pressure trans-
ducer, and radiation was measured with an Eppley PSP pyranometer
in the plane of the collector. The photovoltaic panel was a
glass covered, sealed module of 36 silicon quarter circle segments
each with an active area of 11.0 cm^2. The gross area of the panel
was 858 cm^2. The PV panel tests were made at approximately ($\pm$10
minutes) solar noon. Current and voltage values were measured
with calibrated instruments over the range from the open circuit
to short circuit conditions. From this data the peak power output, corrected for ambient temperature variations, was determined
and the panel conversion efficiency was calculated by dividing
this value by the total radiation on the plane of the module.
Both the thermal and the photovoltaic panels were oriented due
south for the tests and tilted at an angle of 26 degrees, the
latitude of Dhahran.

The testing of the thermal collectors was done in two steps.
The panels were cleaned and tested as part of the collector testing program at the Research Institute. Each thermal panel was
then left to accumulate dust over an extended period of time
(about four months in the case of the double glazed collector and
about seven months in the case of the evacuated tube collector).
Although the original plan was to leave the panels for the same
period and then retest, a number of factors precluded this. In
the case of the photovoltaic panel, testing was done with the
panel clean then approximately once a week for a period of about
six months.

3. RESULTS AND DISCUSSION

3.1. Thermal Collectors

The results for the evacuated tube type collector are shown
in Figures 1 and 2. Figure 1 is used to establish the credibility
of the testing procedures and to compare measurements taken on a
clean collector before and after the period of dust accumulation.
All curves represent linear regression best fit of the actual experimental points. The dashed curve presents the results of a
test made in Toledo, Ohio, of an identical collector [8] . The
agreement between this curve and the results obtained in Dhahran
is seen to be quite good and well within the range one expects for
collector performance tests [9] . Lumsdaine [10] has discussed the
differences in collector performance results one might expect even
when testing is done with the commonly recommended procedures.
The ——·—— curve represents results obtained with the original test
at Dhahran of the collector when it was clean and the solid
curve represents the test results obtained when the collector was
again cleaned after the tests with dust accumulated were performed.

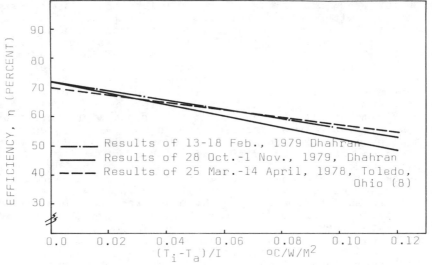

Fig.1 Comparison of thermal efficiency curves for clean
       evacuated tube collector with cylindrical reflec-
       tors.

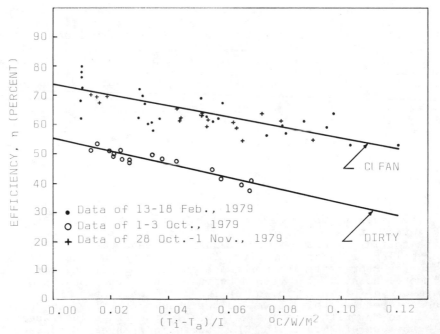

Fig.2 Effect of dust on the thermal performance of an
      evacuated tube collector with cylindrical reflector

Again the results show good agreement and indicate there was negligible change in the collector itself over the seven month period
of exposure.

Figure 2 shows the results obtained in Dhahran for the clean
collector, including both before and after clean test data, and
the collector with seven months dust accumulation. The actual experimental points are shown in this Figure. The intercept,
$\Gamma F_R(\tau\alpha)d/D$, has been reduced from 0.74 to 0.55, a 25 percent reduction. The difference in slope of the two curves is a relatively small 13 percent. Again, this is probably within the testing
accuracy.

Results for the double glazed steel plate collector tests
are shown in Figure 3. Here also a substantial reduction in
efficiency took place as a result of the dust accumulation. The
original intercept, $F_R(\tau\alpha)$, value of 0.51 has been reduced to
0.37, a 27 percent decrease. In both this case and that of the
evacuated tube unit the period of dust accumulation included the
time of heavy sand/dust storm activity typically associated with
late spring and early summer. There was no rain during this
period. However, the average relative humidity ranges from 40 to
65 percent during the months between the clean and dirty tests.
Large daily variations in humidity also commonly occur with a
minimum of perhaps 30 percent and a maximum of over 90 percent.
Thus some amount of dew is assumed to have formed on the collectors and this tends to harden or "cake" on the dust.

3.2. Photovoltaic Panel

Figure 4 shows the results of the measurements on the photovoltaic panel. The ordinate displays the ratio of the dirty panel
efficiency to the initial clean panel efficiency, an approach
which serves to normalize the data and to clearly reveal the percent decrease due to dust accumulation. Unfortunately, to the
author's knowledge, there is no commonly accepted procedure for
the testing of photovoltaic panels comparable to that for thermal
collectors. In the case of our measurements, total radiation in
the horizontal plane was measured a few feet from the panel and
the radiation in the plane of the panel was determined from
calculations as described in reference 1. Errors introduced
by this approach should be minimized by using the ratio of the
dirty to clean efficiencies. Temperature effect compensation
was based on equations supplied by the manufacturer. However,
results were corrected for the ambient temperature referenced
to the original clean panel test ambient temperature rather
than the panel (or cell) temperature and one would expect some
error to be introduced as a result. Panel temperatures were
not measured during most of the tests. Some data was taken
towards the end of the tests, however, to get some idea of the
panel temperatures and their relation to the air temperature. The
panel sensor was a spring loaded "banjo" type surface temperature
thermistor fixed against the middle of the back side of the panel
and in the shade. Air temperatures were measured with a mercury
in glass thermometer, also shaded. As expected the maximum
difference between the two temperatures occurred about midday, the
time when the panel tests were made. The difference was on the

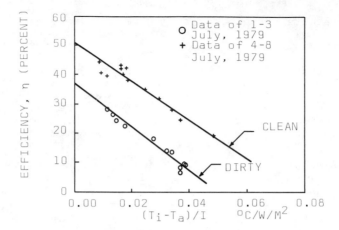

Fig.3   Effect of dust on the thermal
         performance of double glazed
         collector.

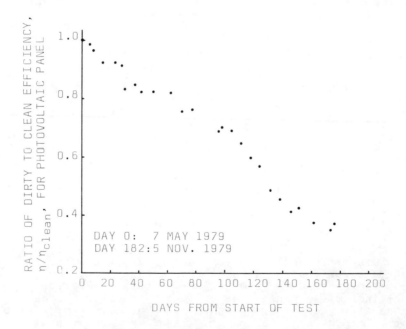

Fig.4   Time dependence of ratio of dirty peak
         power efficiency of photovoltaic panel
         to initial clean efficiency value.

order of 12°C. It is felt that by referencing the panel ambient
test temperature to the original clean panel ambient temperature,
the effect of this 12°C differential will be minimized.

At the end of the test period, the panel was cleaned and
tested. The ratio of this clean test efficiency to the original
clean test efficiency was 1.03, indicating that the degradation
in performance was due to the dust and not a change in the panel
cells themselves. The results of Figure 4 show a more or less
constant decrease in performance of the panel up until the end
of the test where the efficiency ratio had dropped to 40 percent
of its original value.

4. CONCLUSIONS

The present findings lie somewhere between the small corrections usually found in the literature and those suggested more
recently by Garg and Sayigh. Dust conditions in Dhahran are
comparable to those in Riyadh but are aggravated by higher humidity levels which sometimes lead to a hardening of the dust layer
on a cover. The months of testing for the results of reference
[3] and the present work were approximately the same and include
the periods of highest sand storm activity. Continued work in
this area will include air-borne dust sampling, quantitative
measures of dust accumulation and the effect on collector performance, and extension to focusing collectors. The results
indicate that the dust problem is serious. Although no comments
have been made on the means for cleaning collector surfaces, it
is obvious that this topic needs careful consideration. More
specifically, the design of the 350 KW solar village project in
Saudi Arabia must account for the results presented here.

REFERENCES

1. Duffie, J. and Beckman, W., Solar Energy Thermal Processes,
   John Wiley, 1974.

2. Whillier, A., "Design Factors Influencing Solar Collector
   Performance", Chapter III in Low Temperature Engineering
   Application of Solar Energy, ASHRAE 1966.

3. Sayigh, A.A.M., "Effect of Dust on Flat Plate Collectors",
   Proceedings of the International Solar Energy Congress,
   New Delhi, India; Jan. 1978, Vol. II.

4. Hottel, H.C., and Woertz, B.B., "The Performance of Flat Plate
   Solar Heat Collectors", ASME Trans. 64, 1942.

5. Garg, H.P., "Effect of Dirt on Transparent Covers in Flat-
   Plate Solar Energy Collectors", Solar Energy, Vol. 15, No. 4,
   1974.

6. "Solar Villages for Saudi Arabia", Sunworld, Vol. 3, No. 2, 1979.

7. Hill, J.R., and Streed, E.R., "A Method of Testing for Rating Solar Collectors Based on Thermal Performance", Solar Energy, Vol. 18, 1976.

8. Mathur, G.R. Jr., and Beekley, D.C., "Performance of an Evacuated Tubular Collector Using Non-Imaging Reflectors", Owens - Illinois Inc., P.O. Box 1035, Toledo, Ohio, 43666.

9. Streed, E.R. et al., "Results and Analysis of a Round Robin Test Program for Liquid Heating Flat Plate Solar Collectors", Solar Energy, Vol. 22, 1979.

10. Lumsdaine, E., "On the Testing of Solar Collectors to Determine Thermal Performance", Proceedings of 1978 Annual Meeting, Am. Section of ISES, Vol. 2.1, 1978.

ACKNOWLEDGEMENTS

Mr. RATNA NANAYAKKARA was responsible for taking much of the data on the photovoltaic panel.

CONCENTRATING COLLECTORS

Solar Image Characteristics
of Concentrators

PAUL PHILLIPS
I.A. Naman and Associates, Inc.
Houston, Texas 77046, USA

YILDIZ BAYAZITOGLU
Rice University
Houston, Texas 77001, USA

ABSTRACT

A model is developed to study solar image characteristics of solar concentrators. The reflecting and absorbing surfaces of the concentrator are
represented by finite elements. Each element is considered to have nine nodes.
A piecewise ray tracing method is studied such that, instead of representing
the reflected beam from an element with one ray, nine rays are used. At each
node the solar beam is assumed to be reflected with the same solid angle of
the solar disc, which provides a diverging reflected image of the element.
This model accommodates the size of the solar disk; can study complex reflector and absorber geometries and can consider ill-defined or broad incident
fields. The computer implimentation of the present model is used to study a
conical reflector with its base as an absorber. The reflector is chosen such
that a ray reflected from the top edge strikes the outer edge of the absorber
on the opposite side.

1. INTRODUCTION

To analyze the intensity distribution, the incident radiation field and
the reflector geometry must be known. Three dimensional reflectors are usually
designed with surfaces of revolution that can be described with simple mathematical expressions. This makes the analysis of concentrating characteristics
relatively simple. However, in some designs it may be desirable to evaluate
the radiation reflected from the secondary reflector of a geometry which is
not expressible with a simple mathematical expression. Furthermore, effect of
the size of the solar disk can be significant.

1.1 Related Works

Burkhard and Shealy [1] worked with both two and three dimensional concentrators and obtained an equation for the shape of a reflecting surface which
distributes light in a specified manner. This equation was a differential
energy balance equation which was subsequently solved numerically assuming a
uniform distribution. Although this is a relatively simple means of calculating the proper reflector shape, the method becomes complex if the absorber
shape is not simply expressible or the distribution becomes non-uniform.

Wijeysundera [2] examined various three dimensional concentrators for misorientation effects using a finite element technique. Since he did not consider the size of the solar disk, only one node per element was used. The areas of these elements, when projected in the direction of the incident beam, are the same. This makes the method used a true raytracing technique. Though the intensity distribution is not considered, the results are very helpful for examining concentrators that might need piecewise tracing. Evans [3] studied the intensity distribution on a flat absorber in the focal plane of a perfect parabolic trough concentrator. He used an empirical equation for the Sun's intensity distribution and included surface slope errors. The intensity distribution over the absorber of the FMDF (Fixed Mirror/Distributed Focus) concentrator was studied by O'Niel [4]. A closed-form solution based on cone optics was used to find this distribution. The method involves tracing cones instead of rays so the size of the solar disk may be accounted for.

2. ANALYSIS

Although there are several models, they all have several things in common. The most important of these is the basic method for analysis of specular reflections. This method uses simple analytic geometry. Consider three systems as shown in Figure 1. The first, System 1, is arbitrary and will be the reference system. System 2 is defined with respect to one of the reflecting elements such that the z axis is normal to the element plane. System 3 is defined with respect to an absorbing element, and its z axis is normal to the plane of that element. The origins of Systems 2 and 3 are located by the position vectors R_2 and R_3 respectively, which are written in terms of System 1.

Several transformations will be used here. T_{12} transforms a vector of System 1 to one in terms of System 2. The subscripts of the transformations signify the systems considered. Note that $\bar{T}_{21} = \bar{T}_{12}^{-1} = \bar{T}_{12}^{-T}$, since these are orthogonal transformations.

An incident beam strikes the reflector at the point P on the reflector plane. P is located by r in terms of System 2. The reflection of this beam strikes the absorbing plane at Q which is located by s in terms of System 3. The vector s is the only unknown. The incident beam $\tilde{I}_1$ is written in terms of System 1. $\tilde{J}_1$, the reflected beam, may be written

$$\tilde{J}_1 = \bar{T}_{21} \bar{A} \bar{T}_{12} \tilde{I}_1 \quad \text{and} \quad \bar{A} = \begin{bmatrix} 1 & 0 & 0 \\ 0 & 1 & 0 \\ 0 & 0 & -1 \end{bmatrix} \quad (1)$$

where $\tilde{J}_1$ is also in System 1, and $\bar{A}$ is the reflection transformation.

If λ is the distance between P and Q, then simple vector addition yields

$$\tilde{R}_3 + \bar{T}_{31}\tilde{s} = \tilde{R}_2 + \bar{T}_{21}\tilde{r} + \lambda \tilde{J}_1, \text{ or } \tilde{s} = \bar{T}_{13}[(\tilde{R}_2 + \bar{T}_{21}\tilde{r}) - \tilde{R}_3] + \lambda \bar{T}_{13}\tilde{J}_1 \quad (2)$$

Let

$$\underset{\sim}{c} = \bar{T}_{13} [(\underset{\sim}{R}_2 + \bar{T}_{21}\underset{\sim}{r}) - \underset{\sim}{R}_3] \text{ , and } \underset{\sim}{b} = T_{13} \underset{\sim}{J}_1 \tag{3}$$

The above expression can be reformed in terms of $\underset{\sim}{s}$, $\underset{\sim}{c}$, λ, and $\underset{\sim}{b}$ as

$$\underset{\sim}{s} = \underset{\sim}{c} + \lambda \underset{\sim}{b} \tag{4}$$

Because Q lies on the absorbing plane $s_3 = 0$ and $\lambda = - c_3/b_3$, we have

$$s_1 = c_1 - b_1(c_3/b_3), \ s_2 = c_2 - b_2(c_3/b_3) \tag{5}$$

The components of $\underset{\sim}{s}$ may easily be found by these expressions.

Consider now a line passing through P such that J, is included in the line. In all but two cases, this line must intersect the absorbing element. The first of these cases occurs when the planes are parallel. If the angle of incidence is 90°, then the reflection stays on the reflecting plane. Because of this, and if the angle of incidence is 90°, the particular problem is ignored. This case then will not arise. The second case, which is also ignored, occurs when the reflected beam is parallel to the absorbing plane. In this instance the point Q is either non-existent or becomes a line.

In the cases to be considered the line through P must intersect the absorbing element. The case solved above assumed that the intersection was in the positive $\underset{\sim}{J}_1$ direction. Suppose that the intersection occurs in the negative

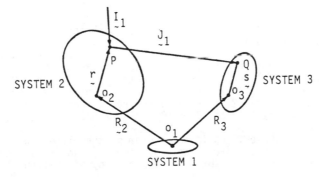

Fig. 1 A schematic model representation of absorber, reflector, and reference systems.

J_1 direction. This case must be automatically disregarded. The problem can-

not be avoided when the model is set up, and it is not a viable solution. Since λ represents the length of the vector $-\lambda J_1$, it must be a positive number.

If Q is reached by traveling in the positive directin, s is found by

$$s = c + (+b) \tag{6}$$

where $\lambda = c_3/b_3$. On the other hand, if Q is reached by traveling in the neg- ative J_1 direction, s is found by

$$s = c + (-b) \tag{7}$$

where $x = c_3/b_3$. In both cases must be positive. If λ is calculated as for the first case, the following will occur; for Q in the positive J_1 direction λ is greater than zero, but for Q in the negative J_1 direction λ will be less than zero. If λ is calculated as a negative number the absorbing plane is behind the reflecting plane and the case will be disregarded.

It should be noted that in some cases the incident beam may be used in- stead of the reflected beam. This would occur if P was being examined for shading. In this case the incident beam is simply retraced. The reflection transformation becomes the retrace transformation.

$\bar{A}$ is written

$$\bar{A} = \begin{bmatrix} -1 & 0 & 0 \\ 0 & -1 & 0 \\ 0 & 0 & -1 \end{bmatrix} \tag{8}$$

Once $\bar{A}$ has been changed, the problem continues as if a reflected ray was being traced.

The above analysis is used with any model that employs planar elements. In all these models it is important to determine whether or not the point of intersection Q falls within the absorbing element being examined. A method for this determination depends on how the absorbing elements are defined. In the models developed here the elements will be defined by four points. The points define four lines which form the sides of the element. Line 1 is de- fined by points 1 and 2. Line 2 is defined by points 3 and 4. Line 3 and 4 are defined by points 2 and 3 and 1 and 4 respectively.

Examining these lines in the system of the element with the x axis hor- izontal, lines 1 and 2 are roughly horizontal, and lines 3 and 4 are roughly vertical. The equations for lines 1 and 2 may be written

$$y_1 = m_1 x_1 + b_1, \; y_2 = m_2 x_2 + b_2 \tag{9}$$

where m_1 and m_2 are the slopes defined as $\Delta y/\Delta x$ and b_1 and b_2 are y axis in- tercepts.

These equations may be slightly modified to get

$$t_1 = m_1x - y + b, \quad t_2 = m_2s - y + b_2 \tag{10}$$

The coordinates of the point Q are applied, and a value for t is easily found. If Q is above a line, t will be negative; if it is below, t will be positive. Of course, if Q is on a line, then t will be zero. So if t_1 and t_2 have different signs, then Q lies between the lines 1 and 2.

Considering lines 3 and 4, m_3 and m_4 are the slopes defined by $\Delta x/\Delta y$ and b_3 and b_4 are the x axis intercepts; changing the method of defining lines 3 and 4 eliminates the confusion caused when m_3 and m_4 have opposite signs. After modifying these equations

$$t_3 = m_3y - x + b_3, \quad t_4 = m_4y - x + b_3 \tag{11}$$

the coordinates of Q may be applied to obtain its position relative to lines 3 and 4.

If Q lies between lines 1 and 2 and also between lines 3 and 4 then it is within the element. Of course, Q is considered inside the element if it lies on any of the sides of the element.

All models using four sided elements use a method similar to this one. If the four points describe an element that does not conform to the pattern described earlier, some rotations within the element system may be necessary. In this case the method becomes much more complex.

Not much can be done to improve the accuracy of planar element assumption for absorbing elements. This assumption can however be improved upon for reflecting elements in axially symmetric reflectors. When the model deals with these three dimensional concentractors, the use of planar elements with constant specular properties becomes inaccurate. Radiation falling normal to the aperture should be reflected radially from all parts of the reflector. A planar reflector element only reflects radially from the azimuthal (with respect to the reference system) center. Only small errors arise because of this when normal radiation is encountered. When the radiation is not normal to the aperture, the amount of error increases.

The specular properties of the reflector elements must be changed according to the azimuthal position. First, consider the element system. The projection of the x axis onto the reference system's horizontal plane points radially outward. The z axis of the element system is normal to the plant of the element. The incident beam I may be written in terms of this system. The rules of specular reflection show that only the z component of I is changed.

These rules cause radiation falling on the azimuthal center of the element to be reflected radially. Radiation falling elsewhere on the element is not reflected properly.

This results from the element system z axis being in the radial direction at the azimuthal center of the element. For proper reflections, the z axis used in specular reflection should be in the radial direction at all positions in the element. Specular properties will remain constant for radial changes in position.

To correct for azimuthal position the element system is rotated such that the z axis is parallel to the reference vertical. The element system z axis has an angle of β_1 with the reference vertical. The second system is rotated by $\pm\gamma_x$ so projection of the x axis on the reference horizontal points in the radial direction. The element has an azimuthal span of 2γ. γ_x is a fraction of γ and will be determined later. This third system is rotated back until the x axis is parallel to the element plane again. Therefore, the entire rotation is

$$\overline{T}^{2} = \begin{bmatrix} \cos\gamma_x\cos\beta_1\cos\beta_2+\sin\beta_1\sin\beta_2 & \sin\gamma_x\cos\beta_2 & \cos\gamma_x\sin\beta_1\cos\beta_2-\cos\beta_1\sin\beta_2 \\ -\sin\gamma_x\cos\beta_1 & \cos\gamma_x & -\sin\gamma_x\sin\beta_1 \\ \cos\gamma_x\cos\beta_1\sin\beta_2-\sin\beta_1\cos\beta_2 & \sin\gamma_x\sin\beta_2 & \cos\gamma_x\sin\beta_1\sin\beta_2+\cos\beta_1\cos\beta_2 \end{bmatrix} \qquad (12)$$

For the x axis of the new system to be parallel to the element plane the 1, 3 element of $\overline{T}$ must be zero.

$$\cos\gamma_x\sin\beta_1\cos\beta_2-\sin\beta_2\cos\beta_1 = 0, \text{ or } \tan\beta_2 = \cos\gamma_x\tan\beta_1 \qquad (13)$$

Notice that the angle will be larger at the center of the element than at the corners. Since the planar element approximation places the center of the element closer to the central vertical axis, the effects of these two errors will cancel each other out.

$\overline{T}$ may be used to interpolate the specular reflection properties of the element depending on where the reflection takes place. To implement this interpolation the incident beam is written in terms of the system after interpolation. The z component is reversed, and the properly reflected beam results. Any other vectors may also be changed into this new system.

2.1 Piecewise Ray Tracing

Ray tracing is a method in which the incident field of radiation is divided into equal parts. Each part is represented by a ray. The reflected rays can then be traced and intersections with the absorber can be found. This method may also be used by dividing the reflector into elements. In order that true ray tracing be established, the area of each element projected in the direction of the radiation must be the same. The division of the reflector into elements then must be done during the execution of the problem. This leads to many difficulties, especially in the computer application of this method. Piecewise ray tracing is a variation of ray tracing. It is not true ray tracing because the elements of the reflector are chosen before the problem is begun. This means that the incident field is not divided into equal parts. The amount of energy assigned to each ray is dependent on the size of the element involved. The problem is simplified because there is no need to separate the reflector into elements during the problem solution. The computer application is more general because it does not require an equation for the shape of the reflector.

One-node model. The piecewise ray tracing one-node model is the simplest possible model. The energy falling on the element under consideration is assigned to a ray which strikes the center of the element. The direction of the reflected ray is found by applying the previously discussed analysis.

Recall that the position of the intersection point on the absorbing plane, s is found by equation (2). For single reflections the origin of the reflecting system is at the node of the element being considered. Since the incident beam strikes there, r = 0 and s is calculated. The position within an element for multiple reflections is known. s will be calculated from the original equation in multiply reflecting cases. Assuming the position of each corner of the elements are known and the transformations into each element system are known, the piecewise ray tracing one-node model can be easily implemented.

The advantage of this model is its simplicity. It may be used very quickly and easily. The concentrator may be divided into a large number of elements to obtain accurate results. The only possible way to account for the solar size is to integrate over the entire solar disk. This integration must be done numerically so the model will have to be executed several times. Each execution will use a different incident field. If the integration is not done, then the sizes of the absorber elements are dictated by the reflector element size and the solar disk size.

Nine-node model. The piecewise ray tracing nine-node model is similar to but more complex than the one-node model. Instead of representing the reflected beam from an element with one ray, nine rays are used. On the original element, each of the nine-nodes serves as an origin. For multiply reflected rays the position with the element is given with respect to the first node. An element will be divided as shown in Figure 2.

Since each node represents a different area, the energy allotted to each node will be different. One quarter of the energy falling on the element is allotted to node 9; one eighth of the energy falls on nodes 5, 6, 7, and 8. The remaining one quarter of the energy is divided among the four corners.

The solar disk is easily accounted for by adjusting the angle of incidence at nodes 1 to 8. The angle of incidence is adjusted to give the largest amount of divergence of reflected energy. Each incident beam is adjusted by the amount of the solar half angle towards the center node, node 9. The resulting reflected beams will diverge away from the node 9.

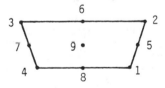

Fig. 2 A schematic shape of the Nine-Node element.

Assuming the position of the corners of all the elements are known along with the proper transformations, an outline for implementation of this model would be as follows:

I.   Reflected Energy

Repeat for each reflector element.
A.   Preliminary
Find the amount of energy that should fall on the element.
Calculate the position vector of each node.
Correct the incident beam for each node.

B.   Shading
For each node, N, of the reflector element, retrace the incident beam to find $\underset{\sim}{s}$.

$$\underset{\sim}{s} = \underset{\sim}{c} + \lambda\underset{\sim}{b}, \; \underset{\sim}{c} = \bar{T}_{13}(R_{\underset{\sim}{2}}^N - R_{\underset{\sim}{3}}), \; \underset{\sim}{b} = -\bar{T}_{13} \, \underset{\sim}{I}_1 \tag{14}$$

i.   Absorber
Examine the absorber elements for an intersection. R_3 refers to corner number 1 of an absorber element, and $\bar{T}_{13}$ is the transformation into the system of that element. If an intersection is found, then the node under consideration is shaded, and it should be disregarded. The energy given to its incident beam is lost.
If, after examining all nine nodes, each one is shaded then the entire element is shaded, and a new element should be examined. (Return to A.)

ii. Reflector
Examine the reflector elements for an intersection. R_3 refers to corner 1 of a reflector element and $\bar{T}_{13}$ is the transformation into the system of that element.

If an intersection is found, then the node under consideration is shaded and should be exempt from further calculations. Again if all nine nodes are shaded, the element should be disregarded, and a new element should be examined.

C.   Single Reflections
For each unshaded node, N, of the reflector element trace the reflected beam to find the point of intersection with the plane of another element $\underset{\sim}{s}$.

$$\underset{\sim}{s} = \underset{\sim}{c} + \lambda\underset{\sim}{b}, \; \underset{\sim}{c} = \bar{T}_{13} \, (R_{\underset{\sim}{2}}^N - R_{\underset{\sim}{3}}), \; \underset{\sim}{b} = \bar{T}_{13} \, \bar{T}_{21} \bar{A}\bar{T}_{12} \underset{\sim}{I}_1 \tag{15}$$

i. Absorber
    Examine each absorber element for an intersection. R_3 refers
    to the corner of an absorber element, and $\bar{T}_{13}$ also refers to
    that element.
    If one intersection is found, the energy from that node is
    added to the amount of energy falling on the absorber
    element.
    If more than one intersection is found, the length of the
    reflected beam, λ, must be examined. The intersection with
    the smallest λ will be counted, and the energy of the re-
    flected beam will fall on the intersected element. If no
    intersection is found, then continue.

ii. Reflector
    If a ray has not intersected an absorber element, examine
    all reflector elements for an intersection. R_3 refers to
    corner 1 or node 1 of a reflector element, and $\bar{T}_{13}$ refers
    to that element.
    The intersection of shortest length should be chosen if more
    than one intersection is found.
    The value of s and b should be retained for the chosen element
    so they may be used to calculate multiple reflections.
    If no intersection is found, the reflected beam passes
    out of the concentrator, and the node should be disregarded
    since the reflected energy is lost.

D. Multiple Reflections
   For each node, N, trace the multiply reflected beam starting with
   the second reflection. Find the point $\underset{\sim}{s}$ by

$$\underset{\sim}{s} = \underset{\sim}{c} + \lambda\underset{\sim}{b}, \quad \underset{\sim}{c} = \bar{T}_{13} \ [(R_{\underset{\sim}{2}}^{N^1} + \bar{T}_{21}\underset{\sim}{r}) - R_3], \quad \underset{\sim}{b} = \bar{T}_{13}\bar{T}_{21}\bar{A}\bar{T}_{12}\underset{\sim}{I}^1 \qquad (16)$$

System 2 in this case refers to an intersected element which is
found from a previous calculation. $R_{\underset{\sim}{2}}^{N^1}$ is the position of corner
1 of a multiply reflecting element. The superscript N shows the
number of the node on the original element whose ray is being
traced. $\bar{T}_{13}$ refers to this same element. I^1 and $\underset{\sim}{r}$ are actually $\underset{\sim}{b}$
and $\underset{\sim}{s}$ determined in the previous calculations.

1. Absorber
   The absorber is inspected for any intersection by the multiply
   reflected beam. R_3 and $\bar{T}_{13}$ refer to an absorber element.
   If an intersection is found, the energy of the beam is added
   to the amount of energy falling on the intersected element.
   If more than one intersection is found, then the intersection
   with the smallest is chosen.
   If no intersection is found, then continue.

ii. Reflector
    The reflector elements are inspected for an intersection. R_3
    and $\bar{T}_{13}$ refer to a reflector element.
    If an intersection is found, an additional reflection is in-
    dicated. The amount of energy left in the ray after imperfect

reflections must be examined. If it is too low, the energy
can be assumed lost. Again the intersection with the smallest
λ is chosen if more than one intersection is found.
If no intersection is found, the ray is lost, and the node
may be disregarded.

II. Direct Energy

   Repeat for each absorber element.
   A. Preliminary
       Calculate the amount of energy that should fall on the element.
       Find the position vectors of each of the nodes in the same manner
       they were found for reflector elements.

   B. Shading
       Repeat for each node, N, of the reflector element.
       Find the position of intersection of the retraced incident ray
       with the plane of another element, $\underset{\sim}{s}$.

$$\underset{\sim}{s} = \underset{\sim}{c} + \lambda \underset{\sim}{b}, \quad \underset{\sim}{c} = \bar{T}_{13} (R^N_{\underset{\sim}{2}abs} + \underset{\sim}{R}_3), \quad \underset{\sim}{b} = -\bar{T}_{13}\underset{\sim}{I}_1 \tag{17}$$

   i. Absorber
       Examine the absorber elements for an intersection. $\underset{\sim}{R}_3$ and
       $\bar{T}_{13}$ refer to an absorber element.
       If an intersection is found, then the node is shaded. No
       energy falls on the node under consideration.
       If no intersection is found, then continue.

   ii. Reflector
       Examine the reflector elements for an intersection. $\underset{\sim}{R}_3$ and
       T_{13} refer to a reflector element.
       If an intersection is found, then the node is shaded, and no
       direct energy hits it.
       If no intersection is found, then the node is not shaded, and
       direct energy does strike it.

   Recall that for any reflected energy the reflecting system must be interpolated according to the position of the reflection point. Of course
the value of X mentioned previously will be -1 or 1 at nodes 1, 2, 3, 4, 5,
and 7 and 0 at nodes 6, 8, and 9. X must be calculated for multiple reflections.

   The solar disk size may be easily accounted for in this model which is its
main advantage. The model is, however, significantly more complex than the
single node approach if the solar disk is considered to be infinitely small.
Using this model, broad fields of incident radiation may be considered.

2.2 The Incident Ray Correction For the Finite Size Of the Sun

   The correction of the incident beam for size of the solar disk is an integral part of the previous outline. This correction is accomplished through
vector analysis.

Two coordinate systems are important in this case. The element system is the first, and the system with its z axis pointing back along the incident ray is the second. Figure 3 shows the latter system. The vector A_1 extends to the center node of the element. A_2 is the projection of this vector onto x-y plane of the system shown. The unit vector $\underset{\sim}{a}$ then points toward the center node while remaining in the x-y plane of this system. The unit vector $\underset{\sim}{b}$ is opposite the direction of the corrected beam. This beam, then could be written

$$
\underset{\sim}{I}_c = \begin{Bmatrix} -a_1 \sin \alpha \\ -a_2 \sin \alpha \\ -\cos \alpha \end{Bmatrix} \tag{18}
$$

where α is the angular half size of the solar disk.

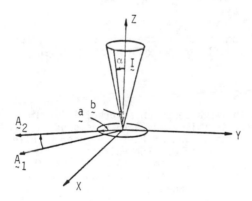

Fig. 3 A schematic diagram of system used in correcting Incident Beam: Z-axis alligned with original Incident Beam.

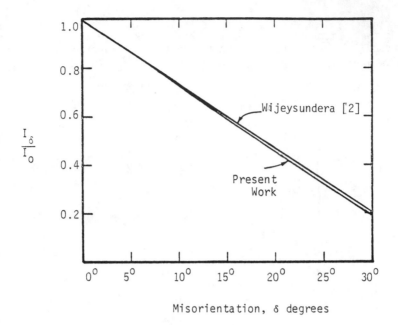

Fig. 4   A graphic comparison of the effects of misorientation
for Conical Reflector with Base as Absorber.

3.   RESULTS AND DISCUSSIONS

A conical reflector with its base as an absorber is studied.   Figure 4
compares the results obtained in this study with Wijeysundera's [2] work.   The
effects of misorientation on the amount of reflected energy collected, using a
collimated incident field, are compared.   The reflectors were ideal and the
size of the solar disk is not accounted for. I_0 and I_δ are the reflected
energy collected without misorientation and with misorientation respectively.
When misorientation is increased to exceed 10 degrees, there is a slight difference between the results of the two works.   However, the difference is very
small and indicates that the two studies compare well.

The surface error characteristics of a conical reflector are also studied.
The introduced errors were in the form of sinusoidal curves as shown in Figure
5.   For an absorber of radius, R, reflector errors of wave length 2R, R, and
approximately 0.6 R were examined.   The most obvious difference between ideal
collectors and the ones with surface errors is that the amount of collected
energy decreases with errors.

The concentration profile of an ideal collector is shown in Figure 6.
This profile is obtained for a collimated incident field normal to the conical
reflector aperture and it compares well with the closed form solution.   When
errors are introduced the most noticeable change in the profile is in the
center of the absorber.   Cencentration gradients introduce a peak which is
sharper and closer to the center when the surface errors are present.   As
the number of cycles increases concentration falls below 2.5 at the outer
edge of the absorber.   It is evident that the energy normally striking these

outer absorber areas is either diverted to the inner rings or is reflected
back out of the absorber.

   Note that, as the number of cycles increases, the shape of the reflector
is a closer approximation to the ideal reflector. The concentration profile
is also a close approximation of the ideal profile, except the concentration
levels are lowered.

   The concentration profiles of an ideal reflector under misorientation are
shown in Figure 7. After 10 degrees of misorientation the profile begins to
become incoherent. Note that the misorientation and also the surface errors
make the area of high concentration smaller. As the misorientation increases
the concentration gradient increases in the direction of misorientation and
because of this severe temperature gradients may result. Figure 8 shows the
concentration profiles with no misorientation for the two types of surface
errors. Note that only the orientation of the sinusoidal error is involved.
The two profiles are very similar. In one case the concentration spike at
the center is sharper, but in general, the effects of the orientation of
sinusoidal error on concentration profiles are minimal.

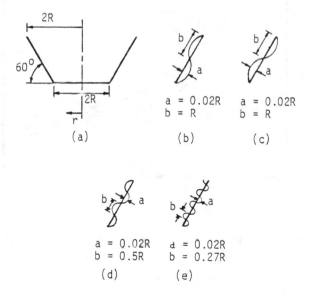

Fig. 5 Sinusoidal surface distortions shown on Conical Reflector
        with Base as Absorber; (b) Distortion 1A, (c) Distortion
        1B, (d) Distortion 2A, (e) Distortion 3A.

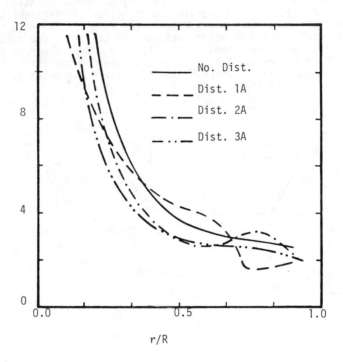

Fig. 6 Concentration distribution for Conical Reflector
       with Base as Absorber.

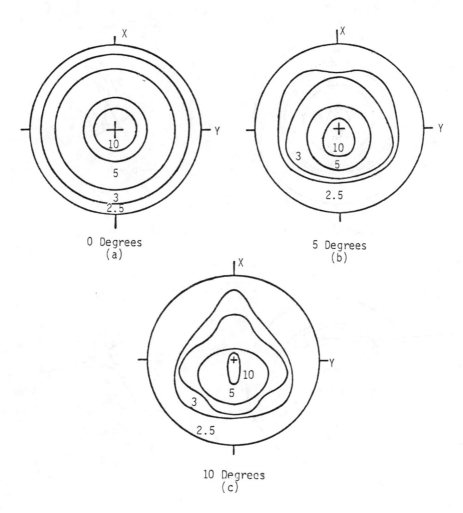

Fig. 7 Effects of misorientation on concentration distribution for an ideal Conical Reflector with its Base as an Absorber.

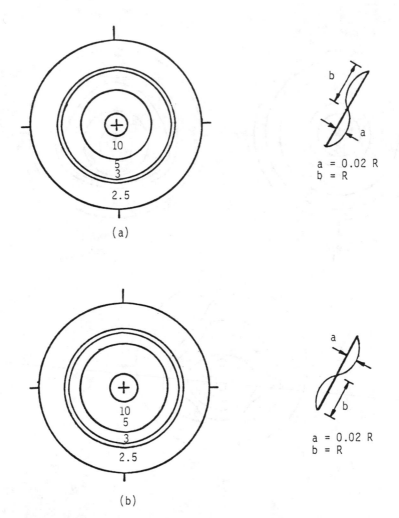

Fig. 8   A graphic comparison of the effects of two symmetrical
         sinusoidal surface distortions on concentration distri-
         bution.

4.   CONCLUSIONS

The computer implimentation of the present ray tracing nine node model is used to study a conical reflector with its base as an absorber. The reflector is chosen such that a ray reflected from the top edge strikes the outer edge of absorber on the opposite side. For the purpose of demonstration a collimated incident field is considered.

The effect of the finite size of the sun for the collimated incident field without any misorientation slightly rounds of the local concentration ratio peak that exists at the middle of the circular absorber.

The results obtained to study the effect of angular misorientation are compared well with Wijeysundera [2]. Since the model can accommodate the actual shape of the reflector and the absorber, the reflector surface errors are studied. A sinusoidal variation of the reflector surface errors of one, two and four cycles are considered. The most obvious difference between the characteristics of an ideal reflector and the reflectors with errors is that, the amount of energy collected decreased. The area of high concentration at the center of the absorber is distorted.

Since the larger number of cycles has closer behaviour to the ideal concentrator, the profile is more like the profile given by the ideal concentrator. If the amplitude of one of these sinusoidal surface errors is increased, the distortion in the profile shape is also increased.

The intensity profile of the ideal reflector remains fairly coherent when angle of misorientation is small. As the angle of misorientation past ten degrees, the profile begins to lose its coherency.

The size of the element is an important consideration. Although this factor is not critical to the actual operation of the program, the accuracy of the results depends on the sizes of absorber elements relative to those of the reflector. Since this method is a ray tracing analysis, considerable error can be incurred simply by the improper choice of absorber elements. The angular size of the sun is compensated for in the program, so no adjustment is necessary on that account. Each reflector node covers the area of approximately half the element area. The minimum size of the absorber elements should be half the size of the reflector elements. The maximum size is arbitrary and depends on the individual application. Note, however, that if the minimum size of the absorber elements is chosen, considerable error may still occur. For reasons of accuracy the absorber elements should be about the same size as the reflector elements. This restriction is relaxed somewhat for axis symmetric concentrators. If any error is incurred in the azimuthal direction, it will be compensated for by the other elements.

ACKNOWLEDGEMENTS

The authors would like to acknowledge the support of Mechanical Engineering and Materials Science Department of Rice University and I. A. Naman + Associates, Inc.

REFERENCES

1. Burkhard, D. G. and Shealy, D. L., "Design of Reflectors which
    will Distribute Sunlight in a Specified Manner", Solar Energy,
    Vol. 17 (1975), pp. 221-227.
2. Wijeysundera, N. E., "Effects of Angular Misorientation on the
    Performance of Conical, Spherical and Parabolic Solar Concentrator",
    Solar Energy, Vol. 19, (1977), pp. 583-588.
3. Evans, D. L., "An Abstract on the Ultimate Performance of Cylindrical
    Parabolic Concentrators with Flat Absorber", Proc. of Am. Sec. of
    ISES and SESC, Winnipeg, Manitoba, Vol. 2, (1976), pp. 253-263.
4. O'Neil, M. T., "Optical Analysis of the Fixed Mirror/Distributed
    Focus (FMDF) Solar Energy Collector", Proc. of Am. Sec. of ISES,
    Orlando, Florida, Vol. I, (1977), pp. (35-4) - (35-28).
5. Goodman, N. B., Rabl, A. and Winston, R., "Optical and Thermal
    Design Considerations for Ideal Light Collectors", Proc. of Am.
    Sec. of ISES and SESC, Winnipeg, Manitoba, Vol. 2, (1976),
    pp. 336-350.

Analysis of a High-Performance Tubular Solar Collector

F.L. LANSING and C.S. YUNG
Jet Propulsion Laboratory
Pasadena, California 91103, USA

ABSTRACT

This article analyzes the thermal performance of the new General Electric vacuum tube solar collector. The assumptions and mathematical modeling are presented. The problem is reduced to the formulation of two simultaneous linear differential equations characterizing the collector thermal behavior. After applying the boundary conditions, a general solution is obtained which is found similar to the general Hottel, Whillier and Bliss form, but with a complex flow factor. The details of the two-dimensional thermal model of the solar collector at steady state is also presented to include the computer simulation and the performance parameterization. Comparison of the simulated performance with the manufacturer's test data showed good agreement at wide ranges of operating conditions. The effects of nine major design and performance variables on the performance sensitivity were presented. The results of this parameterization study were supportive in detecting the areas of design modifications for future performance optimization and improvement.

NOMENCLATURE

A	thermodynamic availability		I	solar flux, W/m^2
$a, \bar{a}$	glass absorption coefficient		K	thermal conductivity, W/m^oC
b	"absorber" reflection coefficient			
$B_o\text{-}B_8$	thermal conductance, $W/m^{2o}C$		L	collector-unit length, m
C	specific heat, W/kg^oC		M	fluid mass flow rate, kg/h
C_o, C_1	constants		N	number of collector units per module
D	diameter			
$E_1\text{-}E_6$	energy flux, W/m^2		n	characteristic constant m^{-1}
F	flow factor		Q	heat rate, W
G	heat capacity $= MC_w/D_{f,o}$			
H	convective heat transfer coefficient, $W/m^{2o}C$			

R   equivalent radiation heat transfer coefficient,
            $W/m^2 °C$

r   glass reflection coefficient

S   spacing between 2 consecutive collector units, m

T   temperature, K

t   thickness, m

x   distance, m

W   wind speed, m/s

α   absorptivity

ρ   reflectivity

τ   transmissivity

λ   augmented radiation factor

η   collector efficiency

δ   parameter, $°C/m$

ε   emissivity

μ   viscosity

φ   extinction coefficient

Subscripts

A   ambient air   v   V-shape reflector

a   "absorber" metallic shell   w   working fluid (hot
                                                          or cold)
c   cold fluid

e   effective

f   first (outer) glass tube

h   hot fluid

i   inside

o   outside

s   second (inner) glass tube

t   serpentine tube

u   insulation

INTRODUCTION

Several solar-powered heating and cooling facility modifications in the Deep Space Network ground stations are planned for future implementation as part of the DSN Energy Conservation Project. In order to support the relevant feasibility and advanced engineering studies, special attention is given to new technologies in low-concentration, nontracking solar collectors. These non-imaging low-concentration types (with intensity concentration between 1 and 5) possess several advantages compared to the high-temperature, high-concentration ones. Examples are:

(1) The ability to harness diffuse and direct portions of sunlight.

(2) Low cost due to less precision requirements in manufacturing, no sun-tracking mechanisms and no sophisticated optics controls.

(3) Good collection efficiency in the range of heating/cooling interest from 80 to 140°C.

One of the new designs that emerged in this field is the tubular and evacuated collector manufactured by General Electric. The collector resembles, but is not identical to, a hybrid system combining: (1) the serpentine tube on a flat absorber enclosed in an evacuated glass cylinder manufactured by Corning Glass Works, and (2) the all-glass concentric tubes manufactured by Owens-Illinois (Refs. 1-3). The semiproprietary GE collector performance data given by the manufacturer (Refs. 4,5) claim that the collector is able to provide about double the energy collection capability of flat plate collectors employing double glazing and selective coating. Although the GE collector does possess the best of each of the Owens-Illinois and the Corning types, it is suggested that its performance superiority be investigated at a wide range of operating conditions. This article is intended to provide the details of collector thermal analysis, the relevant equations needed for a full para-meterization study and the results of performance sensitivity.

Only few experimental data were supplied by the manufacturer (Refs. 4 and 5), and therefore the results of the performance simulation will be compared against these test data only. Some unknown material properties, physical di-mensions and boundary conditions were assumed in this work to complete the modeling process as will be described later. Appendixes A and B give the details of the optical and thermal governing equations for the components while Appendices C and D give the computational sequence of the computer program written for the performance simulation.

COLLECTOR DESCRIPTION

Two versions of the collector design have been manufactured by G.E. Both versions have the same basic features, with the exception of a few differences explained as follows. The collector module consists of a number of heat collection units: 10 in the first version and 8 in the second. The units are mounted in parallel with a highly reflective back reflector. The back re-flector is a V-shaped surface in the first version as shown in Fig. 1. In the second version, the back reflector is a double cusp (parabolic shape) as shown also in Fig. 1b. Each unit contains a U-shaped copper tube and the tubes of the units are connected in series to form a serpentine. Each collector unit con-sists of two coaxial cylindrical glass tubes with evacuated annular space in between.

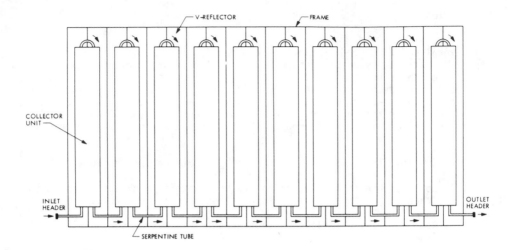

Fig. 1a. A single collector module composed of 10 units

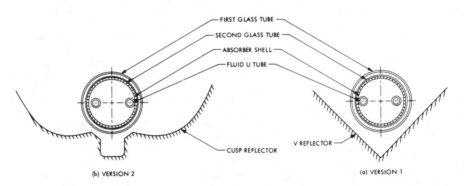

Fig. 1b. Two versions of a collector unit

The first (outer) cylinder serves as a "window," and the second (inner) cylinder is selectively coated on the outer surface to serve as the "absorber." The heat is transferred through the second glass tube to a conforming cylindrical metallic shell made of copper. The latter transfers the heat to the working fluid passing through the U-tube. To allow for thermal expansion, one side of the U-tube is attached to the copper shell while the other side is left free to move as shown in Fig. 1b. The thermal model, however, considers that both sides of the U-tube are in contact with the copper shell.

Although the second version of the collector design was made public after the mathematical model was established, the first design will still be analyzed. The differences between the two designs, namely, the shape of the back re-flector and the number of units per module, will not be changed in the thermal model.

The annular space between the glass tubes is evacuated and sealed to form

a "thermos-bottle" effect. The collector manufactured by G.E. has a selective
coating on the outer surface of the inner glass tube, thus making the inner
glass tube serve as the absorber for transmitting heat to the copper shell by
conduction. Another design is visualized by using a selective coating on the
outer surface of the copper shell instead, while keeping the inner glass tube
transparent since glass is a poor conductor. In this case, the double-walled
glass bottle serves as the'window'and the thin copper shell with its selective
outer surface acts as the'absorber.' Both designs will be investigated in the
analysis.

Besides employing:(1) a vacuum technique to reduce convection losses, (2)
a selective coating to reduce outward infrared radiation losses, and (3) a
back augmenting reflector, the collector has unique features compared to other
tubular collectors. Examples are:(1) a glass tube which, if damaged, will not
discontinue the collector service, and (2) the collector is lightweight with a
low thermal inertia which can be translated to ease in installation, connection
and structural requirement plus a fast temperature response that increases a
full-day performance.

THERMAL ANALYSIS

Before we proceed with the analytical energy expression for each collector
tube in a collector unit, the following assumptions were made to simplify the
simulation process.

(1)   The collector is assumed at steady state, located in an environment
      with uniform ambient temperature and solar irradiation.

(2)   The problem is treated as a two-dimensional heat transfer in the
      axial and radial directions. Collector tubes are assumed to be of
      uniform temperature in the tangential direction, even though the
      solar flux distribution on the outer glass tube may not be uniform
      due to the space allowed between the collector units and the effect
      of the back V-reflector. In the radial direction, the temperature
      distribution is assumed to be in steps with negligible conduction
      thermal resistance for all thin tubes.

(3)   Axial conduction heat transfer from one end to another is neglected.

(4)   Material optical properties are assumed uniform and independent of
      temperature and direction. Physical properties for solids and
      liquids are also assumed uniform and independent of working tempera-
      ture and pressure.

(5)   Sky and ambient temperatures are assumed approximately the same to
      simplify computations.

(6)   The metallic absorber shell, the hot water tube wall and the cold
      water tube wall are all assumed at a single temperature, which is
      the average of these three surfaces. The absorber temperature varies
      only axially. The metallic absorber actually acts as a fin stretched
      from both sides of the fluid tubes in a circular shape. This
      assumption is supported by the observation that the difference in
      temperature between inlet and outlet fluids is small at each
      collector unit.

(7)   The convective heat transfer coefficient between the serpentine tube

and either the hot or cold fluid sides is assumed the same since its variation with temperature is insignificant. The convective co-efficient is a dominant function of tube diameter, length and fluid mass flow in the laminar range.

(8) The deformation, due to lateral thermal expansion of the U-shape tubing, is assumed to be insignificant and not to cause any glass breakage. Also, the present slit suggested by the manufacturer in the metallic shell is assumed to be narrow enough to keep the cold and hot fluid tubing always in contact with the shell.

A segment of the collector unit whose thickness is dx, as shown in Fig. 1c and located at a distance x from the open end of the fluid tubes, is analyzed. Appendix A gives the effective optical properties of the double concentric glass cylinders. In Appendix B, the details of the heat balance equations are given for reference. The collector and thermal behavior is characterized by the following two linear simultaneous differential equations for the hot fluid temperature T_h (x) and the cold fluid temperature T_c (x):

$$\frac{dT_h}{dx} = -\delta - C_0\,T_c + C_1\,T_h \tag{1}$$

$$\frac{dT_c}{dx} = \delta + C_0\,T_h - C_1\,T_c \tag{2}$$

where δ, C_0 and C_1 are collector characteristic constants given in Appendix B. The following subsections describe some important points in solving the above two differential equations.

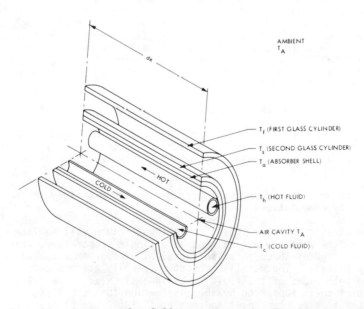

Fig. 1C. Segment of a Collector Unit with Thickness dx

A. Boundary Conditions

The differential equations (1) and (2) are subject to the following two boundary conditions:

(1) At the inlet fluid section (x=0), the colid fluid temperature $T_c(0)$ is given.

(2) At the closed end section (x=L), the cold fluid temperature is equal to the hot fluid temperature.

The temperature distribution $T_c(x)$ and $T_h(x)$ is given by Eqs. C-28 and C-29 in Appendix C.

B. Net Heat Collection Rate

The temperature difference between cold and hot fluids at any location (x) is given from Eqs. C-28 and C-29 by

$$[T_h(x) - T_c(x)] = \left[\frac{\delta}{(C_1 - C_0)} - T_c(0)\right]$$

$$\left[\frac{2(C_1 - C_0)\sinh\, n\,(L - x)}{n \cosh nL + (C_1 - C_0)\sinh nl}\right] \tag{3}$$

where n = $\sqrt{C_1^2 - C_0^2}$

Particularly, at the open end (x=0), both the temperature difference and the net heat collected per unit collector area are determined using Eq. (3) as

$$[T_h(0) - T_c(0)] = \left[\frac{\delta}{(C_1 - C_0)} - T_c(0)\right]\left[\frac{2(C_1 - C_0)\sinh nL}{n \cosh nL + (C_1 - C_0)\sinh nL}\right] \tag{4}$$

and

$$Q''_{coll} = \frac{MC_w}{SL}\, [T_h(0) - T_c(0)] \tag{5}$$

where S is the spacing between any two collector units.

Using Eq. 4 and the conductance coefficients (B's) defined in Appendix B and Eq. B-36, the extracted energy by the fluid is rewritten as

$$Q''_{coll} = \left[\left(E_1 + E_2\frac{B_4}{B_7} + E_3\frac{B_6}{B_7}\right) - B_0\,(T_c(0) - T_A)\right]\cdot F \tag{6}$$

$$\underbrace{\phantom{\left(E_1 + E_2\frac{B_4}{B_7} + E_3\frac{B_6}{B_7}\right)}}_{\text{Energy absorbed}} \quad \underbrace{}_{\substack{\text{Energy lost to}\\ \text{ambient air}}}$$

where F is a dimensionless "flow-factor" defined by

$$F = \frac{GD_{f,0}}{SL\, B_0}\left[\frac{2\sinh\, nL}{\left(\dfrac{B_3}{nG}\right)\cosh nL + \sinh nL}\right] \tag{7}$$

where B_0 is the overall heat transfer coefficient given by Eq. B-37. If the collector glass tubes were made such that the glass absorptivity $\alpha_{f,e}$ and $\alpha_{s,e}$ are negligible, Eq. (6) will become similar to the general Hottel, Whillier and Bliss form

$$Q''_{coll} \cong \{E_1 - B_0 [T_c(0) - T_A]\} \cdot F$$

C. Collector Efficiency

The collector efficiency based on the solar radiancy on the projected idea is defined by

$$\eta = \frac{Q''_{coll}}{I}$$

or, using Eq. (6)

$$\eta = F \left\{ \lambda \left[\alpha_{a,e} + \alpha_{f,e} \frac{B_4}{B_7} + \alpha_{s,e} \frac{B_6}{B_7} \right] - B_0 \frac{[T_c(0) - T_A]}{I} \right\} \tag{8}$$

Eq. (8) suggests that if the collector efficiency is plotted vs $(T_c(0) - T_a)/I$, the results would fit approximately a straight line whose slope $(B_0^c F)$ is an indication of the heat losses to the ambient and the intercept is an indication of the optical characteristics.

D. Highest Temperature at "No-Flow"

The temperature of the collector with "no-flow" or stagnant condition is an important value needed for coating stability and temperature control. Setting the temperature difference $[T_h(0) - T_c(0)]$ from Eq. 4 to the limit as G approaches zero (or n approaches ∞) one can prove that

$$\underset{\text{as } n \to \infty}{\text{Limit}} [T_h(0) - T_c(0)] = \frac{2 \left[\left(E_1 + E_2 \frac{B_4}{B_7} + E_3 \frac{B_6}{B_7} \right) - (T_c(0) - T_A) \right]}{B_0 \left[1 + \sqrt{1 + (2B_3/B_0)} \right]} \tag{9}$$

where $\bar{B}_3$ is the free-convection heat transfer coefficient between absorber tubes and fluid. If the collector was left with a very small flow rate under the sun with an inlet temperature equal to ambient temperature T_A, the simulated maximum temperature of the leaving fluid will be

$$T_{h_{max}}(0) \cong T_A + \frac{E_1}{B_0} \cdot \frac{2}{1 + \sqrt{(2\bar{B}_3/B_0) + 1}} \tag{10}$$

To support the parameterization study and the numerical evaluation of the above findings, a short computer program is written in Appendix C using the optical properties of Appendix A, the heat balance equations of Appendix B. The results of the second phase of this study will be discussed next.

COMPARISON WITH EXPERIMENTAL TESTS

In order to provide a cross-checking on the computer program validity, a comparison is made of some simulated performance results against the manufacturer test data. The dotted line in Fig. 2 was provided by the manufacturer (Ref. 4) based on the experimental tests made in 1978 at Desert Sunshine

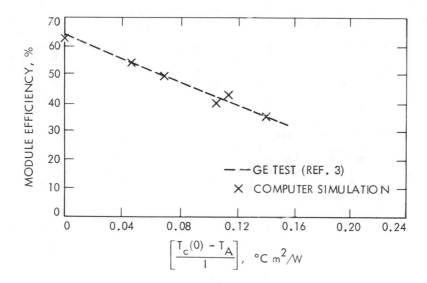

Fig. 2. Performance Comparison Between Some Simulated and
         Experimental Data.

Exposure Tests, Inc., and Florida Solar Energy Center, using a collector module
of the first design version. The X points in Fig. 2 are the simulated results
of some arbitrarily selected operating conditions having different inlet fluid
temperature $T_c(0)$, ambient temperature T_A, and solar intensity I. The abscissa
in Fig. 2, $[T_c(0) - T_A]/I$, is commonly chosen to compare solar collectors of
different optical and thermal characteristics. The coordinates of Fig. 2 also
fit Eq. 6, where the ordinate intercept represents the optical efficiency and
the curve slope is an indication of the thermal losses to the ambient.

The collector characteristic equation can be expressed, from Fig. 2, by
the approximate formula

$$\eta_{module} = 0.640 - 2.0669 \frac{T_c(0) - T_A}{I} \tag{11}$$

where the temperatures T_A and $T_C(0)$ are in degrees Celsius and I is in W/m^2.

The good agreement between the computer simulation results and the experimental tests, indicated in Fig. 2, provided the validation needed for the computer program. Consequently, the performance sensitivity to nine major variables was done next using a baseline set of operating conditions. The
numerical example presented in Appendix D gives the magnitude of this set of
variables used as a baseline. The flow rate of a collector module (10
collector units) was chosen to be 50 kg/h (instead of 5 kg/h used in Appendix B)
to follow the manufacturer's specification. The simualted module efficiency,
at the reference conditions, is 49.47%, corresponding to:(1) a solar radiancy I
of 630.7 W/m^2, (200 Btu/h ft^2), (2) a wind speed W of 4.47 m/s, (3) an ambient
temperature T_A of 4.4°C (40°F), (4) a reflectivity of absorber shell b of 0.5,
(5) a second glass tube emissivity, ε_g of 0.2, (6) a reflectivity of back reflector ρ_v of 0.9, (7) a U-tube size of 6 mm (1/4 in.) nominal diameter, (8) a
fluid mass flow rate M of 50 kg/h, and (9) an inlet fluid temperature to the
first collector unit $T_c(0)$, of 48.89°C (120°F). The selection of these reference conditions was made using explicit and implicit data provided by the
manufacturer.

PERFORMANCE PARAMETERIZATION

The performance of the collector is mainly determined by the above nine major variables. Each variable is set to change in value around the pre - selected reference state, and the results of one collector module (10 units) are discussed in the following subsections.

A. Effect of Solar Radiancy Variations

The performance sensitivity to the solar radiancy I is plotted as shown in Fig. 3. A nonlinear relationship is evident between the radiancy vs. the fluid temperature gain or the collector efficiency. An increase of the solar intensity by 50%, for example, will improve the fluid temperature gain by 63.8% and the collector efficiency by 9.21%. On the other hand, a 50% reduction of

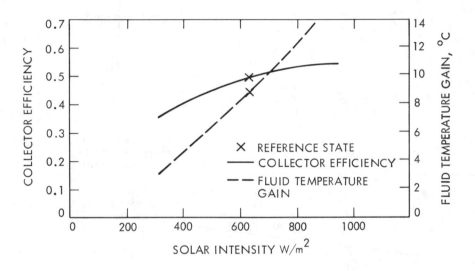

Fig. 3. Effect of Solar Intensity Variations

the solar intensity will lower the fluid temperature gain by 60.9% and the collector efficiency by 27.9%. This behavior can be best explained by using the efficiency expression Eq. 8 . The efficiency of one collector unit given in Eq. 8 is divided into two terms. The first term

$$\left[F\lambda \left(\alpha_{a,e} + \alpha_{f,e} \frac{B_4}{B_7} + \alpha_{s,e} \frac{B_6}{B_7} \right) \right]$$

which is independent of the intensity I, represents the optical efficiency of the collector at zero thermal losses. The second term, namely,

$$FB_0 \left[T_c(0) - T_A \right] / I$$

herein called the thermal loss factor, is proportional to the temperature difference $(T_c(0) - T_A)$ and inversely proportional to the intensity I. This explains the nonlinear relationship of the efficiency vs the intensity as illustrated in Fig. 3. In addition, Eq. 8 can be used to interpret the increase of collector efficiency when the solar intensity increases.

B. Effect of Wind Speed Variations

The convective loss coefficient, H_{fA} between the first (outer) glass tube
and ambient air is solely a function of the wind speed as given by Eq. (C-12)
in Appendix C. Two extreme values of the wind speed were assumed to take place
around the reference state. The first is a no-wind condition and the second is
a wind speed of 8.94m/s (20 mph), which is double the reference speed of 4.47m/
s (10 mph). The results are plotted in Fig. 4. It is evident from Fig. 4 that
the effect of wind speed variations on efficiency and fluid temperature gain
is small. At the no-wind condition, for example, the wind velocity decreased
by -100% compared to the reference point and caused an increase in the efficiency by only 1.8%. On the other hand, at double the reference wind speed, an
efficiency decrease of 0.4% was found. These findings lead to the conclusion
that the collector performance has a very small sensitivity to variations in
wind speed.

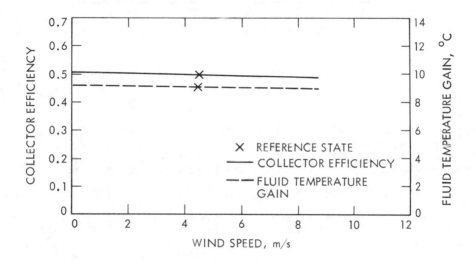

Fig. 4. Effect of Wind Speed Variations

C. Effect of Ambient Temperature Variations

The effect of ambient temperature was investigated by varying the ambient
temperature from -23.33°C (-10°F) to 48.89°C (120°F) around the reference state,
which is 4.4°C (40°F), keeping all other parameters the same. The results are
plotted in Fig. 5, showing the effects on the collector efficiency and the fluid
temperature gain.

Increasing the ambient temperature causes an increase in the collector
efficiency due to the reduction of the thermal losses and vice versa. These
thermal losses are porportional to the temperature difference $(T_c(0) - T_A)$ as
given by Eq. 8. An increase of the ambient temperature from 4.4°C (40°F) to
48.89°C (120°F), i.e., an increase of the absolute temperature by 16%, caused
an increase in the thermal efficiency by about 28%. This is equivalent to a
sensitivity of about 1.75. On the other hand, a decrease of the ambient tem-

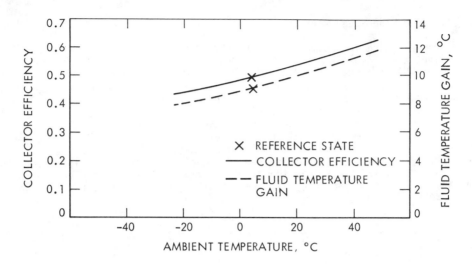

Fig. 5.   Collector Performance vs. Ambient Temperature

perature from 4.4°C (40°F) to -23.33°C (-10°F), i.e., a decrease of the absolute temperature by 10%, caused a decrease in the thermal efficiency by about
12%.   Again, the sensitivity to ambient temperature is 1.2, which shows the
important role that the ambient temperature plays in the performance.

D.   Effect of Absorber Shell Reflectivity b

     The optical properties of the metallic shell and the second (inner) glass
tube should be carefully selected in the design in order to yield a good
collector performance.   The mechanism by which the solar energy is absorbed,
converted into heat, and transmitted to the working fluid can be one of two
types.

     The first heat transfer mechanism could be achieved by adopting a heat-
absorbing glass material for the second glass tube to act as the "absorber"
from which the net absorbed heat is conducted to the metallic shell that holds
the fluid tubing.   This mechanism is already used by the manufacturer, and it
requires that the first (outer) glass tube function only as a "window" for
minimizing the outward infrared radiation losses.

     Another heat transfer mechanism is envisioned in which both the first and
second glass tubes act as a double-paned "window" made of common clear glass
with negligible heat absorbing capability.   The major portion of solar energy
will be absorbed at the outer surface of the metallic shell, thus acting as the
"absorber".   In these two mechanisms, the reflection coefficient r for the first
glass tube outer and inner surfaces, and that for the outer surface of the
second tube, do not play a significant role.   The reflection coefficient r is
known to be a function only of the incident angle and the refraction index of
glass.

     The absorption coefficients a (for the first glass tube) and $\bar{a}$ (for the
second glass tube) depend on the glass extinction coefficient ϕ and thickness t

(Ref. 8) such that

$$\left.\begin{array}{c} a = e^{-\phi t_f} \\[2mm] \bar{a} = e^{-\bar{\phi} t_s} \end{array}\right\} \tag{12}$$

or

     In general, the percentage of ferrous oxide (F_2O_3) in glass is important since iron accounts for most of the absorption. Reference 8 gives the extinction coefficient ϕ for the three different types of clear, medium-heat absorbing and high-heat absorbing glass panels as 6.85 m⁻¹, 129.92 m⁻¹, and 271.26 m⁻¹, respectively.

     The reflectivity of the metallic shell b, on the other hand, affects the balance of heat absorbed, reflected, or transmitted to and from the second glass tube. The metallic shell outer surface can be either polished or coated to change the value of b, which will be shown next to be an important factor.

     Several variations of the reflection coefficient b were made in the program using the above three different types of second glass tube material. Both $T_c(0)$ and T_A were set equal to isolate the effects of thermal losses and to focus on the collector optical efficiency alone. The results were plotted as shown in Fig. 6. Since the details of the optical properties of the collector components were not given by the manufacturer, the baseline collec-

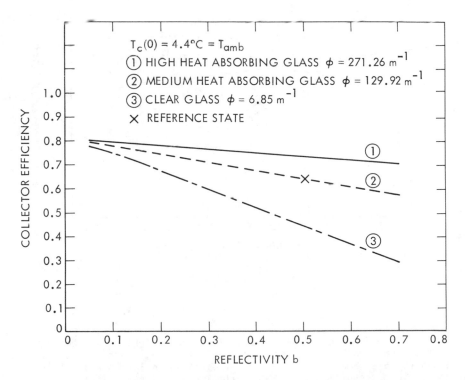

Fig. 6. Effect of Metallic Shell Reflectivity, b, on Efficiency

tor properties were selected arbitrarily to be having a medium-heat absorbing
second glass tube ($\bar{\phi}$ = 129.92 m^{-1}), a polished copper shell (b = 0.5), and a
clear first glass tube (ϕ = 6.85 m^{-1}).

It can be observed from Fig. 6. that reducing the shell reflectivity b
always improves the optical efficiency with any type of second glass-material.
The percentage improvement is large at small extinction coefficients. The
collector efficiency for the first mechanism of heat transfer was found always
higher than that for the second mechanism when both mechanisms have the same
coefficient b. At small values of b, or at high shell absorptivity, the effect
of ϕ becomes diminishing, and the collector optical efficiency reaches about
80%. This last result leads to the recommendation that high absorptivity or
black coating is necessary for the metallic shell in order to achieve the highest performance, whether or not a heat-absorbing glass is used.

E.   Effect of Second Glass Tube Outer Surface Emissivity

If a "selective" coating is used on the outside surface of the second
(inner) glass tube, it will reduce the outward infrared radiation losses, thus
improving the performance. Different emissivity values were tested in the
parameterization study, ranging from 0.05 to 0.9, with a reference value at 0.2
which is also "selective." The results are plotted as shown in Fig. 7.

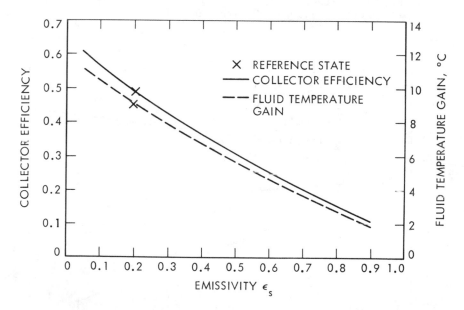

Fig. 7.   Effect of Second Glass Surface Emissivity, ε_s

The strong relationship between the efficiency and the emissivity ε_s is clearly
indicated in Fig. 7. The overall thermal loss coefficient B_o is dominantly
dependent on the "equivalent" radiation coefficient B_4, which is given in Eqs.
(C-26) and (C-35). Smaller emissivity values cause smaller overall thermal
loss coefficient, resulting in higher collector efficiency. If the infrared
emissivity drops from the reference "selective" value (ε_s = 0.2) to a lesser

emissivity value of 0.05, for example, i.e, a decrease of 75%, the overall
thermal loss coefficient B_0 for the first collector unit will be decreased by
72% and the collector module efficiency will be improved by 24.1%. This is
equivalent to a sensitivity of -0.32 for the collector module efficiency and
+0.96 for the coefficient B_0. On the other hand, an increase of the emissivity
(ϵ_s) from 0.2 (selective) to 0.9 (flat black), i.e., an increase of 350% causes
an increase in the loss coefficient B_0 for the first collector unit by 269%
i.e., a loss coefficient sensitivity of 0.77. The corresponding module efficiency will drop to a low value of 10.4% i.e., a decrease of 78.9% compared to
the reference state. The efficiency sensitivity in the latter case is equivalent to -0.22.

It can be concluded from the above discussion that "selective" coatings
having an infrared emissivity in the order of 0.2 or less are recommended to
achieve higher performance. Coating instability due to temperature recycling,
aging, or operation at high temperatures and the associated increase in
collector operation and maintenance cost, should be traded off against the improvement in collector performance.

F. Effect of the Back Panel Reflectivity

The back reflector used, whether it is a V-shape or a cusp shape, is
necessary in order to enhance concentration of the solar flux on the glass
tubes. Equation (C-11) gives the relationship between the augmentation factor
λ and the units spacing S, outer glass tube diameter $D_{r,o}$, and the back surface reflectivity (ρ_v). Equation (C-11) assumes that for both the V and cusp
reflector types, the solar energy falling on the unshaded areas of the back reflector is reflected totally, with no loss, upon the external surface of the
first glass tube. The higher the reflectivity ρ_v is, the higher the augmentation
factor, and the higher the efficiency will be. The results of varying ρ_v are
plotted in Fig. 8.. The baseline design assumes a highly reflective mirrorlike
material that is used for the back panel with ρ_v of 0.9. If a polished aluminum
rack with ρ_v of 0.5, for example, is used, it means a drop in the reflectivity

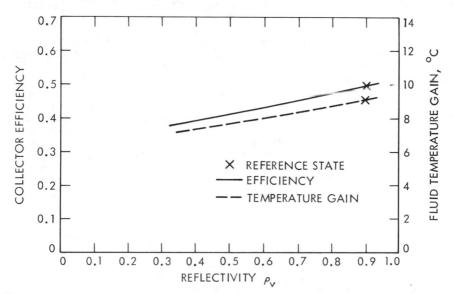

Fig. 8. Effect of the Reflectivity, ρ_v, of the back V-shape Reflector

by 44.4%. The resulting drop in module efficiency is found to be 16.7%. This is equivalent to an efficiency sensitivity of about 0.38, which is not insignificant.

G. Effect of Tubing Size

The size of the copper serpentine tube carrying the fluid is set using the manufacturer data to be a 6 mm (1/4 in.) nominal diameter. With a fluid flow of 50 kg/h, an inside diameter $D_{t,i}$ of 10 mm, and an outside diameter $D_{t,o}$ of 14mm, the Reynolds number is computed as 6121, which lies in the transition region. Given a fixed mass flow rate, the effect of varying the tube diameter on the efficiency was studied for two different tube sizes. The first tubing has a nominal diameter of 1/8 in. ($D_{t,i}$ = 7.2 mm, $D_{t,o}$ = 10.2 mm) and the second tubing, has a nominal diameter of 3/8 in. ($D_{t,i}$ = 12.6 mm, $D_{t,o}$ = 17.2 mm). The resulting efficiencies were 49.50, 49.47, and 49.41%, corresponding to the nominal diameters of 1/8, 1/4, and 3/8 in., respectively.

The effect of tubing size could be considered, therefore, practically negligible. The slight improvement noticed above when a small tubing is used contributed by the increased convective heat transfer coefficient H_{ah} caused by the higher fluid velocities attained. The effect of the latter on performance was somewhat counterbalanced by the corresponding smaller heat transfer surface area.

H. Effect of Fluid Mass Flow Rate

The choice of the operating mass flow rate is important if the collector efficiency needs to be improved and the pump horsepower to be decreased. Several references in the literature, including Ref. 9 have indicated a practical range from 24.4 kg/h-m^2 (5 lb/h-ft^2) to 97.7 kg/h-m^2 (20 lb/h-ft^2) for flat-plate collectors to trade off between collector efficiency and pump horsepower. A recommended rate of 48.8 kg/h-m^2 (10 lb/h-ft^2) was given in Ref. 9. The flow rate recommended by the manufacturer for the 10-unit collector module is 50 kg/h or 34.2 kg/h-m^2 based on 1.464 m^2/module, which lies in the above practical range.

Increasing the fluid mass flow rate always increases the heat transfer between the copper tubing and the fluid, therefore increasing the extracted heat rate and the collector efficiency. On the other hand, for a given inlet fluid temperature, increasing the flow rate reduces the fluid exit temperature. Apart from the fact that a tradeoff analysis needs to be done with the feed pump horsepower, the "quality" of the extracted heat should be investigated from a thermodynamic availability viewpoint. If a reversible engine is connected to the collector and made to operate utilizing the collector extracted heat as if the latter is taken from a finite heat reservoir at the exit temperature $T_b(0)$, the availability A will be sketched in Fig. 9(a). A is defined as the maximum useful mechanical work that could be obtained from the above collector-reversible engine system. A is written for a constant specific heat fluid as

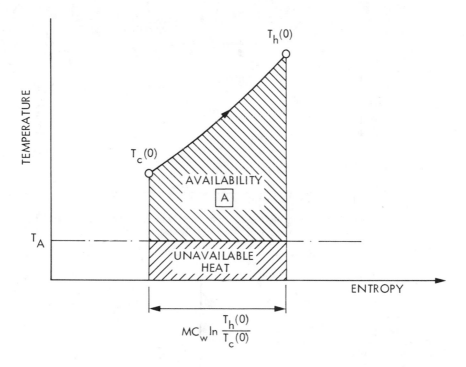

Fig. 9a. Thermodynamic Availability for the Collector Extracted Heat

$$A = Q_{ext} - T_A \, MC_W \ln \frac{T_h(0)}{T_c(0)} \tag{13}$$

where

$$Q_{ext} = MC_W \, [T_h(0) - T_c(0)] \tag{14}$$

For a module with 10 units, the temperature $T_c(0)$ is taken at the entrance of the first unit and $T_h(0)$ is taken at the exit of the 10th unit. The relationship between the availability A and the mass flow rate at given fluid inlet temperature $T_c(0)$ and ambient temperature T_A has the same trend as sketched in Fig. 9(b). The values of peak availability A^*_m at the optimum mass flow rate M* are plotted in Fig. 10 at different inlet fluid temperatures. Figure 10 indicates that for given inlet fluid temperature and ambient temperature, there exists an optimum mass flow rate M* that corresponds to a maximum thermodynamic availability A^*_m. This result is highly important in optimizing the collector operating conditions for solar-thermal-electric applications.

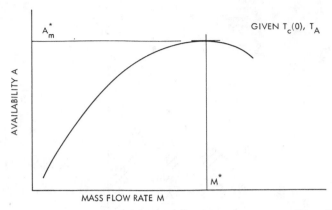

Fig. 9b.   Sketch of Thermodynamic Availability vs. the Mass Flow Rate
           for a Given Inlet and Ambient Temperature

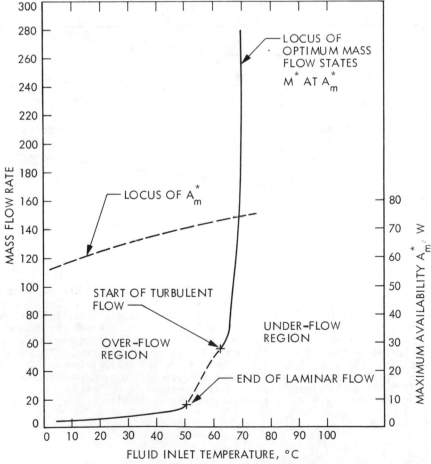

Fig. 10.   Fluid Inlet Temperature Effect on Optimum Mass Flow Rate

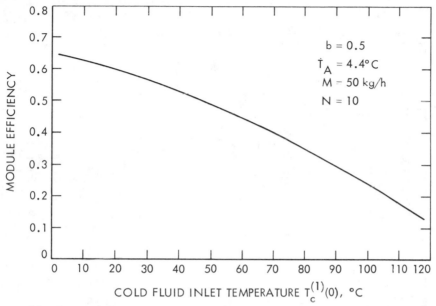

Fig. 11.   Effect of Inlet Fluid Temperature on Module Efficiency

I.   Effect of Inlet Fluid Temperature

The results of changing the cold fluid temperature at the entrance of the first collector unit are plotted as shown in Fig. 11. Increasing the temperature $T_c(0)$ causes an increase of the thermal losses from the collector surface since these losses are proportional to the temperature difference between the collector operating temperature and ambient air. The correlation between $T_c(0)$ and the module efficiency tends to deviate from the approximate straight line form due to the increasing effect of the radiation losses at higher fluid temperatures. Although increasing, the inlet fluid temperature causes a setback in the collector efficiency; the resulting increase of the outlet fluid temperature may be favored, especially when the heat is converted to mechanical work via engines. In order to find the optimum inlet fluid temperature at which the production of mechanical work is maximum, the thermodynamic availability A is introduced as in Eqs. (13) and (14). For a given mass flow rate, the availability A was computed at different inlet temperatures, keeping all other variables unchanged. The peak availability A^*_T and the optimum temperature $T^*_c(0)$ were computed at the given mass flow rate. The results were plotted as shown in Fig. 12. It is indicated from Fig. 12 that the maximum availability A^*_T corresponding to the optimum inlet temperature, stays approximately constant if the flow rate is beyond 50 kg/h. The latter matches the flow rate recommended by the manufacturer.

From Fig. 12, at a flow rate of 50 kg/h, the availability A^*_T is maximum at an inlet fluid temperature of 69^0C (156.2^{0}F) and equal to 74.33 watts/module. On the other hand, entering the optimum fluid temperature (69^0C) into Fig. 10, the optimum flow rate corresponding to a maximum availability A^*_m is about 140 kg/h where A^*_m is about 74 watts/module. In other words, maximizing the availability using the optimum inlet fluid temperature approach rather than the optimum mass flow rate approach is found convenient since the first gives more practical values of flow rates compared to the second approach.

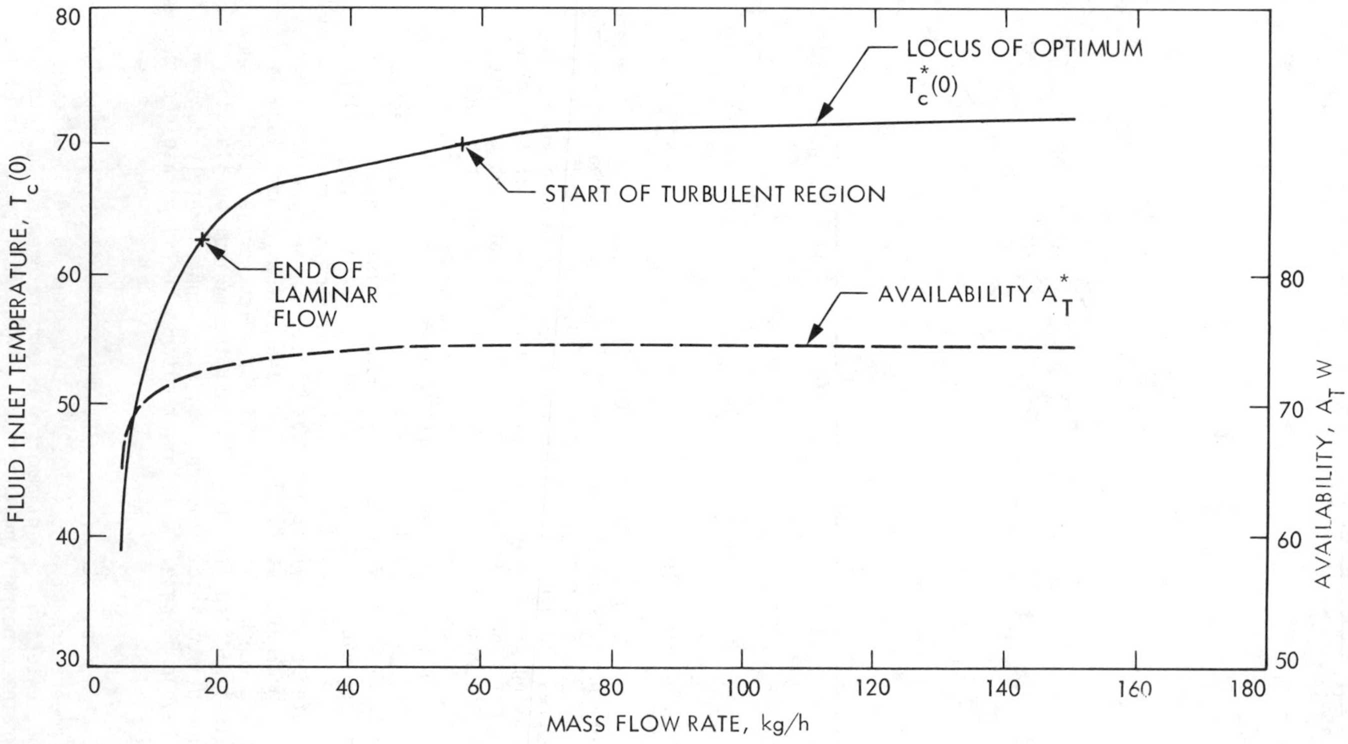

Fig. 12. Optimum Fluid Inlet Temperature at Various Mass Flow Rates

SUMMARY

In order to evaluate, in detail, the high-performance tubular collector recently manufactured by General Electric, a parameterization analysis was made. An in-house computer program was written for this purpose following the optical and thermodynamic analysis presented in Appendicies A and B and the computational sequence in Appendices C and D. Comparison of simulated results and manufacturer's test data showed good agreement at a wide range of operating conditions. The comparison is considered also a validation method for the computer program. Nine design and performance parameters were investigated to evaluate the performance sensitivity to their changes. The parameters considered were:(1) solar radiancy, (2) wind speed, (3) ambient temperature, (4) reflectivity of metallic shell, (5) second tube outer surface emissivity, (6) reflectivity of back reflector, (7) fluid tubing size, (8) fluid mass flow rate, and (9) inlet fluid temperature to first collector unit. The results of this parameterization study shed some light onto variables of insignificant effects and others that need to be modified in the design in order to yield a higher performance than the present one.

REFERENCES

1. Lansing, F.L., "Heat Transfer Criteria of a Tubular Solar Collector - The Effect of Reversing the Flow Pattern on Collector Performance", The Deep Space Network Progress Report 42-31, pp. 108-114, Jet Propulsion Laboratory, Pasadena, California, Feb. 1976.

2. Lansing, F.L., The Transient Thermal Response of a Tubular Solar Collector, Technical Memorandum 33-781, NASA, July 1976.

3. Lansing, F.L., "A Two-Dimensional Finite Difference Solution for the Transient Thermal Behaviour of a Tubular Solar Collector", The Deep Space Network Progress Report 42-35, pp. 110-127, Jet Propulsion Laboratory, Pasadena, California, October 1976.

4. "SOLARTRON TC-100 Vacuum Tube Solar Collector- Commercial and Industrial Installation Manual", Document No. 78DS4214A, General Electric Advanced Energy Programs, Philadelphia, Pennsylvania, November 1978.

5. "SOLARTRON TC-100 Vacuum Tube Solar Collector- Commercial and Industrial Application Guide", Document No. 78DS4215A, General Electric Advanced Energy Programs, Philadelphia, Pennsylvania, November 1978.

6. Duffie, J.A. and Beckman, W.A., Solar Energy Thermal Processes, Wiley Interscience Publication, New York, New York, 1974.

7. Kreith, F., Principles of Heat Transfer, 3rd ed., Intext Educational Publishers, New York, 1973.

8. Threlkeld, J.L., Thermal Environmental Engineering, Prentice-Hall, Inc., Englewood Cliffs, New Jersey, 1962.

9. Hewitt, H.C. and Griggs, E.I., "Optimal Mass Flow Rates through Flat Plate Solar Collector Panels", ASME Publication 76-WA/SOL-19, presented at ASME Winter Annual Meeting, New York, New York, December 1976.

APPENDIX A

DERIVATION OF OPTICAL PROPERTIES

The effect of multiple reflections, absorption, and refraction of direct solar rays incident upon the collector tubes is discussed next. The derivation of the effective absorptivity, reflectivity, and transmissivity will be made for each of the exterior glass tubes, the interior glass tube and the metallic (absorber) shell.

Since the radial spacing between the two coaxial glass tubes is small compared to their diameter, the optical properties derivation will be carried out assuming two parallel glass plates instead.

A-1. OPTICAL PROPERTIES OF A SINGLE SEE-THROUGH SHEET OF GLAZING

Figure A-1 shows the paths of a single beam of light when it falls on a single see-through sheet of glass. The intensity is assumed unity since the optical properties are dimensionless. The coefficients of absorption a and reflection r are applied both to the top and the bottom surfaces of the single glazing. The net transmissivity of a single glazing will be given by

$$\tau_f = a\,(1 - r)^2 + a^3 r^2\,(1 - r)^2 + a^5 r^4 (1 - r)^2 + \cdots$$

Using the infinite geometric series sum, τ_f is written as

$$\tau_f = \frac{a\,(1 - r)^2}{1 - a^2 r^2} \tag{A-1}$$

Similarly, the absorptivity α_f is written as

$$\alpha_f = (1 - a)(1 - r) + ar(1 - a)(1 - r) + a^2 r^2\,(1 - a)(1 - r) + \cdots$$

or

$$\alpha_f = \frac{(1 - r)(1 - a)}{1 - ra} \tag{A-2}$$

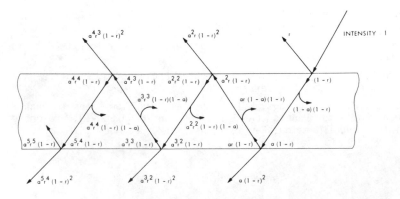

Fig. A-1. Paths and Intensities of a Light Beam on a Single Glazing.

Since the first law of thermodynamics states that

$$\alpha_f + \tau_f + \rho_f = 1$$

then the reflectivity ρ_f can be expressed as

$$\rho_f = r + \frac{ra^2(1-r)^2}{1-r^2a^2} \tag{A-3}$$

The above optical properties are given the subscript f since they represent the properties of the first (outer) glass tube of the collector unit.

A-2. OPTICAL PROPERTIES OF A SINGLE SHEET OF GLAZING WITH OPAQUE BOTTOM SURFACE

Figure A-2 illustrates the paths of a single light beam on a single sheet of glass whose bottom surface is opaque. The optical properties for this glazing type will be given a subscript S since it represents the second (inner) glass tube that surrounds the metallic absorber shell. In this case, the reflection coefficient r at the top surface and that at the bottom opaque surface b will be different. The absorption coefficient a will be the same as in case A-1, if all glazing have the same thickness and material. The optical properties can be obtained by summing the infinite geometric series of intensity taken from Fig. A-2. Accordingly, for the bottom surface

$$\alpha_B = a(1-r)(1-b) + a^3br(1-r)(1-b)$$

$$+ a^5b^2r^2(1-r)(1-b) + \cdots$$

or

$$\alpha_B = \frac{a(1-b)(1-r)}{1-a^2br} \tag{A-4}$$

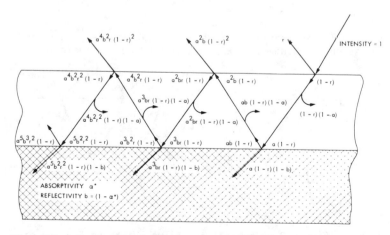

Fig. A-2. Paths and Intensities of a Light Beam on a Single Glazing with Opaque Bottom Surface

Also,

$$\alpha_S = (1-r)(1-a) + ab(1-a)(1-r) + a^2 br(1-r)(1-a) + \cdots$$

$$\alpha_S = \frac{(1-r)(1-a)(1+ab)}{1-a^2 br} \tag{A-5}$$

Since the energy conservation law can be written as

$$\alpha_B + \alpha_S + \rho_S = 1$$

then one can prove that

$$\rho_S = r + \frac{a^2 b (1-r)^2}{1-a^2 br} \tag{A-6}$$

A-3.   COMBINING TWO SHEETS OF GLAZING WITH AN OPAQUE BOTTOM SURFACE FOR THE SECOND SHEET

Combining the above two separate cases in Sections A-1 and A-2, the effective optical properties of each glazing will be derived using the light paths as illustrated in Fig. A-3. The properties of the first glazing will be τ_f, α_f, and ρ_f, as expressed by Eqs. (A-1), (A-2), and (A-3), respectively. For the second glazing, the properties α_B, α_S, and ρ_S are given by Eqs. (A-4), (A-5), and (A-6), respectively. The effective absorptivity $\alpha_{a,e}$ of the metallic absorber shell surface next to the second glazing will be found by summing the infinite series.

$$\alpha_{a,e} = \alpha_B \tau_f + \alpha_B \tau_f \rho_f \rho_S + \alpha_B \tau_f \rho_f^2 \rho_S^2 + \cdots$$

or

$$\alpha_{a,e} = \frac{\alpha_B \tau_f}{1 - \rho_f \rho_S} \tag{A-7}$$

Also, the effective absorptivity of the first glazing $\alpha_{f,e}$ when it is placed next to the second glazing is determined by summing the infinite series

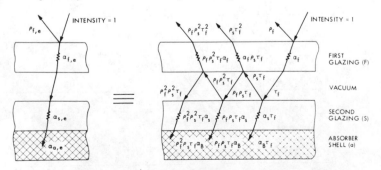

Fig. A-3.   Combination of Two Sheets of Glazing with an Opaque Bottom Surface for the Second

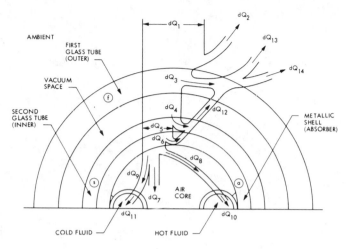

Fig. B-1. Sankey Diagram for Tubular Collector

dQ_2 = outward reflection loss from the collector unit

$$dQ_2 = \rho_{f,e}\, dQ_1 \tag{B-3}$$

where $\rho_{f,e}$ is given by Eq. (A-10)

dQ_3 = effective absorbed energy by the outer glass tube

$$dQ_3 = \alpha_{f,e}\, dQ_1 \tag{B-4}$$

where $\alpha_{f,e}$ is given by Eq. (A-8)

dQ_4 = effective absorbed energy by the second (inner) glass tube

$$dQ_4 = \alpha_{s,e}\, dQ_1 \tag{B-5}$$

where $\alpha_{s,e}$ is given by Eq. (A-9)

dQ_5 = effective absorbed energy by the metallic absorber shell

$$dQ_5 = \alpha_{a,e}\, dQ_1 \tag{B-6}$$

where $\alpha_{a,e}$ is given by Eq. (A-7)

dQ_6 = conduction heat transfer between the second (inner) glass tube and the absorber shell

$$dQ_6 = \frac{2\pi\,(T_a - T_s)\,dx}{\left[\dfrac{\ell n\,(D_{s,o}/D_{s,i})}{K_s} + \dfrac{\ell n\,(D_{a,o}/D_{a,i})}{K_a}\right]} \tag{B-7}$$

$$\alpha_{f,e} = \alpha_f + \alpha_f \tau_f \rho_S + \alpha_f \tau_f \rho_f \rho_S^2 + \alpha_f \tau_f \rho_f^2 \rho_S^3 + \cdots$$

or

$$\alpha_{f,e} = \alpha_f + \frac{\alpha_f \tau_f \rho_S}{1 - \rho_f \rho_S} \tag{A-8}$$

Similarly, for the second glazing, the effective absorptivity $\alpha_{S,e}$ is expressed as:

$$\alpha_{S,e} = \alpha_S \tau_f + \alpha_S \tau_f \rho_f \rho_S + \alpha_S \tau_f \rho_f^2 \rho_S^2 + \cdots$$

or

$$\alpha_{S,e} = \frac{\alpha_S \tau_f}{1 - \rho_f \rho_S} \tag{A-9}$$

On the other hand, the sum of reflections from the first glazing will be written as:

$$\rho_{f,e} = \rho_f + \rho_S \tau_f^2 + \rho_f \rho_S^2 \tau_f^2 + \rho_f^2 \rho_S^3 \tau_f^2 + \cdots$$

or

$$\rho_{f,e} = \rho_f + \frac{\rho_S \tau_f^2}{1 - \rho_f \rho_S} \tag{A-10}$$

Equations (A-7) through (A-10) are the effective properties to be used for the thermal analysis. As a cross-checking, the properties also satisfy the energy conservation law where

$$\alpha_{f,e} + \alpha_{s,e} + \alpha_{a,e} + \rho_{f,e} = 1 \tag{A-11}$$

APPENDIX B

HEAT TRANSFER RATE EQUATIONS

Following the assumptions made in the text, a segment of the collector unit whose thickness is dx and located at a distance x from the inlet fluid section is as shown in Fig. 3. The differential rates of heat flux are divided as shown by Sankey diagram in Fig. B-1, whereby

dQ_1 = total solar irradiation (direct and diffuse) on the outer glass tube from all sides including the irradiation from the back V-reflector

$$dQ_1 = \lambda I D_{f,o} dx \tag{B-1}$$

where λ is given by

$$\lambda = 1 + \rho_v \left(\frac{S}{D_{f,o}} - 1 \right) \tag{B-2}$$

dQ_7 = convection heat transfer between the absorber shell and the air trapped in the absorber core with the ambient air (see Appendix C, Eq. C-24 for the equivalent coefficient B2)

$$dQ_7 = D_{f,o} \, B_2 \, (T_a - T_A) \, dx \qquad \text{(B-8)}$$

dQ_8 = convection heat transfer between the absorber tubes and the hot fluid

$$dQ_8 = H_{ah} \, (T_a - T_h) \pi D_{t,i} \, dx \qquad \text{(B-9)}$$

dQ_9 = convection heat transfer between the absorber tubes and the cold fluid

$$dQ_9 = H_{ac} \, (T_a - T_c) \pi D_{t,i} \, dx \qquad \text{(B-10)}$$

dQ_{10} = sensible heat gain by the hot fluid

$$dQ_{10} = -MC_W \cdot \left(\frac{dT_h}{dx}\right) dx \qquad \text{(B-11)}$$

where the (-) sign was introduced since the hot fluid flow is in the opposite direction to the positive x direction.

dQ_{11} = sensible heat gain by the cold fluid

$$dQ_{11} = MC_W \left(\frac{dT_c}{dx}\right) dx \qquad \text{(B-12)}$$

dQ_{12} = radiation heat transfer between the second (inner) glass tube and the first (outer) glass tube

$$dQ_{12} = \frac{\sigma (T_S^4 - T_f^4) \pi D_{S,o} \, dx}{\dfrac{1}{\epsilon_S} + \left(\dfrac{1}{\epsilon_f} - 1\right) \left(\dfrac{D_{S,o}}{D_{f,i}}\right)} \qquad \text{(B-13)}$$

dQ_{13} = radiation heat losses from the outer surface of the first (outer) glass tube to the ambient air

$$dQ_{13} = \sigma \epsilon_f (T_f^4 - T_A^4) \pi D_{f,o} \, dx \qquad \text{(B-14)}$$

dQ_{14} = convection heat transfer from the outer surface of the first (outer) glass tube to the ambient air

$$dQ_{14} = H_{fA} \, (T_f - T_A) \pi D_{f,o} \, dx \qquad \text{(B-15)}$$

Applying the first law of thermodynamics at steady state to the collector components will yield the following heat balance equations (see Refs. 3 and 7).

For the absorber tube:

$$dQ_5 - (dQ_6 + dQ_7 + dQ_8 + dQ_9) = 0 \qquad \text{(B-16)}$$

For the hot fluid:

$$dQ_8 - dQ_{10} = 0 \qquad \text{(B-17)}$$

For the cold fluid:

$$dQ_9 - dQ_{11} = 0 \qquad \text{(B-18)}$$

For the first (outer) glass cover:

$$dQ_3 + dQ_{12} - dQ_{13} - dQ_{14} = 0 \qquad \text{(B-19)}$$

For the second (inner) glass cover:

$$dQ_4 + dQ_6 - dQ_{12} = 0 \qquad \text{(B-20)}$$

For the incident solar energy;

$$dQ_1 - dQ_2 - dQ_3 - dQ_4 - dQ_5 = 0 \qquad \text{(B-21)}$$

Summing Eqs. (B-16) through (B-21) yields

$$dQ_1 - (dQ_2 + dQ_7 + dQ_{13} + dQ_{14}) = 0 \qquad \text{(B-22)}$$

Equation (B-22) shows that the thermal losses from the collector will be only the summation of:(1) the outward light reflection from the first (outer) glass tube, (2) the convection and radiation heat transfer to the ambient air and sky from the first (outer) glass, and (3) the convection heat transfer from the inner absorber walls to the air core and ambient air.

The elementary radiation heat transfer dQ_{12} and dQ_{13} are further linearized by defining the radiation heat transfer coefficients R_{sf} and R_{fAm} such that

$$\left. \begin{array}{c} R_{sf} = \dfrac{\sigma (T_s^4 - T_f^4)}{(T_s - T_f)\left[\dfrac{1}{\epsilon_s} + \left(\dfrac{1}{\epsilon_f} - 1\right)\left(\dfrac{D_{s,o}}{D_{f,i}}\right)\right]} \\[4em] R_{fA} = \epsilon_f \dfrac{\sigma (T_f^4 - T_A^4)}{(T_f - T_A)} \end{array} \right\} \qquad \text{(B-23)}$$

Equations (B-13) and (B-14) are reduced to

$$dQ_{12} = R_{sf}\pi D_{s,o} (T_S - T_f) dx$$

(B-13)

$$dQ_{13} = R_{fA} \ \pi D_{f,o} (T_f - T_A) dx$$

(B-14)

The five linearized heat balance equations, Eqs. (B-16) through (B-20), will be grouped after division by $(D_{f,o}dx)$ as follows:

For the absorber shell:

$$E_1 - B_1 (T_a - T_s) - B_2 (T_a - T_A) - B_3 (T_a - T_c)$$

$$- B_3 (T_a - T_h) = 0$$

(B-24)

For the hot fluid:

$$G \frac{dT_h}{dx} = -B_3 (T_a - T_h)$$

(B-25)

For the cold fluid:

$$G \frac{dT_c}{dx} = B_3 (T_a - T_c)$$

(B-26)

For the first (outer) glass tube:

$$E_2 + B_4 (T_s - T_f) - B_5 (T_f - T_A) = 0$$

(B-27)

For the second (inner) glass tube:

$$E_3 + B_1 (T_a - T_s) - B_4 (T_s - T_f) = 0$$

(B-28)

where

$$B_1 = \frac{2\pi/D_{f,o}}{\left[\dfrac{\ell n (D_{s,o}/D_{s,i})}{K_s} + \dfrac{\ell n (D_{a,o}/D_{a,i})}{K_a} \right]}$$

$$B_3 = \pi (D_{t,i}/D_{f,o}) H_{ah}$$

$$B_4 = \pi R_{sf} (D_{s,o}/D_{f,o})$$

$$B_5 = \pi (H_{fA} + R_{fA})$$

(B-29)

$$E_1 = \alpha_{a,e} \lambda I$$

$$E_2 = \alpha_{f,e} \lambda I$$

$$E_3 = \alpha_{s,e} \lambda I \Bigg\}$$ (B-30)

$$G = M C_w / D_{f,o}$$

Expressing the temperatures T_f and T_s in terms of T_a using Eqs. (B-24), (B-27), and (B-28) yields

$$T_f = (E_4 + B_4 T_s)/B_6$$

$$T_s = (E_5 + B_6 T_a)/B_7 \Bigg\}$$ (B-31)

$$T_a = [E_6 + B_3 (T_h + T_c)]/B_8$$

where

$$E_4 = E_2 + B_5 T_A \qquad , \quad B_6 = (B_4 + B_5)$$

$$E_5 = E_3 \frac{B_6}{B_1} + E_4 \frac{B_4}{B_1} \qquad , \quad B_7 = B_6 + \frac{B_4 B_5}{B_1} \Bigg\}$$ (B-32)

$$E_6 = E_1 + B_2 T_A + \frac{B_1 E_5}{B_7}, \quad B_8 = B_2 + 2 B_3 + \frac{B_4 B_5}{B_7}$$

Substituting in Eqs. (B-25) and (B-26), the two differential equations for the hot and cold fluid streams at any position x will be

$$\frac{dT_h}{dx} = -\delta - C_0 T_c + C_1 T_h$$ (B-33)

and

$$\frac{dT_c}{dx} = \delta + C_0 T_h - C_1 T_c$$ (B-34)

where

$$
\left.\begin{array}{l}
\delta = \dfrac{B_3}{B_8} , \dfrac{E_6}{G} \\[2em]
C_0 = \dfrac{B_3^2}{GB_8} \\[2em]
C_1 = \dfrac{B_3}{GB_8}(B_8 - B_3)
\end{array}\right\}
\tag{B-35}
$$

The term $\delta/(C_1 - C_0)$ appears in solving Eqs. (B-33) and (B-34) and can be expressed as:

$$
\frac{\delta}{(C_1 - C_0)} = T_A + \frac{\left(E_1 + E_2 \dfrac{B_4}{B_7} + E_3 \dfrac{B_6}{B_7}\right)}{\left(B_2 + \dfrac{B_4 B_5}{B_7}\right)}
\tag{B-36}
$$

The overall heat loss coefficient B_0 follows from Eq. (B-36), to be

$$
B_0 = \left(B_2 + \frac{B_4 B_5}{B_7}\right)
\tag{B-37}
$$

Eqs. (B-36) and (B-37) are useful in giving the physical meaning needed to the collector efficiency expression as shown in the text.

APPENDIX C

DETAILS OF THE COMPUTER MODEL

I. INTRODUCTION

A computer program is written using the thermal model and analytic equations given in Appendices A and B. The program is divided into several parts described in detail next.

II. PROGRAM INPUT DATA

The following data need to be provided by the user in order to complete the program execution.

A. Optical Properties

Glass reflection coefficient	r
First glass tube absorption (or extinction coefficient)	a (or ϕ)
Second glass tube absorption (or extinction coefficient)	$\bar{a}$ (or $\bar{\phi}$)

Reflectivity of absorber shell   b

Reflectivity of back reflector ρ_v

Emissivity of first glass tube surfaces ε_f

Emissivity of second glass tube outer surface ε_s

B. Collector Dimensions

Spacing between collector units   s

Inner diameter of first glass tube $D_{f,i}$

Outer diameter of first glass tube $D_{f,o}$

Inner diameter of second glass tube $D_{s,i}$

Outer diameter of second glass tube $D_{s,o}$

Inner diameter of metallic shell $D_{a,i}$

Outer diameter of metallic shell $D_{a,o}$

Inner diameter of fluid tubing $D_{t,i}$

Outer diameter of fluid tubing $D_{t,o}$

Length of one side of the U-tube of a collector unit   L

Thickness of insulation at the open end t_u

Number of collector units per module   N

C. Thermodynamic Properties

Working fluid viscosity μ_w

Working fluid specific heat C_w

Working fluid thermal conductivity K_w

Thermal conductivity of second glass tube K_s

Thermal conductivity of absorber metallic shell K_a

Thermal conductivity of open-end insulation K_u

D. Operating Conditions

Fluid flow rate   M

Fluid temperature at the entrance of the first collector unit $T_c(0)$

E. Weather

Solar radiancy   I

Ambient temperature (dry bulb) T_A

Wind Speed   W

III. PROGRAM SEQUENCE

The following equations are listed in the same order of calculations sequence.

A. Optical Properties

1. From Appendix A, the transmissivity, absorptivity, and reflectivity of the first glass tube are τ_f, α_f and ρ_f, respectively, where

$$\tau_f = \frac{a\,(1-r)^2}{1-a^2 r^2} \tag{C-1}$$

$$\alpha_f = \frac{(1-r)\,(1-a)}{1-ar} \tag{C-2}$$

$$\rho_f = r + \frac{ra^2\,(1-r)^2}{1-a^2 r^2} \tag{C-3}$$

where r and a are the reflection and absorption coefficients, respectively, for the first glass tube.

2. The absorptivity (α_B) of the bottom surface of the second glass tube is given by

$$\alpha_B = \frac{\bar{a}\,(1-b)\,(1-r)}{1-\bar{a}^2 br} \tag{C-4}$$

Also, the absorptivity α_s and reflectivity ρ_s of the second glass tube are given by

$$\alpha_s = \frac{(1-r)\,(1-\bar{a})\,(1+\bar{a}b)}{1-\bar{a}^2 br} \tag{C-5}$$

$$\rho_s = r + \frac{\bar{a}^2 b\,(1-r)^2}{1-\bar{a}^2 br} \tag{C-6}$$

where $\bar{a}$ and r are the absorption and reflection coefficients of the second glass tube, and b is the reflectivity of the absorber shell surface.

3. The effective properties $\alpha_{a,e}$ of the metallic absorber shell $\alpha_{f,e}$ of the first glass tube and $\alpha_{s,e}$ of the second glass tube, are given by

$$\alpha_{a,e} = \frac{\alpha_B \tau_f}{1 - \rho_f \rho_s} \tag{C-7}$$

$$\alpha_{f,e} = \alpha_f + \frac{\alpha_f \tau_f \rho_s}{1 - \rho_f \rho_s} \tag{C-8}$$

$$\alpha_{s,e} = \frac{\alpha_s \tau_f}{1 - \rho_f \rho_s} \tag{C-9}$$

$$\rho_{f,e} = \rho_f + \frac{\rho_s \tau_f^2}{1 - \rho_f \rho_s} \tag{C-10}$$

4. Augmentation factor λ is given by

$$\lambda = \left[1 + \rho_v \left(\frac{S}{D_{f,o}} - 1 \right) \right] \tag{C-11}$$

B. Heat Transfer Coefficients

1. Convection coefficient between tubes and ambient. The convective heat transfer coefficient between the outer glass tube and surrounding air, H_{fA} in W/m²°C, is given approximately in Ref. 6 as a linear function of the wind speed W

$$H_{fA} = 5.7 + 3.8W \tag{C-12}$$

where W is in m/sec. Equation (C-12) is also used to determine the film coefficients between the inner surface of the "absorber" shell and the still air core H_{aA}, also between the end insulation and the air core H_{uA}. The coefficients are obtained by the setting W equal to zero in Eq. (C-12). Accordingly,

$$H_{aA} = H_{uA} \cong 5.7 \tag{C-13}$$

2. Radiation coefficient. Two "effective" radiation heat transfer coefficients are calculated in Appendix B namely, R_{sf} and R_{fA}. The coefficient R_{sf} represents the radiation exchange between the inner surface of the first (outer) glass tube and the outer surface of the second (inner) glass tube. Hence,

$$R_{sf} = \frac{\alpha \left(T_s{}^4 - T_f{}^4 \right)}{\left[(T_s - T_f) \left(\frac{1}{\epsilon_s} + \left(\frac{1}{\epsilon_f} - 1 \right) \frac{D_{s,o}}{D_{f,i}} \right) \right]} \tag{C-14}$$

The coefficient R_{fA} represents the radiation exchange between the first (outer) glass tube and the ambient air and is given by

$$R_{fA} = \frac{c_f \alpha \left(T_f{}^4 - T_A{}^4 \right)}{(T_f - T_A)} \tag{C-15}$$

A convergent iterative process is used for each collector unit whereby average temperatures of each of the first glass tubes are assumed. These averages give the first estimate of the coefficients R_{sf} and R_{fA} to determine the temperature distribution $T_f(x)$ and $T_s(x)$ which in turn are used to modify the radiation coefficients. The iteration process is completed for a given collector unit before proceeding to the next unit in the module and so on.

3. Convection between the working fluid and tubes. The present model incorporates all the equations needed to calculate the forced convection heat transfer coefficient between the working fluid and the copper tubing at any one of the three flow regions: laminar, transition, and turbulent. The laminar region is characterized by a Reynolds number (Re) less than 2100 for circular tubes. For the turbulent flow case, the Reynolds number is greater than 7000. References 6 and 7 were used to determine the Nusselt number (Nu) for laminar and turbulent flow regions. For laminar flow

$$Nu = 4.64 + \frac{0.067 \, D_{t,i} \, Re \, Pr/l}{1 + 0.04 \left(\frac{D_{t,i}}{l} \, Re \, Pr \right)^{2/3}} \tag{C-16}$$

where $D_{t,i}$ is the inner tube diameter, l is the total tube length and Pr, Re and Nu are Prandtl, Reynolds, and Nusselt numbers, respectively. The dimensionless numbers are written as follows:

$$\left. \begin{array}{l} Re = \dfrac{4M}{\pi D_{t,i} \mu_w} \\[2em] Pr = \dfrac{C_w \mu_w}{K_w} \\[2em] Nu = \dfrac{H_{ah} D_{t,i}}{K_w} \end{array} \right\} \tag{C-17}$$

For turbulent flow inside tubes, the Nusselt number Nu, is given by

$$Nu = 0.023 \, Re^{0.8} \, Pr^{0.3} \tag{C-18}$$

In the metastable transition region, no reliable expression has been found in the literature to express the Nusselt number. Accordingly, a linear interpolation is used in this work as a first approximation of the fluid transition from laminar to turbulent regions, i.e., in the range where 2100 < Re < 7000. The approximate equation used is

$$Nu_{(transition)} = Nu_{2100} + (Re - 2100) \left(\frac{Nu_{7000} - Nu_{2100}}{7000 - 2100} \right) \tag{C-19}$$

where Nu_{2100} is the Nusselt number computed from Eq. (C-16) at Re equal to 2100 and Nu_{7000} is the Nusselt number computed from Eq. (C-18) at Re equal to 7000.

The fluid thermal properties used in the above equations were taken at some preselected average bulk temerperature and were assumed to be constant during operation. The working fluid used in the modelling exercise was the ethylene glycol-water solution with a volumetric ratio of 50/50 as recommended by the manufacturer. Pure water was not selected because of its inadequacy at working temperatures below 0°C (32°F) or above 100°C (212°F).

C. Absorbed Energy Flux E Coefficients

The energy flux terms (E's), discussed in Appendix B, Eq. (B-30), are rewritten here to complete the program sequence.

1. The fraction of solar energy that is absorbed by the "absorber" shell E_1 is given by

$$E_1 = \alpha_{a,e} \lambda I \tag{C-20}$$

2. The fraction of solar energy that is absorbed by the first (outer) glass tube E_2 is expressed by

$$E_2 = \alpha_{f,e} \lambda I \tag{C-21}$$

3. The fraction of solar energy that is absorbed by the second (inner) glass tube E_3 is written as

$$E_3 = \alpha_{s,e} \lambda I \tag{C-22}$$

D. "Equivalent" Heat Transfer Coefficients (B's)

The B coefficients are herein called "equivalent" since they represent the "equivalent" heat transfer coefficients for a plate-to-plate heat exchange giving the same heat transfer of the present circular geometry. The B coefficients given in Appendix B, Eq. (B-29), are rewritten here, briefly, for completion.

1. The "equivalent" conduction coefficient B_1 between the "absorber" shell and second (inner) glass tube is given by

$$B_1 = \frac{2\pi/D_{f,o}}{\left[\dfrac{\ln\left(D_{s,o}/D_{s,i}\right)}{K_s} + \dfrac{\ln\left(D_{a,o}/D_{a,i}\right)}{K_a}\right]} \tag{C-23}$$

2.   The coefficient B_2 represents the "equivalent" heat loss coefficient from the central air core to the ambient air through the open and closed ends of the collector. Figure C-1 shows a sketch of the location of film coefficients used in deriving the B_2 expression for one collector unit having one end insulated and the other using double hemispheres with vacuum in between. One can prove that

$$B_2 = \frac{1}{\left[\dfrac{D_{f,o}}{\pi h_{aA}\left(D_{a,i}+2\,D_{t,o}\right)} + \dfrac{1}{PA_c}\right]} \tag{C-24}$$

where P and A_c are given by

$$P = \frac{2}{\dfrac{1}{H_{aA}} + \dfrac{1}{R_{sf}} + \dfrac{2\,t_s}{K_s} + \dfrac{1}{H_{f_A}}} + \frac{1}{\dfrac{1}{H_{uA}} + \dfrac{t_u}{K_u} + \dfrac{1}{H_{fA}}}$$

$$A_c = \frac{\pi}{4}D_{a,i}^2$$

3.   The "equivalent" convection coefficient between the absorber shell and the fluid (B_3) is given by

$$B_3 = \pi\left(D_{t,i}/D_{f,o}\right) H_{ah} \tag{C-25}$$

4.   The "equivalent" radiation coefficient between the first glass and second glass tube is given by

$$B_4 = \pi R_{sf}\left(D_{s,o}/D_{f,o}\right) \tag{C-26}$$

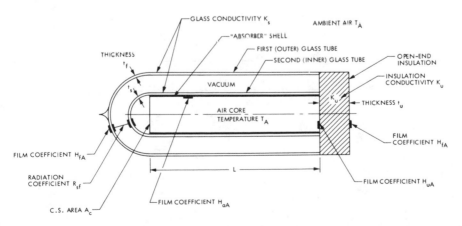

Fig. C-1.   Film Coefficients Used for the Determination of the 'Equivalent' Coefficient B_2

5. The "equivalent" combined radiation and convection coefficient between the first glass tube and ambient is written as

$$B_s = \pi (H_{fA} + R_{fA})$$

(C-27)

E. Overall Heat Loss Coefficient B_0

The coefficients B_1, B_2, B_4, and B_5 represent, by analogy to electric circuits, the thermal conductances between the "absorber" shell, glass tubes and ambient air as shown in Fig. C-2. The thermal resistances $1/B_1$, $1/B_4$, and $1/B_5$ are connected in series and their resultant is connected in parallel with the resistance $1/B_2$. The overall thermal conductance of this circuit is herein called the overall heat loss coefficient B_0 given by Eq. (B-37) as

$$B_0 = B_2 + \cfrac{1}{\cfrac{1}{B_1} + \cfrac{1}{B_4} + \cfrac{1}{B_5}}$$

F. Temperature Distribution

The hot and cold temperatures $T_h(x)$ and $T_c(x)$, respectively, at any position x from the open end of one collector unit, are obtained from solving the differential equation and substituting the boundary conditions.

$$T_h(x) = \frac{\delta}{C_1 - C_0} - \left[\frac{\delta}{C_1 - C_0} - T_c(0) \right]$$

(C-28)

$$\left[\frac{n \cosh n(L-x) - (C_1 - C_0) \sinh n(L-x)}{n \cosh nL + (C_1 - C_0) \sinh nL} \right]$$

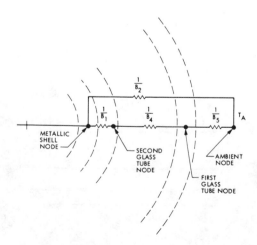

Fig. C-2.   Overall Heat Loss Coefficient, B_0

$$T_c(x) = \frac{\delta}{C_1 - C_0} - \left[\frac{\delta}{C_1 - C_0} - T_c(0) \right]$$

$$\left[\frac{n \cosh n(L-x) + (C_1 - C_0) \sinh n(L-x)}{n \cosh nL + (C_1 - C_0) \sinh nL} \right] \qquad (C-29)$$

where the constants n, c_0, c_1 and δ are computed in sequence from the following equation:

$$\left. \begin{aligned} n &= \sqrt{C_1^{\,2} - C_0^{\,2}} \\[6pt] C_0 &= B_3^{\,2}/(GB_8) \\[6pt] C_1 &= \frac{B_3}{GB_8}(B_8 - B_3) \\[6pt] \delta &= B_3 E_6/GB_8 \end{aligned} \right\} \qquad (C-30a)$$

where

$$\left. \begin{aligned} E_6 &= E_1 + B_2 T_A + \frac{B_1 E_5}{B_7} \\[6pt] E_5 &= E_3 \frac{B_6}{B_1} + E_4 \frac{B_4}{B_1} \\[6pt] E_4 &= E_2 + B_5 T_A \end{aligned} \right\} \qquad (C-30b)$$

$$\left. \begin{aligned} B_8 &= B_2 + 2B_3 + \frac{B_4 B_5}{B_7} \\[6pt] B_7 &= B_6 + \frac{B_4 B_5}{B_1} \\[6pt] B_6 &= B_4 + B_5 \end{aligned} \right\} \qquad (C-30c)$$

and

$$G = \frac{M C_W}{D_{f,o}} \qquad (C-30d)$$

The "absorber" shell temperature $T_a(x)$ is given by Eq. (B-31) as

$$T_a(x) = \frac{\{E_6 + B_3 [T_h(x) + T_c(x)]\}}{B_8} \qquad (C-31)$$

The second glass tube temprature $T_s(x)$ is given by Eq. (B-31) as

$$T_s(x) = \frac{[E_5 + B_6 \, T_a(x)]}{B_7} \qquad (C\text{-}32)$$

Also, the first glass tube temperature $T_f(x)$ is given by Eq. (B-31) as

$$T_f(x) = \frac{[E_4 + B_4 \, T_s(x)]}{B_6} \qquad (C\text{-}33)$$

G.   Performance Factors

The flow factor F is defined by

$$F = \frac{GD_{f,o}}{SL\,B_0} \left[\frac{2 \sinh nL}{\dfrac{B_3}{nG} \cosh nL + \sinh nL} \right] \qquad (C\text{-}34)$$

where B_0 is the overall heat loss coefficient given by Eq. (B-37) as:

$$B_0 = B_2 + \frac{B_4 \, B_5}{B_7} = B_2 + \frac{1}{\dfrac{1}{B_1} + \dfrac{1}{B_4} + \dfrac{1}{B_5}} \qquad (C\text{-}35)$$

The unit collector instantaneous efficiency, based on the solar radiancy falling on the projected area SL is given by

$$\eta_{unit} = \frac{MC_w \, [T_h(0) - T_c(0)]}{ISL}$$

or

$$\eta_{unit} = F \left[\lambda \left(\alpha_{a,e} + \alpha_{f,e} \frac{B_4}{B_7} + \alpha_{s,e} \frac{B_6}{B_7} \right) - B_0 \frac{T_c(0) - T_A}{I} \right] \qquad (C\text{-}36)$$

where SL is the projected area of a collector unit. For a collector module that consists of N collector units in series, the module efficiency is given by

$$\eta_{module} = \frac{MC_w \, [T_h^{(N)}(0) - T_c^{(1)}(0)]}{ISLN} \qquad (C\text{-}37)$$

where $T_h^{(N)}(0)$ is the fluid temperature at the exit of the nth unit and $T_c^{(1)}(0)$ is the fluid temperature at the entrance of the first collector unit.

APPENDIX D

NUMERICAL EXAMPLE

As an illustration of the use of the computer program described in Appendix C, a numerical example is given next to show the sequence followed. For convenience, the performance of only one collector unit will be computed. The performance of a module with N units in series can be computed by repetition, following the same sequence of calculations of a single unit.

I. INPUT VARIABLES

The following input variables are entered in the program where some were based on information provided by the manufacturer and the rest were estimated from past experience. The input data are grouped in order similar to Section II of Appendix C.

A. Optical Properties

Glass reflection[1] coefficient r	0.043
First glass tube extinction[2] coefficient ϕ	6.85 m^{-1}
Second glass tube extinction[3] coefficient ϕ	129.92 m^{-1}
Reflectivity[4] of metallic shell b	0.5
Reflectivity[5] of back V-reflector ρ_v	0.9
Emissivity of first glass tube surfaces ε_f	0.9
Emissivity[5] of second glass tube surface ε_s	0.2

B. Collector Dimensions

Spacing[7] between two consecutive collector units S	0.12 m
Inner diameter[8] of first glass tube $D_{f,i}$	0.076 m
Outer diameter[8] of first glass tube $D_{f,o}$	0.082 m
Inner diameter[8] of second glass tube $D_{s,i}$	0.058 m
Outer diameter[8] of second glass tube $D_{s,o}$	0.064 m

---

[1]Taken at zero incidence angle and a glass refraction index of 1.526.
[2]Assumed made of clear glass (Refs. 4 and 5).
[3]Assumed made of a medium heat absorbing glass.
[4]For polished copper surface.
[5]For silvered aluminum surface.
[6]Assumed for a selective coating.
[7]See Refs. 4 and 5.
[8]Estimated from sketches in Refs. 4 and 5

Inner diameter[1] of metallic shell $D_{a,i}$ 0.056 m

Outer diameter[1] of metallic shell $D_{a,o}$ 0.058 m

Length of one[1] side of U-tube L   1.22 m

Thickness[1] of insulation at the open end t_u 0.004 m

C. Thermodynamic Properties

Type of working fluid: ethylene glycol-water (50/50) by volume

Fluid viscosity[2] μ_w 1.04 kg/m h

Fluid specific heat[2] c_w 1.0122 Wh/kg°C

Fluid thermal conductivity[2] K_w 0.4 W/m°C

Glass thermal conductivity K_s 0.7 W/m°C

Thermal conductivity of shell K_a 380.16 W/m°C

Thermal conductivity of open-end insulation K_u 0.319 W/m°C

D. Operating Conditions

Fluid flow rate, m   5.0 kg/h

Inlet fluid temperature $T_c(0)$ 48.89°C (120°F)

E. Weather

Solar radiancy I   630.7 W/m^2 (200 Btu/h ft^2)

Ambient temperature T_A 4.4°C (40°F)

Wind speed W   4.47 m/s (10 mph)

II. OUTPUT RESULTS

The following is a partial list of the output results obtained from running the computer program using Sec. D.1 data.

A. Optical Properties

α_f (Eq. C-2)   0.0203

$\alpha_{f,e}$ (Eq. C-8)   0.0251

α_s (Eq. C-5)   0.4176

---

[1]Estimated from sketches in Refs. 4 and 5.

[2]Taken as an average from 4°C to 120°C.

$\alpha_{s,e}$ (Eq. C-9)   0.3833

$\alpha_{a,e}$ (Eq. C-7)   0.3004

λ (Eq. C-11)   1.4171

B. Heat Transfer Coefficients

    Reynolds number Re   612.1 (laminar)

    Prandtl number Pr   2.632

    Nusselt number Nu   4.748

    Fluid heat transfer coefficient H_{ah} 189.9 $W/m^2 \, ^\circ C$

    "Equivalent" coefficient B_1 544.5117 $W/m^2 \, ^\circ C$

    "Equivalent" coefficient B_2 0.1336 $W/m^2 \, ^\circ C$

    "Equivalent" coefficient B_3 72.7588 $W/m^2 \, ^\circ C$

    "Equivalent" coefficient B_4 3.1424 $W/m^2 \, ^\circ C$

    "Equivalent" coefficient B_5 85.1437 $W/m^2 \, ^\circ C$

    Overall loss coefficient B_0 3.1473 $W/m^2 \, ^\circ C$

C. Temperature Distribution

    The temperature profile along each collector tube is plotted as shown in Fig. D-1 at the above input conditions. The arithmetic average of the tube temperatures are as follows: 6.58°C for the first glass tube; 58.57°C for the second glass tube; 58.24°C for the metallic shell; 53.05°C for the cold fluid and 57.38°C for the hot fluid. The actual fluid temperature gain $[T_h(0) - T_c(0)]$ is computed as 8.6996°C using Eqs. (C-28) and (C-29).

    In addition to the above results, the flow factor F is computed as 0.6401 and the collector efficiency is found to be 47.68%.

III. TEMPERATURE TRENDS FOR HOT AND COLD FLUIDS

    The hot and cold fluid temperatures $T_h(x)$ and $T_c(x)$ were given in Appendix C by Eqs. (C-28) and C-29), respectively. The useful temperature gain $[T_h(x) - T_c(x)]$ at any position x from the open end can be written as

$$[T_h(x) - T_c(x)] = \left[\frac{\delta}{C_1 - C_0} - T_c(0) \right]$$

$$\left[\frac{2(C_1 - C_0) \sinh n(L - x)}{n \cosh nL + (C_1 - C_0) \sinh nL} \right]$$

                                                             (D-1)

To find the location of any maximum or minimum fluid temperatures along the collector length, Eqs. (C-28), (C-29), and (D-1) are differentiated with respect to (x) keeping all other parameters constant. One can prove mathematically that neither the cold temperature $T_c(x)$ nor the useful temperature

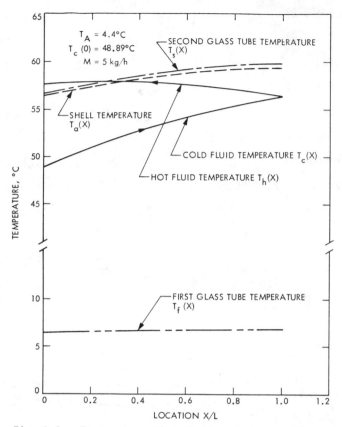

Fig. D-1.   Temperature Distribution along a Collector
            Unit

gain $[T_h(x) - T_c(x)]$ can ever possess any maximum or minimum value.   In spite
of this finding, the hot fluid temperature $T_h(x)$ possesses one peak value at
some location $\bar{X}$ (measured from the open end) given by

$$\bar{X} = L - \frac{1}{n} \tanh^{-1} \left(\frac{C_1 - C_0}{n} \right) \tag{D-2}$$

$\bar{X}$ can be either within the collector tube or theoretically outside $0 \leqslant x \leqslant L$
depending on the constants C_1, C_0.   The corresponding temperature difference
$[T_h(\bar{X}) - T_c(\bar{X})]$ can be proven to be smaller than that at the open end $(x = 0)$
at all times.   The conclusion is that the rate of energy extracted from the
collector is maximum at the open end and is unaffected by the internal temperature peaks.   The optical stability of selective coatings on the other hand,
is related to the highest temperature attained, and therefore needs this
investigation.

Development and Study of a Flat Mirror Multivalent Concentrator

L.M. SCHWARTZ, R. LOUAT, and G. MENGUY
Universite Claude Bernard
69622 Villeurbanne-Cedex, France

ABSTRACT

The concentrator described is made up of square flat mirrors each measuring 25 cm x 25 cm. The total reflecting area is 4 m^2. The advantage of plane mirrors is their low cost. The theoretical concentration ratio is 60, so that the temperature reaches 300°C when the intensity of the solar radiation is 850 W.m^{-2}. The power received by the system is about 3 kW. The slope of each mirror to the horizontal plane is $\frac{\theta}{2}$, if θ is the angle of the axis of the concentrator to the axis of each mirror.

The shape of the spot is a four branched star with an inner part, nearly circular, in which the concentration ratio is maximal. The mounting of the captor is equatorial. Within a cradle, a parallelepipedic frame turns whose bottom bears the mirrors and the superior part, the furnace, centered on the axis of the system and fitting a square covering the four central mirrors. The frame is adjusted with the declination.

A "blind" tracking type has been chosen. The captor is oriented by data directly calculated from the equations of the Earth's movement around the Sun. A D.C. motor is used, mounted on a big reductor with a pulley system whose speed output is about 1/30 r.p.m. The control is made by means of a feedback loop with an angular detector and a potentiometer. The automatism is activated by numerical means with a hit - or - miss governor. The principle of a 1° angular increment for the captor imposes a $\pm$ 0.50 precision of the captor position. The feedback loop which checks the instructions works with 8 bit words.

This sort of concentrator could be useful in a lot of studies involving the use of direct solar radiation, particularly in not very well insolated areas.

NOMENCLATURE

ℓ	concentration ratio
F	furnace ratio

$\mathscr{C}_e$ effective concentration ratio

$\mathscr{P}$ power (W)

T   temperature (K)

t   temperature (^oC)

θ angle of the concentrator axis to the axis of each mirror

x_i, y_i, z_i coordinates of a mirror summit

α, β coordinates of the mirror center

X_i, Y_i coordinates of the intersection of the reflected ray with the focus
       plane

α_s sun's angle width

h   distance between the focal plane and the plane of the mirrors

1. INTRODUCTION

The concentrator here described has been designed as a laboratory appara-
ratus rather than an industrial product. Its price, when operating, is about
$ 4 000. It would be able to provide a temperature of 300°C with a geometrical
concentration ratio $\mathscr{C}$ = 60. Its possibilities would be as follows :

1°) Study of the behavior of a receiver with a mean concentration ratio,
under random conditions of insolation (mean yearly insolation ratio in Lyon :
0.4).

2°) Substituting the existing mirrors by smaller ones, one can easily increase the concentration ratio at least cost and bring it, for example, to
$\mathscr{C}$ = 240, an intermediate value between the maximum concentration of a parabolic
cylindrical mirror and a spherical one, so that the temperature might reach 500°C.

3°) With testing different boilers, one can find the best shape giving the
highest yield.

4°) Study of thermoelectrical generators.

5°) Study of hot air machines or steam machines.

6°) Study of the behavior of materials under high energetical flux and
thermal shocks.

7°) Experiments on drying materials by hot air or overheated steam.

8°) Experiments on pyrolysis (e.g. wood, to change it into charcoal) or
distillation.

9°) Production of cold by using heat pumps.

2. DESCRIPTION

2.1. The Concentrator

The square reflecting area measuring 2 m x 2 m is made of 64 plane square

mirrors, 2.5 cm on side. The interest of plane mirrors is their low cost
($\simeq$ \$ 2 per unit). The four central mirrors are hidden by the furnace and are
not useful for concentration. The effective concentration ratio is $\mathcal{C}_e = F\mathcal{C}$.
F is the "furnace ratio" whose value usually is 0.6. So, $\mathcal{C}_e \simeq 36$. The power
received by the furnace is $\mathcal{P} = \mathcal{C}_e.I = \sigma T^4$ (σ is the Stefan contant). If
$I = 850$ W.m^{-2}, $T = 857$ K and t = 584°C. In practice, the expected temperature
300°C should be reached, since, during preliminary experiences, paper was igni-
ted very quickly in open and windy air.

To have a focusing in a given point 0 for all the pencils reflected by the
mirrors, it is easy to show that the slope of each mirror to the horizontal
plane must be $\frac{\theta}{2}$ if θ is the angle of the axis of the concentrator to the axis
of each mirror (fig. 1). Each mirror is fastened by three nylon rods on a metallic treillis and the regulating of the three points in altitude allows the suitable orientation of the mirror.

If three rectangular axis O'x, O'y, O'z are chosen on the mirror mosaic,
with origin in O', the center of the system, and if ω is the center of the mirror A B C D (fig. 2), the mirror, to be properly oriented, must turn, with an
angle $\frac{\theta}{2}$ about the straight line EF which is perpendicular to O'ω. If x_i, y_i, z_i
are the coordinates of a summit of the mirror, α and β the coordinates of ω
and a, the half-side of the mirror, we find :

$$x \simeq \alpha \pm a$$
$$y \simeq \beta \pm a$$
$$z \simeq \pm a \frac{(\alpha \pm \beta)}{\sqrt{\alpha^2 + \beta^2}} \sin \frac{\theta}{2} ,$$

the sign depending on the considered summit.

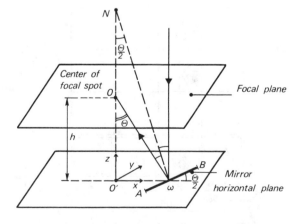

Fig. 1. Position of an elementary mirror.

The focal spot is situated on a plane which is parallel to the plane x O' y
whose altitude is h. If X_i, Y_i are the coordinates of the intersections of the
reflected rays with the focal plane, it may be prooved that :

$$X_i = x_i + \alpha \left(\frac{z_i}{h} - 1 \right)$$

$$Y_i = y_i + \beta \left(\frac{z_i}{h} - 1 \right),$$

or more explicitely :

$$X_A = -a \left[1 + \frac{\alpha}{h} \frac{(\alpha + \beta)}{\sqrt{\alpha^2 + \beta^2}} \sin \frac{\theta}{2} \right] = -X_C$$

$$Y_A = -a \left[1 + \frac{\beta}{h} \frac{(\alpha + \beta)}{\sqrt{\alpha^2 + \beta^2}} \sin \frac{\theta}{2} \right] = -Y_C$$

$$X_B = a \left[1 + \frac{\alpha}{h} \frac{(\alpha - \beta)}{\sqrt{\alpha^2 + \beta^2}} \sin \frac{\theta}{2} \right] = -X_D$$

$$Y_B = a \left[-1 + \frac{\beta}{h} \frac{(\alpha - \beta)}{\sqrt{\alpha^2 + \beta^2}} \sin \frac{\theta}{2} \right] = -Y_D.$$

The form of the focal spot is given by fig. 3. The concentration is the highest
in the inner area and the lowest and equal to 1 in the farthest corners.
Because the angular width of the Sun is $\alpha = \frac{1}{100}$, the maximum concentration area
is a little reduced by a band whose width is $\frac{h\alpha}{2}$ (here h = 180 cm, therefore
$\frac{h\alpha}{2}$ = 0.9 cm) and the area of lowest concentration is enlarged by the same width.

The power received by all the mirrors is about 3 kW. One may hope to reco-
ver more than the half of it on the furnace.

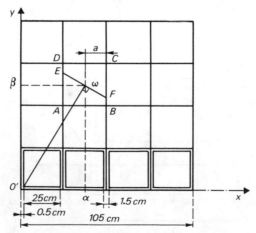

Fig. 2. Arrangement of the mirrors in the first quadrant
of the concentrating system.

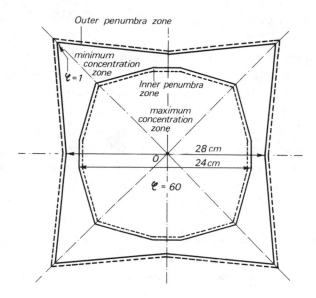

Fig. 3. Approximative shape of the focal spot. Scale : ¼.

2.2. The Furnace.

The first boiler was a motor car radiator, set in a plexiglass box covered
with a glass and isolated by glass wool. The radiator, soldered with tin, was
deteriorated and the glass broken by thermal shock. A new tubular boiler with a
thermal screen is on building. It enters within a square covering the four central mirrors exactly. Thermocouples are attached in several points allowing the
estimation of thermal losses.

2.3. Mounting of the Captor.

An equatorial mounting was chosen (fig. 4). The solar receiver is made of
three distinguished parts : a fixed part named the chassis ; a first movable set
tilted according to the latitude of Lyon and named the cradle ; a second movable
system supporting the mirrors and the furnace and named the receiver. The deta-
chable chassis has a mass of 150 kg and involves a pyramidal basis and a tripod
which allows the adjustment in latitude. A complete study of the cradle was
achieved, particularly concerning the behavior in wind. Its axis has an angle
to the horizontal plane equal to the latitude. Within it, a parallelepipedic
frame is turning, whose bottom bears the mirrors and the superior part, the
furnace. The frame may be adjusted with the declination so that the axis of
symmetry of the system always remains pointed towards the Sun.

3. TRACKING POSITION CONTROL.

We have chosen a "blind" tracking type, that is to say that the system is oriented owing to data directly calculated from the equations of the movement of the Earth around the Sun. The disadvantage of a system watching the Sun is that it is rather poorly adapted to intermittent sunny countries : in this case, the automatism, each time, has to proceed to a new seek of the Sun ; this work, again, reduces the time of insolation of the receiver.

A "blind" system needs to be oriented precisely since its automatism does not react to outer data. Its advantage is to be tilted always to the direction of the Sun, so that the time of insolation is maximum, even if the apparition of the Sun is intermittent.

In blind tracking, the position of the Sun must be calculated. The precision depends on the right orientation of the chassis of the concentrator (North-South and latitude) and the position of the concentrator in connection with the chassis ; this regulating needs an accurate angular detector.

The motor which is used is a D.C. one mounted on an important reductor whose output speed is 1/3 round per minute. An automatic command and a manual one are foreseen. The control is made by means of a feedback loop with an angular

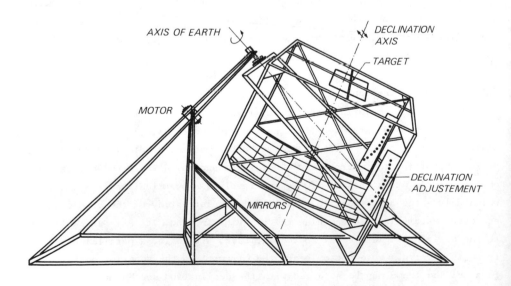

Fig. 4. Mounting of the concentrator.

detector directly mounted on the shaft of the solar concentrator (without any
free motion). The automation is acted by numerical means with a hit-or-miss
governor ; this principle is permitted by the low output speed. The electrical
diagram is given in fig. 5.

The command organs include :

3.1. The Instructions.

The angular variation instructions of the system are assumed by means of a
digital clock working with the electric network. The angular increment is sett-
led to be 1°, this leads to a displacement of the concentrator every fourth
minute. A divisor allows to obtain this signal, which is transmitted then to a
binary counter. At every moment, this one owns the instruction value, translated into an angular one. It may be set at the value corresponding to the starting angle ; likely, the clock is able to ungear the movement owing to the alarm
which is provided with.

3.2. The Feedback Loop.

The feedback loop of the control is assumed by a potentiometer. The potential difference, proportionnal to the angle it delivers, then is converted into
a numerical word by means of a digital-analogical converter. This solution,
though surprising, is less expansive than an incremental coder discriminating
the sense of rotation and associated with a counter-discounter.

The hour angle varies at most of 233° on the summer solstice : 8 bits words
are sufficient to obtain a 1° resolution.

To avoid any risk of relaxation oscillations, the feedback loop may be provided by a phase advance system located between the potentiometer and the digital-analogical converter.

3.3. The Command of the Motor.

The informations issued from the counter (instruction) and the converter
(feedback) are analysed in a numerical comparator which pilots the on-off relays
and those of the sense of rotation. The stop at the end of the day is assumed
by means of a run-end with a mercury contact pointing out the horizontality of
the axis of symmetry of the mirror system. It needs no regulating during the
year, when the declination varies. The run-end stops the alarm signal of the
clock ; this, in automatical mode, prevents the counter to increment the instructions. It puts back the counter to the preselected value and this resets the
captor for the next day. The movement only will begin during the rising up of
the alarm signal of the clock, thus allowing the variation of the instructions.

Also a manual command is foreseen.

3.4. Precision of the System.

The principle of a 1° angular increment for the captor imposes a ± 0.5° precision of the captor position. The feedback loop, assuming the respect of the instructions works with 8 bits words, i.e. in words written in degrees. The precision of the digital-analogical converter is ± 1/2 less significant bit.

4. CONCLUSION

Such a concentrator will be able to be used in a lot of studies involving the utilization of the direct solar radiation, particularly in countries which are not very favorable as for the insolation, like the Lyonese one, and to test the profitability of such concentration systems.

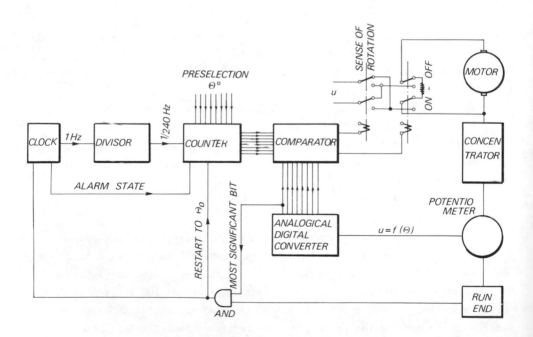

Fig. 5. PRINCIPLE OF THE ELECTRONICAL COMMAND

Investigations on the Prediction of Thermal Performance of Compound Parabolic Concentrators

C.R. HARIPRASAD, R. NATARAJAN, and M.C. GUPTA
Department of Mechanical Engineering
Indian Institute of Technology
Madras 600 036, India

ABSTRACT

The present paper focuses attention on the prediction of thermal performance of a compound parabolic concentrator for different values of insolation and mass flow rate of collector medium (water), under steady-state conditions. The analysis involves an iterative scheme and a method is proposed by which the absorber temperature, outlet temperature and glazing temperature can be predicted for given insolation and mass flow rate. The convective and radiative losses are then computed to evaluate the collector efficiency. The results are presented for different concentration ratios of a single-glazed CPC collector.

NOMENCLATURE

A	area (m^2)
C	concentration ratio
C_p	specific heat of water $J/(kg)(^\circ C)$
H	insolation W/m^2
h	heat transfer coefficient $W/(m^2)(^\circ C)$
k	thermal conductivity $W/(m)(^\circ C)$
l	aperture width (m)
L	collector length (m)
$\dot{m}$	mass flow rate of water (kg/s)
Nu	Nusselt number
Pr	Prandtl number
Q	heat transfer rate (W)
Re	Reynolds number
T	temperature (K)
T_{air}	ambient air temperature (K)
T_{sky}	sky temperature (K)

$U_{ambient}$ heat transfer coefficient from cover to ambient
air $W/(m^2)(^{\circ}C)$

V_w flow velocity of water (m/s)

V wind speed (m/s)

σ Stefan Boltzmann constant $W/(m^2)(K^4)$

α absorptivity

ρ reflectivity

ε emissivity

γ kinematic viscosity (m^2/s)

Subscripts

a   air

L   cover

R   reflector

s   absorber

w   water

1. INTRODUCTION

The type of solar collector to be employed in solar energy
utilization depends upon the intended application. Flat plate
collectors would suffice for relatively low-temperature applications such as, for example, in domestic water heating. To
achieve higher temperatures, as encountered in power plant and
metallurgical applications, focussing collectors are required;
this however, imposes the disadvantage of sophisticated tracking devices to follow the sun. Under these circumstances, it
would be worth-while to develop collectors that are not too
complex like concentrators, yet provide higher temperatures
than flat plate collectors. Compound parabolic concentrators
(CPC) belong to this class of collectors. CPC collectors have
the following important properties [1]:

1) The CPC is a non-focussing light funnel that concentrates light by the maximum amount permitted by physical principles.

2) The CPC has a uniform response over a large acceptance
angle (θ).

3) The concentration ratio of the CPC is given by [2]:

$$C = \frac{n}{\sin\theta} \quad \text{(trough type)}$$

$$C = \frac{n}{\sin^2\theta} \quad \text{(cone type)}$$

where n is the refractive index of CPC medium (n=1, for air).

The above properties imply the following advantages in using the CPC collectors for thermal applications over other types of concentrators:

1) The CPC collector achieves the highest possible concentration of radiant energy for a given acceptance angle.

2) There is no need for diurnal tracking of the sun [2,3]. By suitably adjusting the concentration ratio, (and in turn, the acceptance angle) it would be possible to construct a totally stationary, yet optically concentrating, solar collector.

3) Being a non-imaging device the reflecting surfaces of a CPC need not be fabricated with high precision.

Figure 1 shows a typical CPC. The CPC collector consists of an absorber (s) surrounded by two parabolic reflectors (R), with their arcs inclined at an angle equal to acceptance angle θ, and focii f_1 and f_2, situated at the edges of the absorber (s), as shown in Fig.1. The top of the CPC is covered by a glazing (L) to minimize the convection and radiation losses.

The heat transfer processes occurring in the CPC are convective and radiative heat exchange between the absorber and the glazing in the presence of the reflector, which renders the analysis rather complex. Not much work appears to have been done on the prediction of thermal performance of CPC collectors.

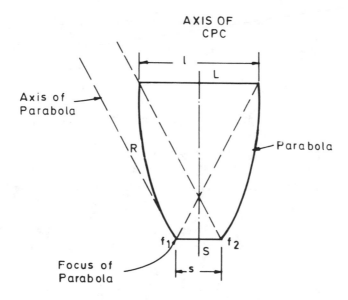

FIG.1. PROFILE OF CPC

2. PREDICTION OF THERMAL PERFORMANCE OF CPC

The present work proposes an iterative scheme to predict the temperatures of the absorber, the glazing, and the outlet fluid, and for the evaluation of radiative and convective losses and the collector efficiency, for given values of mass flow rate of the collector medium (water) and insolation, under steady-state conditions.

The analysis presented in reference 2 for the evaluation of convective and radiative losses forms the basis for the present investigation. That analysis is based on prediction of glazing temperature for an assumed absorber temperature, whereas the present scheme is more general.

The steps involved in the iterative scheme are outlined below:

2.1 Energy Balance for Fluid to Predict Outlet Temperature (T_2) for an Assumed Absorber Temperature (T_s)

Useful heat gain by water, $Q_{u1} = \dot{m} \, C_p \, (T_2 - T_1)$ (1)

where T_1 = inlet temperature = ambient temperature (assumed)

  T_2 = outlet temperature

Rate of heat transfer from absorber to water $= Q_{u2} = h_w \, A(T_s - T_w)$ (2)

where T_w = mean temperature of water = $(T_1 + T_2)/2$

The heat transfer coefficient, h_w is evaluated from:

$$h_w = (\frac{k_w}{L}) \, Nu_w$$ (3)

The Nusselt number is computed from the following relations for the flow over a flat plate [4]:

$$Nu_w = (0.664) \, (Re_w)^{0.5} \, (Pr_w)^{1/3} \text{ for laminar flow}$$ (4)

$$Nu_w = 0.036 \, (Pr_w)^{1/3} \left[(Re_w)^{0.8} - 23,200 \right] \text{ for turbulent flow}$$ (5)

The following property values for water are needed to compute h_w, and are parameterized in the temperature range 303 to 373 K:

(a) For thermal conductivity (k_w)

$$k_w = [-78.72714 + 0.724032 \, T_w - 0.9533993 \times 10^{-3} \, T_w^2]$$
$$\times 1.1603 \times 10^{-2}$$ (6)

(b) For kinematic viscosity (γ_w)

$$\gamma_w = [12.7210 - 0.6566775 \times 10^{-1}\ T_w + 0.860092 \times 10^{-4}\ T_w^2] \times 10^{-6} \quad (7)$$

(c) For Prandtl number

$$Pr_w = \left[98.21904 - 0.51519 \times T_w + 0.6882821 \times 10^{-3}\ T_w^2\right] \quad (8)$$

Equations (1) and (2) are solved iteratively by changing T_2 until

$$Q_{u1} \approx Q_{u2} \quad (9)$$

The value of T_2 corresponding to this condition is the outlet temperature for the assumed absorber temperature (T_s).

2.2 Energy Balance for Cover to Predict the Glazing Temperature (T_L) and Computation of the Losses, for the Assumed T_s

Convective heat transfer from absorber (s) to cover (L) is given by [2];

$$Q_{conv\ sL} = 14.2\ A_s\ \frac{(\Delta T_s)^{5/4}}{s^{1/4}}\left(T_s - \frac{\Delta T_s}{2}\right)^{-0.4} \quad (10)$$

where
$$T_s = \frac{T_s - T_L}{1 + C^{-0.6}}$$

$$C = \text{concentration ratio} = \frac{\text{Aperture area}}{\text{Absorber area}} = \frac{A_L}{A_s}$$

$$s = \text{width of the absorber}$$

Radiative heat transfer from absorber to cover is given by

$$Q_{rad\ sL} = \varepsilon_{eff}\ A_s\ \sigma\left(T_s^4 - T_L^4\right) \quad (11)$$

$$\sigma = \text{Stefan-Boltzmann constant} = 5.67 \times 10^{-8}$$
$$W\ m^{-2}\ (^{\circ}K)^{-4}$$

ε_{eff} is the effective emissivity of the CPC, which depends on the emissivities of the reflector (ε_R), absorber (ε_s), cover (ε_L) and concentration ratio [2].

Net heat flow from absorber s

to cover $L = Q_{conv\ sL} + Q_{Rad\ sL}$

$$= 14.2 \ A_s \ \frac{(\Delta T_s)^{5/4}}{S^{1/4}} \left(T_s - \frac{\Delta T_s}{2} \right)^{0.4}$$

$$+ \ \varepsilon_{eff} \ A_s \ \sigma \left(T_s^4 - T_L^4 \right) \qquad (12)$$

This heat transfer should be equal to the convective and radiative heat transfer from cover to ambient, which is given by

$$Q_{L \ ambient} = \varepsilon_L \ A_L \ \sigma \ (T_L^4 - T_{sky}^4) + U_{ambient} \ A_L(T_L - T_{air}) \quad (13)$$

$U_{ambient}$ is calculated from

$$U_{ambient} \ = \ \frac{k_a}{L} \ Nu_a \qquad (14)$$

To evaluate $U_{ambient}$, the following Nusselt number relations are recommended for forced convection over the flat plate, for wind speeds above 0.5 mph [4].

$$Nu_a \ = \ 0.664 \ (Re_a)^{\frac{1}{2}} \ (Pr_a)^{1/3} \quad \text{for laminar flow} \quad (15)$$

$$Nu_a \ = \ 0.036 \ (Pr_a)^{1/3} \ \left[(Re_a)^{0.8} - 23,200 \right]$$
$$\text{for turbulent flow} \quad (16)$$

The property values for air are evaluated at the bulk mean temperature

$$T \ = \ \frac{(T_L + T_{air})}{2}$$

The following parameterized expressions are used to evaluate the properties of air in the temperature range 250 to 600 K [2]:

(a) For kinematic viscosity (γ_a),

$$\gamma_a \ = \ \gamma_0 T^{1.7} \quad \text{with} \ \gamma_0 = 9.76 \times 10^{-10} \ m^2 \ sec^{-1}(K)^{-1.7} \qquad (17)$$

(b) For thermal conductivity (k_a)

$$k_a \ = \ k_0 \ T^{0.7} \quad \text{with} \ k_0 = 4.86 \times 10^{-4} \ W \ m^{-1}(K)^{-0.7} \qquad (18)$$

Assuming T_{sky} and T_{air}, equations (12) and (13) can be solved iteratively to compute cover temperature (T_L), for the assumed absorber temperature (T_s). After predicting the cover temperature (T_L), the convective and radiative losses are evaluated using equations (10) and (11), respectively.

Total heat losses $= Q_{Loss} = Q_{conv\ sL} + Q_{rad\ sL} + Q_{cond}$

$$\tag{19}$$

where Q_{cond} is the conductive heat loss given by

$$Q_{cond} = U_{back}\ A_L (T_L - T_{air}) \tag{20}$$

U_{back} = effective conductance, which accounts for conductive losses through back, sides, pipe fittings etc. In practice, the value of U_{back} lies in the range 0.3 to 0.6 W m^{-2}($^{\circ}$C)$^{-1}$. Useful heat gain is also given by

$$Q_{out} = \eta_o\ H\ A_L - Q_{Loss} \tag{21}$$

where H is the total insolation on the CPC,

η_o is the optical efficiency, which accounts for the fraction of insolation H that is absorbed by the collector and is given by

$$\eta_o = \gamma\ \tau\ (\rho_R)\ \alpha_s \tag{22}$$

where γ = fraction of total insolation accepted by the CPC, which depends upon the turbidity of the atmosphere and on acceptance angle [2].

From equation (21) the useful heat gain can be evaluated. If the assumed T_s is correct then

$$Q_{out} = Q_{u2} = Q_{u1} \tag{23}$$

Otherwise T_s has to be changed until $Q_{out} \approx Q_{u2} \approx Q_{u1}$. The values of T_s, T_2, and T_L corresponding to this condition are the predicted values, for given mass flow rate and insolation under steady-state conditions. Figure 2 shows the flow chart for the above iterative scheme.

The reflector temperature T_R can be predicted, following Ref.2, from the known values of T_L and T_s by the equations given in the Appendix.

The collector efficiency is given by

$$\eta = \frac{Q_{out}}{H\ A_L} \tag{24}$$

The overall heat transfer coefficient for the CPC is given by

$$U_L = \frac{Q_{Loss}}{A_L (T_s - T_{air})} \tag{25}$$

where U_L is based on the aperture area.

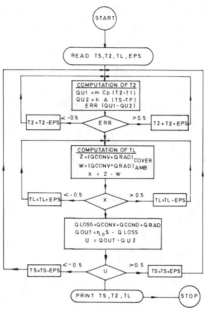

FIG.2. ITERATIVE SCHEME FOR PREDICTION OF ABSORBER TEMP.
OUTLET TEMP AND COVER TEMP. OF CPC FOR GIVEN
INSOLATION AND FLOW RATE

A computer program has been developed for the prediction of thermal performance of CPC using the iterative scheme indicated above.

3. NUMERICAL COMPUTATIONS

The numerical results are presented for three typical concentration ratios, 1.6, 5, and 8. The following inputs have been used in the calculation s.

(a) For the optical efficiency [2]:

$\gamma = 0.95$	for $\theta = 33.7^{\circ} (C = 1.6)$
$\gamma = 0.9$	for $\theta = 11.5^{\circ} (C = 5)$
$\gamma = 0.88$	for $\theta = 5.74^{\circ} (C = 8)$
$\tau = 0.9$	
$\rho_R = 0.85$	
$\alpha_s = 0.9$	

(b) for effective emissivity of CPC, ε_{ff} [2]:

$$\varepsilon_{eff} = 0.83 \qquad \text{for} \quad C = 1.6$$

$$\varepsilon_{eff} = 0.85 \quad \text{for} \quad C = 5$$
$$\varepsilon_{eff} = 0.86 \quad \text{for} \quad C = 8$$

For the total insolation normal to the collector, a sinusoidal function is assumed for predicting the hourly performance:

$$H = H_{max} \sin(\pi\, t_1/12) \tag{26}$$

where $H_{max} = 1000 \text{ W/m}^2$; t_1 = time of the day in hours

wind speed = 4.5 m/s

emissivity of the reflector $\varepsilon_R = 0.5$

emissivity of the cover $\varepsilon_L = 0.94$ (glass)

effective conductive loss factor, $U_{back} = 0.3 \text{ W/m}^2 \, ^{\circ}C$

L = collector length = 1 m

l = aperture width = 0.196 m

C_p = specific heat of water = 4190 J/kg $^{\circ}$C

$T_{sky} = T_{air} = 30^{\circ}C$

4. RESULTS AND DISCUSSION

Figure 3 shows the predicted absorber and outlet temperatures for a CPC of concentration ratio 5, and mass flow rate of 2.5 kg/hr, as a function of time of day. The insolation pattern assumed for predicting the above temperatures is also indicated in the figure. It can be seen that the absorber and outlet temperatures follow the insolation curve over most of the day, and the difference between absorber and outlet temperatures is maximum at noon. The distributions are symmetric about noon.

Figure 4 indicates the effect of mass flow rate on the collector efficiency for the three different concentration ratios. It can be seen that the collector efficiency increases more rapidly at the smaller flow rates.

Figure 5 shows the plot of efficiency vs $(t_s - t_a)/H$, depicting the performance characteristics of the CPC collectors investigated. For a given flow rate, as concentration ratio increases, the collector performance improves initially, but shows a decline with further increase, suggesting thereby that there exists an optimal concentration ratio. For the range of parameters investigated a concentration ratio of around 5 is found to be optimal from the standpoint of collector performance.

Figure 6 shows the effect of wind velocity on the performance of CPC collectors, for a concentration ratio of 5. It is observed that in the range of wind velocities investigated, namely, 4.5 to 15 m/s, there is no significant change in the collector performance.

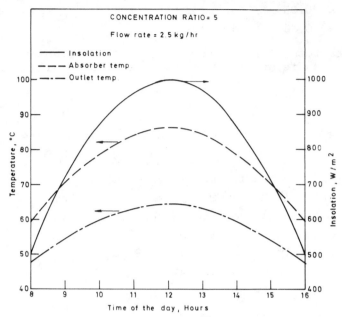

FIG.3. TRANSIENT VARIATION OF INSOLATION AND ABSORBER
AND OUTLET TEMPERATURE

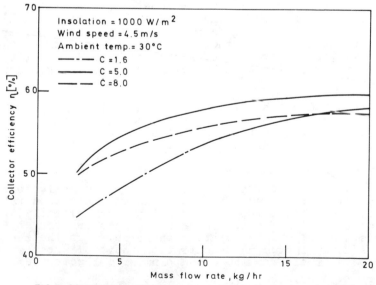

FIG.4. COLLECTOR EFFICIENCY Vs FLOW RATE FOR DIFFERENT
CONCENTRATION RATIOS.

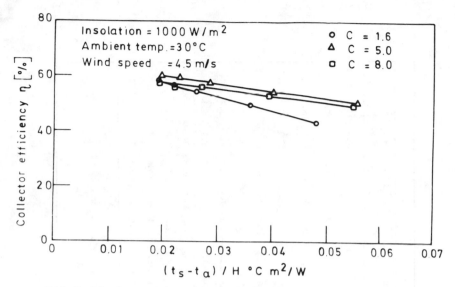

FIG. 5. COMPARISON OF PERFORMANCE FOR DIFFERENT CONCENTRATION RATIOS

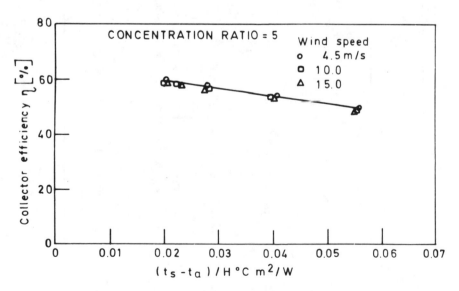

FIG. 6. EFFECT OF WIND VELOCITY ON THE CPC PERFORMANCE C = 5

TABLE I – PREDICTED PERFORMANCE OF CPC COLLECTORS

Concentration Ratio	1.6					5.0					8.0				
Optional Efficiency (%)	65.4					64.0					60.0				
Mass Flow Rate (kg/hr)	2.5	5.0	10.0	15.0	20.0	2.5	5.0	10.0	15.0	20.0	2.5	5.0	10.0	15.0	20.0
$T_{absorber}$ °C	78.0	66.0	56.0	52.0	49.0	85.8	70.1	58.4	5.3	49.9	85.2	69.3	57.0	52.0	49.0
T_{outlet} °C	58.3	46.4	39.1	36.3	34.8	64.2	48.6	39.8	36.6	35.1	63.2	48.3	39.2	36.4	34.9
T_{cover} °C	44.1	39.8	36.7	36.5	35.8	38.6	35.9	34.0	33.2	32.8	35.1	34.4	33.0	32.5	32.2
$T_{reflector}$ °C	49.4	43.8	39.5	38.7	37.6	42.3	38.5	35.8	34.6	34.0	38.3	36.5	34.4	33.6	33.2
Radiative Loss (watts)	29.2	20.9	14.5	11.4	9.6	13.5	8.9	6.0	4.7	4.0	8.9	5.7	3.7	2.9	2.5
Convective Loss (watts)	11.5	8.5	5.8	4.5	3.7	10.0	6.8	4.5	3.5	2.9	7.4	4.8	3.0	2.2	1.9
Conductive Loss (watts)	2.8	2.1	1.5	1.3	1.1	3.3	2.4	1.7	1.3	1.2	3.2	2.3	1.6	1.3	1.1
Total Heat Loss (watts)	43.5	31.5	21.8	17.2	14.4	26.8	18.1	12.2	9.5	8.1	19.5	12.8	8.3	6.4	5.5
U-Value W/(m²)(°C)	4.6	4.5	4.3	4.0	3.8	2.4	2.3	2.2	2.1	2.0	1.8	1.7	1.6	1.5	1.5
Useful Heat Gain (watts)	84.6	96.7	106.3	111.0	113.8	98.6	107.3	113.3	115.9	117.3	98.0	104.8	109.3	111.0	112.0
Efficiency (%)	43.1	49.5	54.2	56.6	58.0	50.3	54.7	57.8	59.1	59.8	50.0	53.4	55.7	56.6	57.2

Inputs used: Insolation = 10^3 W/m²; T_{air} = T_{sky} = 30°C; Wind speed = 4.5 m/s

Emissivity of reflector = 0.5; Emissivity of cover = 0.94; Emissivity of absorber = 0.9

Effective conductive loss factor U_{back} = 0.3 W/(m²)(°C)

Table 1 indicates the predicted radiative, convective and conductive losses, and the collector efficiency for different mass flow rates and for different cocentration ratios.

5. PRINCIPAL CONCLUSIONS

1. The absorber and outlet fluid temperatures follow the insolation curve through most of the day. The difference between the absorber and outlet fluid temperatures is maximum at noon.

2. Collector efficiency increases with flow rate for the concentration ratios investigated.

3. In the range of parameters investigated, a concentration ratio of 5 is found to be optimal from the stand point of collector performance.

4. Wind velocities, from 4.5 to 15 m/s, have negligible influence on the collector performance, for a concentration ratio of 5.

REFERENCES

1. A.J.Gorski, R.L.Cole, R.M.Craven, and W.R.Mcintire, 'Photo-Voltaic Power Generation by use of Compound Parabolic Concentrators', Alternative energy sources, Solar Energy 3, pp.1050.

2. A.Rabl, 'Optical and Thermal Properties of Compound Parabolic Concentrators', Solar Energy, 18, 497 (1976).

3. R.Winston and H.Hinterberger, 'Principles of Cylindrical Concentrators for Solar Energy', Solar Energy, 17, 255 (1975).

4. F.Krieth, 'Principles of Heat Transfer', 3rd Edn., Intext. Educational Publishers, New York (1973).

APPENDIX: PREDICTION OF REFLECTOR TEMPERATURE T_R:

$$T_R^4 = \frac{\varepsilon_{eff\ SR}\ T_S^4 + \varepsilon_{eff\ RL}\ T_L^4}{\varepsilon_{eff\ SR} + \varepsilon_{eff\ RL}} \tag{i}$$

where $\varepsilon_{eff\ SR}$ = effective emissivity for radiation between absorber (S) and reflector (R)

$\varepsilon_{eff\ RL}$ = effective emissivity for radiation from reflector (R) to cover (L)

Effective emissivity of CPC, ε_{eff}, is related to $\varepsilon_{eff\ RL}$ by the relation:

$$\varepsilon_{eff} = \varepsilon_{eff\ SL} + \frac{\varepsilon_{eff\ SR}\ \varepsilon_{eff\ RL}}{\varepsilon_{eff\ SR} + \varepsilon_{eff\ RL}} \tag{ii}$$

where $\varepsilon_{eff\ SL}$ = effective emissivity for radiation between absorber S and cover L,

$$\varepsilon_{eff\ SL} = \frac{\varsigma_{Ri}\ \varepsilon_L\ \varepsilon_S}{1 - \varsigma_{Ri}^2\ \varsigma_L\ \varsigma_S}, \tag{iii}$$

and $$\varepsilon_{eff\ SR} = \frac{\left(1 + \varsigma_{Ri}\ \varsigma_L\right)}{\left(1 - \varsigma_{Ri}^2\ \varsigma_L\ \varsigma_S\right)}\ \varepsilon_{Ri}\ \varepsilon_S, \tag{iv}$$

ς_{Ri} = effective reflectivity of the CPC for radiation within the acceptance angle = shape factor F_{SL} between S and L

ε_{Ri} = effective absorptivity of CPC for radiation within the acceptance angle.

ε_{Ri} = $1 - \varsigma_{Ri} = 1 - F_{SL}$

ς_L = reflectivity of the cover = $1 - \varepsilon_L$

ς_S = reflectivity of the absorber = $1 - \varepsilon_S$

Rearranging terms in equation (ii), $\varepsilon_{eff\ RL}$ is given by

$$\varepsilon_{eff\ RL} = \frac{(\varepsilon_{eff} - \varepsilon_{eff\ SL})\ \varepsilon_{eff\ SR}}{(\varepsilon_{eff\ SL} + \varepsilon_{eff\ SR} - \varepsilon_{eff})} \tag{v}$$

Optical and Thermal Analysis of Linear Solar Receiver for Process Industries

RAVINDRA KUMAR
M.N.R. Engineering College
Allahabad 211004, India

ABSTRACT

An eccentric cylindrical annulus receiver performance is projected for industrial applications and a two-stage CPC is discussed for solar cell applications.

INTRODUCTION

Almost every process industry requires some kind of process heat to yield any chemical or physical change on materials in process. The economic potential of solar thermal energy systems to provide industrial process heat is largely influenced by the temperature level required, the form of heat used, the amount of heat needed, its competative cost and the solar manufacturing techniques available. Basically the success of any solar energy system lies in its ability to catch the insolation by absorbing it at useful temperatures. The requirement for high fluid temperatures to achieve reasonable thermodynamic conversion efficiencies has led to many complexities of the receiver design. The possible use of spectrally selective surfaces, honeycomb materials, vacuum enclosures and convection suppression cavities are often reported.

In this paper, the performance of an eccentric cylindrical annulus receiver is projected for industrial applications. In section II, a two-stage CPC, one over other, is discussed for photovolatic applications where high cost demands that the absorber area be as small as possible.

BACKGROUND

Receivers may have a variety of shapes depending on the optical systems. An interesting example of determination of receiver geometry for a circular cylindrical reflector which led to a receiver of triangular cross section is given by Tabor and Zeimer [1]. Cobble [2] studied the influence of receiver geometry on maximum concentration and concludes that parabolic receiver shape is the optimum when used with a parabolic mirror of relative aperture four. Elliptic cylinder [3] is also argued to be near optimum. The manufacture and use of such a receiver

may pose difficulties whereas 'tube-in-tube' geometry finds its
wide application in any industry. Lumsdaine [4] has observed that
annular receiver yields higher thermal efficiencies especially at
large Nusselt number. The optical analysis of linear receivers
are also investigated and presented very recently [5] .

A compromise between optics and engineering permits a cylindrical black cavity receiver with an opening parallel to axis of
the collector to be near economical. The high flux density of
absorbed energy on a portion of the cavity creates very large
temperature gradient locally which means greater thermal losses
through the aperture. Boyd et al [6] suggest a scattering surface in the receiver cavity (Fig. 1). So that radiations are not

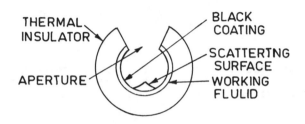

Fig. 1 - Boyd Receiver

absorbed at the point of maximum concentration but after defocu-
ssed and scattered uniformly over a much larger internal area of
the cavity. Thus lower effective temperature of the aperture
results into smaller temperature gradient and hence lower thermal
losses. This design involves (i) high fabrication cost of the
receiver for the need of true scattering surface such that view
factor between deflector and rest absorber surface would approach
unity and (ii) skilled hand.

If appears reasonable, instead of scattering surface in the
cavity, to design the receiver itself in such a way that the heat
transfer rate is enhanced at the point of maximum concentration of
the radiant energy. It can be achieved by an eccentric annulus
receiver (Fig. 2 and Fig. 3) providing maximum eccentricity that
means more fluid bed at the point of maximum concentration. This
design has the advantage of its simple fabrication technology
required with commonly available material, say, G.I. pipes painted
black inside and white outside. In the annular passage fins may
be provided to enhance the heat delivery at the maximum eccentric
point as suggested by McVeigh [7] .

ANALYSIS

Concentrator and receiver efficiency play important role in
maximum heat delivery. The requirement for the concentrating
system is based on delivering a maximum fraction of the incident
radiation into a minimum width aperture. Optimum receiver aperture width is discussed in terms of concentration ratio, tracking
allowance, alignment errors, f-ratio, rim angle and the contour

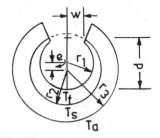

Fig. 2 - Receiver Notations

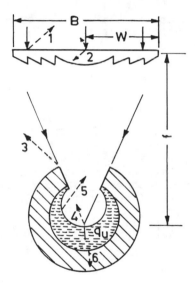

Fig. 3 - Solar Thermal Con-
version, Dotted
Arrows Indicate
Losses

errors. Values of intercept factor, γ , as function of receiver
width and acceptance angle is determined either by analytical
relationship or by numerical method. The rim angle corresponding
to this highest γ is the optimum value. A simplified relationship
between the receiver width, w, inside cavity radius, r_1, and
f-ratio of the optical system can be obtained. Vee-shaped
entrance aperture in the insulation is an important design feature
in limiting the infrared radiation less from the receiver aperture,
the amount of reduction being related to vee angle [6] .

    The conduction through insulator is the dominant loss at
moderate temperature where radiation loss dominats at higher
temperatures. Equating the surface losses to the conducted flux,
the relation between outer annulus radius, r_2, and outermost
insulation radius, r_3, can be obtained of the form

$$r_2 = r_3 \, e^{-\mathcal{F}}$$

where $\mathcal{F}$ is constant for given heat flow conditions and is
defined as

$$\mathcal{F} = \frac{\Delta T_f - \Delta T_{s-a}}{2\,\pi K \times \text{surface loss}}$$

which also is a function of inner cavity radius r_1.

The parameter, eccentricity of the annulus, e, can be associated
with temperature distribution radially and along the length of
the receiver.

It should be emphasized that the study presented here do not
represent perfectly optimized design of the receiver.

EXPERIMENTATION

Fig. 4 shows the experimental set-up consists of Fresnel lens
and linear cylindrical blackbody receiver. The heat absorber was
made of two G.I. pipes of lengths 183 cm each having diameters
90 mm, 50 mm and an axial slot of 4 cm width. The slot was cut on
a planer machine and edges were then welded. The receiver was
suspended with the frame work itself by means of clamping device
so that it could swing at the desired focal distance of the concentrator as and when the concentrator is tilted. Performance
curves for the two types of receivers are shown in Fig. 5.

It appears that proposed receiver design performs satisfac-
torily at moderate temperature, required for many applications in
process industries.

Fig. 4 - Experimental Setup

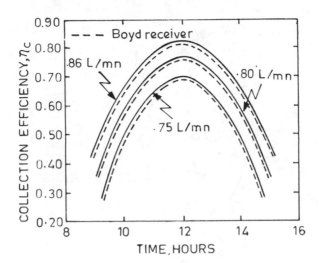

Fig. 5 - Performance curves, η_c, vs hours of
the day for Boyd receiver and the
proposed receiver

LINEAR CONCENTRATOR

Recently CPC has been widely investigated not only as single
stage but also for two stage concentration [8,9] . Ravindra Kumar
et al [10] have found that two CPC's in series require less
material than a single CPC for the same concentration ratio when
filled with a dielectric medium having refractive index $\mu > 1$.
This effect is observed because of the fact that an emitter in a
medium of index μ , radiates μ^2 as much energy as an emitter in
vacuum [11] .

Fig. 6 shows a three-stage CPC. Its theoretical depth is
plotted in Fig. 7. It is interesting to note that more than two-
stage CPC have only limited advantages compared with its complexities in design. Some of the design data for a two-state CPC is
given in Table-1. Fig. 8 compares the deepness of a two-stage
over single stage for the same concentration ratio. The decreased
concentration by individual stages results into increased acceptance angle and such overall increase in acceptance angle of the
collectors are shown in Fig. 9. But this advantage is counter
balanced by multireflection of incident beam radiations.

A critical concentration ratio for which depth of the
collector becomes zero was obtained for each stage which depends
on the ratio of refractive indices of the medium and the number
of stages of the collector.

This type of concentrator is found suitable for solar cells
where cells are cooled while receiving the energy.

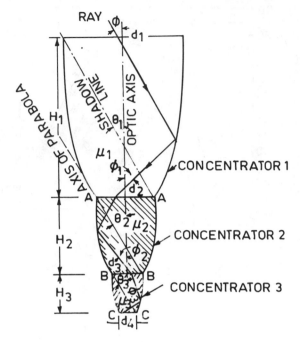

Fig. 6 – Three stage collector

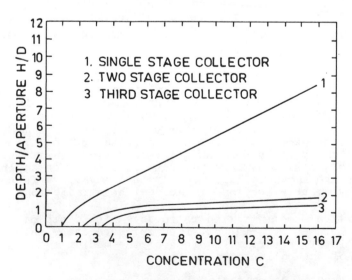

Fig. 7 – Depth variation with concentration for
single and multi-stage collector

TABLE-1 Design Data of Single CPC and Two Stage Collector

SINGLE CPC: Equation $y = \dfrac{x^2}{4f}$, $f = \dfrac{d_e}{2}(1+\sin\theta)$

CONCENTRATION, C	1	2	3	4	5	6	7	8	9	10	11	12	13	14	15	16
HALF ACCEPTANCE ANGLE OF COLLECTOR θ°	90	30	19.45	14.47	11.53	9.55	8.20	7.18	6.37	5.37	5.16	4.76	4.41	4.09	3.78	3.58
DEPTH TO APERTURE RATIO H/D	0.00	1.30	1.89	2.42	2.94	3.46	3.96	4.47	4.97	5.48	6.04	6.50	6.98	7.49	8.07	8.49

TWO STAGE CPC; $C_{cr}=2.25$ C	1	2	3	4	5	6	7	8	9	10	11	12	13	14	15	16
HALF ACCEPTANCE ANGLE OF FIRST COLLECTOR θ_1°	90	45.00	35.26	30.00	26.56	24.09	22.20	20.70	19.65	18.45	17.55	16.78	16.10	15.50	14.96	14.48
HALF ACCEPTANCE ANGLE OF SECOND COLLECTOR θ_2°	—	—	60.11	48.59	42.04	37.75	34.47	32.00	30.00	28.34	26.86	25.69	24.55	23.65	22.81	22.00
DEPTH TO APERTURE RATIO $\dfrac{H_1+H_2}{d_1}$	—	—	.72	.99	1.05	1.28	1.38	1.46	1.54	1.61	1.67	1.73	1.78	1.84	1.89	1.93

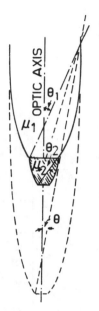

Fig. 8 - Two-stage CPC having c = 6,
$\mu_1/\mu_2 = 1.5$, $\theta_1 = 24°$, $\theta_2 =$
38° and $\theta = 9.5°$

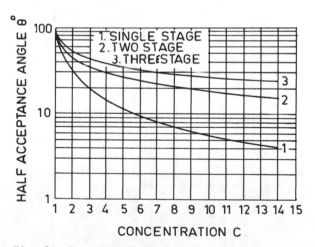

Fig. 9 - Increase in acceptance angle of the CPC

ACKNOWLEDGEMENT

    I should like to thank Dr. B.K. Gupta, Dr. S.B.L. Garg and
Dr. R.K. Bhardwaj for their helpful comments and discussions.

REFERENCES

1. Tabor, H., and Zeimer, H. 1962. Low Cost Focussing Collector
    for Solar Power Units. Solar Energy 6. p. 55.

2. Cobble, M.H. 1961. Theoretical Concentrations for Solar
    Furnaces. Solar Energy 5(2). p. 61-72.

3. Lumsdaine, E., and Cheng, J.C. 1976. On Heat Exchangers used
    with Solar Concentrators. Solar Energy 18(2). p. 157-158.

4. Lumsdaine, E. 1975. Comparison of Solar Heat Exchangers.
    Solar Energy 17(5). p. 269-275.

5. See for example. 1979 ISES Silver Jubilee International
    Congress Proceeding, Atlanta, Georgia. May 28-June 1.

6. Boyd, D.A., et al. 1976. A Cylindrical Blackbody Solar Energy
    Receiver. Solar Energy 18 p. 395-400.

7. McVeigh, J.C. 1978. A Stationary Concentrator for Industry,
    Conference (C14), Solar Energy for Industry, The Royal
    Institute, ISES UK Section. p. 41.

8. Lazor, J.M. and Alexander, C.K. 1976. Yes an Industrial
    Focussing non-tracking Collector will work in northeastern
    OHIO, Solar Cooling and Heating Vol. II edited by T. Nejat
    Veziroglu, Hemisphere Publishing Corporation, Washington.
    p. 367-375.

9. Rabl, A. 1978. Comparison of Solar Concentrators, Solar
    Energy Vol. 18, p. 93-111.

10. Ravindra Kumar et al. 1978. Depth Reduction Technique of
    Compound Parabolic Concentrators Proceeding of Solar Energy
    & Conservation Symposium-Workshop, Miami Beach Florida, USA.

11. Goodman, N.B. et al. 1976. On the use of Solid-dielectric
    Compound Parabolic Concentrators with Photo-Voltaic Devices.
    Proceedings of the Joint Conference of the American Section
    ISES and Solar Energy Society of Canada. Vol. 6, edited by
    Dr. K.W. Böer.

Dynamic Response Analysis of a Solar Powered Heliotropic Fluid-Mechanical Drive System

N.A. COPE, H.A. INGLEY, E.A. FARBER, and C.A. MORRISON
University of Florida
Gainesville, Florida 32611, USA

ABSTRACT

This paper provides a summary of work performed during the design, construction, and subsequent analysis of a solar powered tracking mechanism.[1] This device is a fluid-mechanical system which utilizes solar radiation as its power source.

The mechanism utilizes basic mechanical and thermodynamic principles in its construction and operation. This paper discusses these principles and describes the construction of the tracking device. Data taken during the course of the research and reported in this paper reveal that with a particular combination of system components and working fluid, a high degree of accuracy and wind stability can be achieved with this device when used to drive large concentrating solar collectors. The tracking mechanism was found to be fully self-correcting during normal daily operation and to reorient itself to the morning sun.

NOMENCLATURE

M	moment
A	area
P	pressure
D	diameter
f	friction coefficient
W	load
Wk	work
K	thermal conductivity
σ	Stephen-Boltzman constant
γ	view factor
h_{CR}	combined convection radiation coefficient
I	solar irradiation
C_p	specific heat

Subscripts

D	drive cylinder
C	collector
V	variable

```
gr    gear
b     bearing
pn    piston
f     friction
s     convection surface
amb   ambient
se    sensor element container
g     gas
```

INTRODUCTION

As the uses of solar energy expand into a diversity of areas, the need for high temperature (greater than 100·C) solar collectors strengthens. Whereas flat plate collectors find applications in heating domestic water, space heating and in some cases even comfort cooling, concentrating or focusing collectors are coming into wider use to provide the energy required to drive refrigeration equipment, electricity generating equipment, and to provide heat for industrial processes.

There are many designs for concentrating collectors including those requiring a minimal degree of tracking (vacuum tube types, pyramidal optics systems, and Winston Collectors), those which are set at fixed angles to the horizontal and are rotated about their axes from east to west (parabolic concentrators, linear Fresnel systems, and tracking absorbers) and finally those collectors which require complete tracking capability for both altitude and azimuth angles (dish shaped collectors, heliostats, and tracking absorber elements).

In each case where the collector must be moved in order to maintain the proper orientation, some source of external power is required. Generally this power is in the form of the electricity required to power a small motor. This method requires periodic maintenance and, of course, a source of electricity, precluding the use of these mechanisms in remote locations.

There has been a need for the development of a tracking device that is low cost, simple in construction and self powered. In 1975, the authors and Mr. D.B. Wiggins [2] implemented research to study the use of fluid-mechanical drive systems. The initial designs were simple but proved the concept. The work reported in this paper describes the measures taken to improve the dynamic response of the proposed mechanism. Mechanical tracking systems employ driving mechanisms such as raised weights and compressed springs, bimetallic-expansion springs, and expanding gases. A unique method used in Russia for tracking the sun employed a locomotive engine pulling mirrors mounted on rail cars! Each of these methods has its advantages and disadvantages and in some cases may even be used in remote locations. However, they require either close attention in order to insure proper operation or they lack the precision required in tracking the sun.

METHODOLOGY

The two primary tasks proposed for this study were 1) to increase the sensitivity and accuracy of the tracking device so that it would maintain a collector position within the limits permissible for full solar energy collection and 2) to decrease the time needed for self correction and realignment of the collector after direct solar rays become available.

The parabolic collector used in this study is shown in Figure 1. The solar powered tracking device is attached to this collector as illustrated by the shadow panels and two power cylinders shown in this figure. Even though this mechanism can be used for dual axis tracking, single axis tracking was selected to introduce a minimum of variables in this initial study. The collector shown was set at an angle of 30˙ from the horizontal and allowed to rotate in an east west direction.

Figure 1
Parabolic Collector

An analysis of the collector geometry indicated that an angle of 1.19˙ was the maximum deviation that could be permitted if the maximum energy were to be collected. This corresponds to a response time of 4.7 minutes. An angle of deviation equal to 0.50˙ was selected as a design goal. In order to accomplish these goals, it was necessary to study separately three aspects of the problem: the tracking drive mechanism, the working fluid, and the sensor design.

Figure 2 provides a schematic of the tracking mechanism selected. Other options considered were: 1) four bar linkages
2) linked chains
3) belt drives

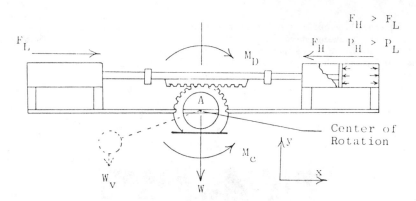

Figure 2
Schematic of Tracking Mechanism

The options were evaluated in view of the following design specifications:
    1) the drive mechanism must be a positive non-slip device
    2) low friction
    3) compact and lightweight
    4) simple in design and construction
    5) low cost.
Ultimately the rack and gear shown in Figure 2 was selected for the study.

    As Figure 2 illustrates, the gear is fixed and the rack rotates around
the gear.   The direction of rotation of the mechanism is clockwise if the
pressure in the right drive cylinder is larger than the pressure in the left
cylinder and vice versa for the reverse conditions.

    The tracking mechanism is set in motion when one cylinder has a pressure
sufficiently above that of the opposing cylinder so that the resultant
pressure force overcomes the frictional resistance of the system.   The
pressure difference required to produce the force necessary for impending
motion can be calculated by summing moments about the center of rotation, A.

$$\Sigma M_A = 0 = M_D - M_C - M_V - M_{gr} - M_b \tag{1}$$

where the drive cylinder torque is given by

$$M_D = (P_H - P_L) (A_{pn}) D_{gr}/2 \tag{2}$$

and the torque due to the balanced weight on the collector shaft (coincident
with the center of rotation is given by

$$M_C = f_b W_b D_b/2 \tag{3}$$

    M_V represents any variable or unbalanced torque produced in the collec-
tor and tracking assembly such as wind loading.

    M_{gr} is due to the friction created between the rack and gear. M_b
is the torque resulting from fractional forces in the drive cylinder.   Substituting into equation 1 the various forms for these torques and rearranging the
equation results in the form:

$$(A_{pn}D_{gr}/2)(P_H - P_L) = (f_bW_bD_b/2 + W_{VD} + 0.02 \ W_g - f_bW_bD_b/2) \tag{4}$$

This equation may then be written as

$$(P_H - P_L) \ A_{pn}D_{gr}/2 = W_f \tag{5}$$

where W_f equals the sum of the system's frictional forces.   These forces were measured experimentally and found to be equal to 7.29 Newton meters (64.5 in.lbf). D_{gr2} was equal to 0.171 m (6.718 in.) and A_{pn} equal to 0.0194 m^2 (30 in^2).   Solving for $P_H - P_L$:

$$P_H - P_L = 4,412.7 \text{ pascals } (0.640 \ lbf/in^2)$$

An analysis of several working fluids was performed in order to select the best fluid for year around applications in Florida.   The criteria observed in the selection were as follows:

1) The absolute pressure rating of the drive cylinders for safe operation (862 kilopascals, 125 psia),
2) the minimum required pressure differentials, a yearly sensor temperature swing of $-6.67 \cdot C$ ($20 \cdot F$) to $51.67 \cdot C$ ($125 \cdot F$) and
3) the compatibility of the working fluid with the diaphragm material.

Using these criteria, two refrigerants, R-12 and R-114, were selected for experimental verification.

The final task was to design sensors which would provide the pressure differential in the drive mechanism.

A first law analysis of the sensors yields the following equation:

$$Q_{in} = Q_{out} + Q_{stored} + Wk_{pn} \tag{6}$$

where $Q_{in} = IA_p$ (7)

$$Q_{stored} = (MC_p \frac{\Delta T}{\Delta T})_{se} + (MC_p \frac{\Delta T}{\Delta T})_g \tag{11}$$

$$Q_{out} = Wk_{pn} + [KA\Delta T + hA_s\Delta T + \sigma\epsilon\gamma A_s \ (T^4 - T_{amb}^4)] \tag{12}$$

Combining terms and setting Wk_{pn} equal to zero (assumes motion is impending), equation 6 can be reduced to:

$$IA_p\alpha = h_{CR} \ A_s \ \Delta T + \Sigma MC_p \frac{\Delta T}{\Delta T} \tag{13}$$

(h_{CR} combines convection and radiation terms)
If the storage term is neglected, equation 13 can be rearranged to provide an expression for the area ratio of thermal convection surface area to the maximum projected irradiated area:

$$\frac{A_s}{A_p} = \frac{I \ \alpha}{h_{CR}\Delta T} \tag{14}$$

For Gainesville, Florida. the terms on the right may be specified:
I   = 93.76W (320 Btuh/ft^2)
α = 0.95
ΔT = $4.44 \cdot C$
h_{CR} = 14.19 W/$m^2 \cdot K$ (2.5 Btuh/$ft^2 \cdot F$)

and

$$\frac{A_s}{A_p} = 3.04$$

The aforementioned analysis led to the development of four sensors (Figures 3 and 4). Each of these sensors was bench tested before selecting Sensor 1 as the design to use in the total tracking system evaluation. The physical characteristics of Sensor 1 are given in Table 1.

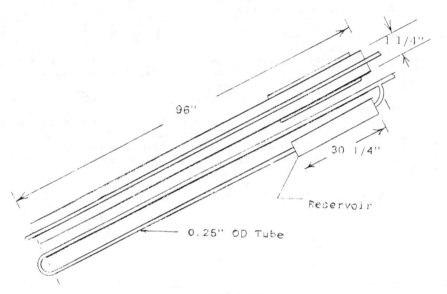

Figure 3a
Type 1 Sensor Element & Fluid Reservoir - Top & Side View

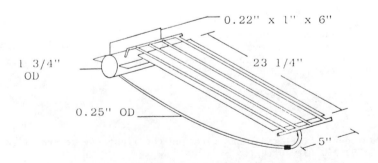

Figure 3b
Type 2 Sensor with Reservoir

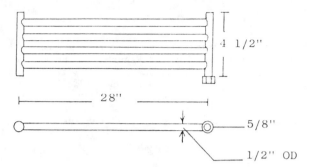

Figure 4a
Type 3 Sensor

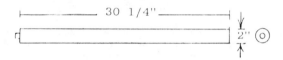

Figure 4b
Type 4 Sensor

Table 1
Physical Characteristics of Sensor 1

Physical Characteristics	#1	
Projected area (m^2)	0.08	(0.83 ft^2)
Total surface area (m^2)	0.20	(2.19 ft^2)
Fluid volume (m^3)	3.99 X 10^{-5}	(14.1 X 10^{-4} ft^3)
Mass of empty element (kg)	1.04	(2.29 1bm)
Internal wetted area (m^2)	0.04	(0.38 ft^2)
Projected area to internal wetted area ratio	2.18	
Total area to internal wetted area ratio	5.76	
Internal wetted area to mass of empty element (m^2/kg)	0.03	(0.17 ft^2/1bm)
Mass of R-12 in full sensor of 38·C (kg)	0.05	(0.11 1bm)
Total area to fluid mass R-12 at 38·C (m^{2}kg)	0.84	(19.91 ft^{2}1bm)
Mass of R-12 at 38·C to mass of empty element	0.05	
Specific heat capacity-filled sensor (R-12 at 38·C (cal/·C)	33.60	(0.24 Btu/·F)

DISCUSSION

The sensor and tracking mechanisms were integrated and mounted on the large parabolic concentrator shown in Figure 1. The collector was instrumented with a pyranometer and an angular position sensor. The straight lines shown in Figures 5 through 8 represent the "ideal" collector position, that is, the position the collector would follow if the tracking mechanism tracked the sun with no deviation. The difference between this straight line and the collector's actual position represents the tracking error.

1. Tracking Accuracy

The use of R-12 provided the tracking accuracy required to meet the design goal of $\pm$ 0.5· deviation. It was concluded that for the existing design R-12 would be best suited for use during late fall, winter and early spring ambient conditions, whereas R-114 would be best for late spring, summer, and early fall conditions. With proper piston design, R-12 could be used for year around applications.

In general, the larger average deviation values resulted from the collector position lagging the ideal position. This was attributed to low levels of irradiation. This suggests the use of wider fins on the sensor, as this would enhance the performance of the tracker during periods of diminished direct radiation.

2. Minimum Solar Irradiation Required

The minimum solar irradiation required for accurate tracking, with both R-12 and R-114, was approximately 583 watts/m^2 (185 Btuh/ft^2).

The minimum insolation required for an accurate tracking also depended upon the amount of direct solar irradiation as compared to the amount of diffuse radiation.

Even though the tracking device required at least 583 w/m^2 (185 Btuh/ft^2) for accurate tracking it would respond to insolation values as low as 252 w/m^2 (80 Btuh/ft^2) but would consistantly lag behind the sun's position.

3. Self Adjustment

The Type 1 sensor responded quickly to changes in the sun's position and to wind loadings and intermittent cloud cover. The wind not only loads the system mechanically but will also increase the convective heat losses from the sensors. Comparing the figures for runs with wind shields on and off illustrates the effects caused by winds. Wind shields on the sensors greatly improved the performance of the system.

4. Early Morning Response

One of the major concerns about the solar powered tracking device was its capability of rotating the collector in the mornings from west back to east. It was found that the collector could rotate from its furthermost western facing position to its furthermost eastern position in less than 7 minutes.

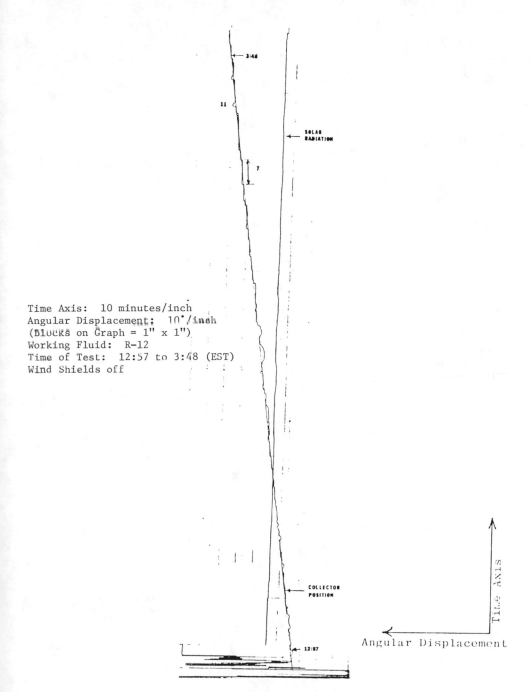

Time Axis: 10 minutes/inch
Angular Displacement: 10°/inch
(Blocks on Graph = 1" x 1")
Working Fluid: R-12
Time of Test: 12:57 to 3:48 (EST)
Wind Shields off

Figure 5
Tracking Performance
22 January 1977

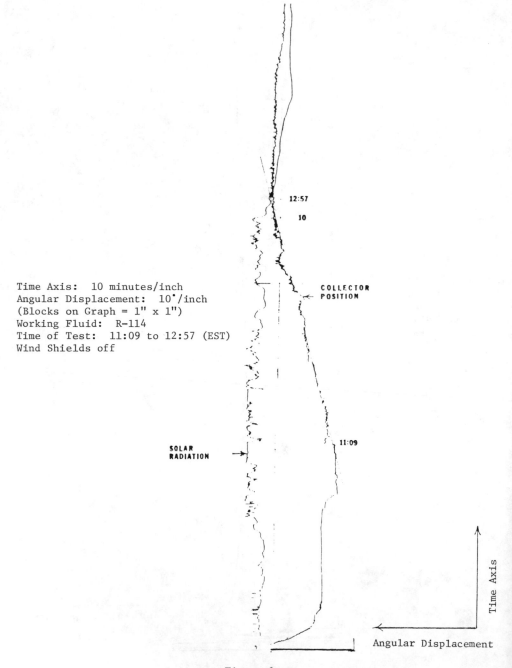

Time Axis: 10 minutes/inch
Angular Displacement: 10°/inch
(Blocks on Graph = 1" x 1")
Working Fluid: R-114
Time of Test: 11:09 to 12:57 (EST)
Wind Shields off

Figure 6
Tracking Performance
24 January 1977

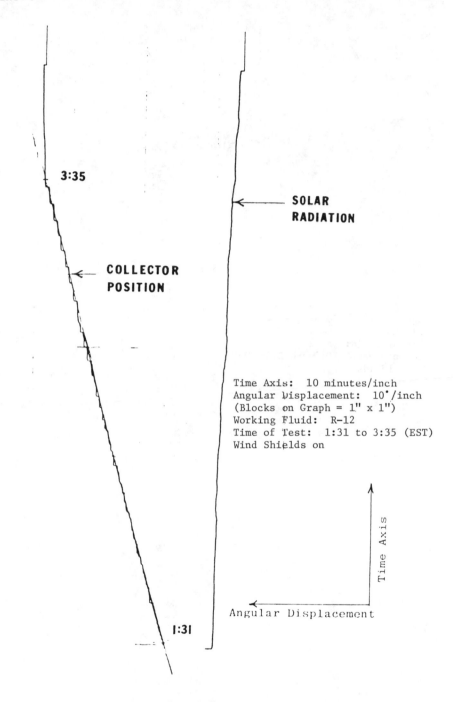

Time Axis: 10 minutes/inch
Angular Displacement: 10°/inch
(Blocks on Graph = 1" x 1")
Working Fluid: R-12
Time of Test: 1:31 to 3:35 (EST)
Wind Shields on

Figure 7
Tracking Performance
8 February 1977

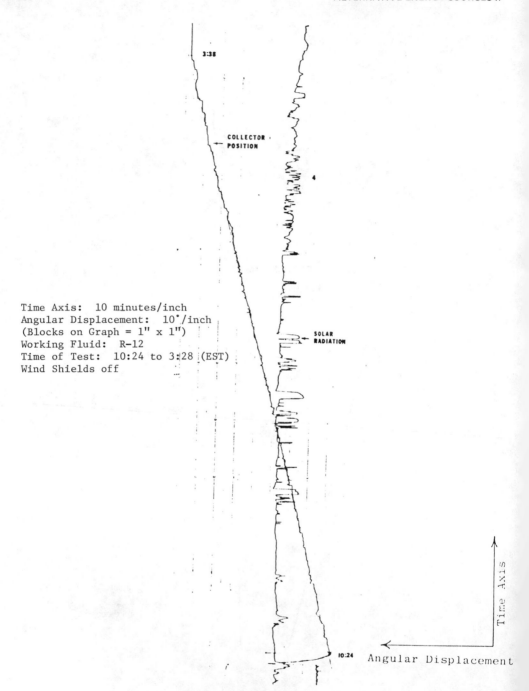

Time Axis: 10 minutes/inch
Angular Displacement: 10°/inch
(Blocks on Graph = 1" x 1")
Working Fluid: R-12
Time of Test: 10:24 to 3:28 (EST)
Wind Shields off

Figure 8
Tracking Performance
12 February 1977

5. Wind Stability

Wind stability was studied in the light of two parameters, wind load and convection heat transfer for the sensor elements. Wind shields opened at the ends protected the sensors from the wind and permitted a higher degree of accuracy. The spikes seen in Figure 6 were due to gusts of wind striking the collector, R-12 was by far less sensitive to wind gusts and the resulting heat losses than R-114.

CONCLUSIONS

It was concluded that the solar-power tracking mechanism investigated by this study demonstrated a viable means of driving concentrating collectors and other equipment which requires moderately precise tracking of the sun.

A design maximum deviation for tracking of $\pm$ 0.50· not only proved to be a realistic design goal, but the data indicate that the maximum deviation of $\pm$ 0.25· is not unrealistic.

The dynamic response of this tracking system was more than sufficient to respond quickly to changing environmental conditions such as insolation, temperature, and wind.

R-12 was found to have an excellent temperature/pressure sensitivity. The tracking system performed well with R-12 as the working fluid. Furthermore, the data taken using R-114 indicate that this refrigerant would best serve its purpose in warmer climates.

REFERENCES

1.   Cope, Norman L. "Dynamic Response Analysis of a Solar Powered Heliotropic Fluid-Mechanical Drive System", a thesis presented to the Graduate Council of the University of Florida in partial fulfillment of the requirements for the degree of Master of Engineering, 1978.

2.   Farber, E.A., Morrison, C.A., Ingley, H.A., and Wiggins, D.B., "Solar Powered Tracking Device", Building Systems Design, (December-January, 1976).

ENERGY STORAGE

Energy Storage Technology— Environmental Implications of Large Scale Utilization

M.C. KRUPKA, J.E. MOORE, and W.E. KELLER
Los Alamos Scientific Laboratory
Los Alamos, New Mexico 87545, USA

G.A. BACA, R.I. BRASIER, and W.S. BENNETT
Los Alamos Technical Associates, Inc.
Los Alamos, New Mexico 87544, USA

ABSTRACT

The Department of Energy has emphasized, in recent years, the development of several advanced energy storage technologies. It is expected that these technologies will have certain environmental impacts. Such impacts must be assessed so that appropriate environmental control technology, where deemed necessary, can be developed on a schedule compatible with the development of the specific energy storage technology. A number of environmental assessment programs have been conducted at the national laboratories including the Los Alamos Scientific Laboratory (LASL) over the past several years. Environmental impacts for several energy storage technologies have been identified. State-of-the-art control technology options were similarly identified. Recommendations for research and development on new control technology were made where present controls were either deemed inadequate or non-existent. Specifically, the energy storage technologies under study included: advanced lead-acid battery, compressed air, underground pumped hydroelectric, flywheel, superconducting magnet and various thermal systems (sensible, latent heat and reversible chemical reaction). In addition, a preliminary study was conducted on fuel cell technology. Although not strictly classified as an energy storage system, fuel cells in conjunction with product recycling units can serve an energy storage function. A very large number of potential environmental impacts can be identified for all of these technologies. However, not all are of primary importance. Detailed discussions of a number of environmental impacts from the latest LASL study as they relate to primarily operational situations are emphasized. In addition, a brief discussion on new applications for energy storage technologies and the additional costs of controls to be used for mitigation of specific impacts are also presented.

1. INTRODUCTION

In one approach to promote resource conservation, minimize foreign resource dependency, increase operational efficiency and reduce costs, consideration has been given by the Department of Energy (DOE) to the development of several advanced energy storage technologies. These have major applications in the areas of transportation, building, heating and cooling, industrial processes and electric power generation. They will permit efficient and continuous usage of otherwise intermittent renewable energy sources such as solar and wind. The general relationship of energy storage technologies to the source, distribution and end use networks is shown in Fig. 1.

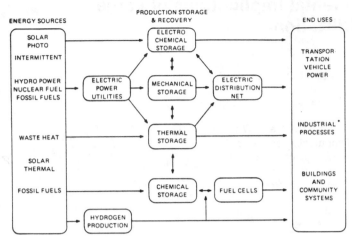

Fig. 1.   Energy storage technology, source, distribution and end
use networks.

As is the case with the prime energy technologies, storage technologies
are projected to have a number of environmental impacts, especially if they
are to be utilized on a large scale. These impacts must be identified and
assessed so that appropriate environmental control technology, where deemed
necessary, can be developed on a schedule compatible with the development of
the specific energy storage technology.

To assist the Office of Environment of DOE in the timely development of
such control technology, a number of assessment programs have been conducted
at the various national laboratories over the past several years. At the Los
Alamos Scientific Laboratory (LASL), two previous programs have been
conducted.   In March 1975, an Electrical Power Transmission and Energy
Storage Systems Working Group was assembled, as part of a larger task-force,
to assist in formulating that part of the DOE (then ERDA) program relating to
environmental control technology.[1]   In 1976, an assessment study in further
detail was made expanding upon the efforts of the Working Group.[2]

This paper presents some results of an additional assessment designed
to update information contained in LA-6979-MS, "Environmental Control Technology R & D Requirements for Energy Storage Systems."[2]   The specific
energy storage technologies under consideration include: advanced lead-acid
battery, underground pumped hydroelectric, superconducting magnet, compressed
air, flywheel and thermal. In addition, a preliminary environmental assessment was made concerning fuel cell technology. Strictly speaking, fuel cells
per se are not energy storage devices. Many of their benefits, however, are
similar to those of energy storage devices. Further, when coupled to certain
recycling units, the total system operates in an energy storage capacity. A
brief study was also included relating to industrial implementation of energy
storage technologies for purposes other than that of centralized electric
power generation by the utility industry. This survey was made in the hope
that new and significant impacts specifically related to a new application
would thus be identified. Finally, wherever possible, an estimate of the
additional cost to the energy storage system or to its product, e.g., elec-
tricity-kWhr, due to the implementation of some form of control technology,
was made.

Certain guidelines and assumptions were used in the course of the latest investigation as follows:

a. Emphasis has been placed primarily upon operational impacts. Previous studies have shown that both preoperational (mostly construction) and decommissioning environmental impacts for most storage technologies have a reasonable degree of commonality (with the possible exception of specific decontamination procedures).[2] Control technologies for these categories of impacts are well known and little or no new control technology is generally required.

b. The general time frame for the investigation covers the period 1985-1990. In view of the many uncertainties involved with funding of research and development programs, this time range for near term energy storage implementation may conceivably be extended. It is also clear that for a number of potential applications, large scale utilization will almost certainly be later than the dates specified above.

c. The environment under consideration is the area located beyond the site boundary and in general would not include activities within the storage technology complex. Thus, those activities regulated by the USEPA and the OSHA were not considered here. However, there does appear to be certain "grey areas" where such activities cause or influence impacts extending beyond the complex perimeter. To this extent, some data has been presented with regard to excursion (accident) situations and certain other operational impacts.

d. Systems concerned with the generation, distribution and storage of hydrogen were not included in this investigation.

e. The methodology used included a computerized reference literature survey, review of present storage programs under development including state-of-the-art control technology, personal contacts with private and governmental organizations and attendance at specific storage technology seminars.

The primary environmental impacts and suggested control technologies associated with the specific energy storage technology will now be discussed. In view of the many technical details and designs associated with these technologies, it is suggested that appropriate engineering references be consulted.

2.   TECHNOLOGIES

Those environmental impacts deemed to be of primary concern as determined in this study and associated with the normal operational phase are summarized in Table 1.[3] Included are suggested mitigation control technologies as well as cost estimates for such control technology. It should be noted, however that an in-depth economic study was not made, hence significant deviation from the values quoted is possible.

2.1. Advanced Lead-Acid Battery

The responsibility for the Department of Energy's battery research, development, and demonstration program lies within its Division of Energy

TABLE 1. SUMMARY OF PRIMARY ENVIRONMENTAL IMPACTS, CONTROL TECHNOLOGY AND COST ESTIMATES

Storage Technology	Operation	Affected Environment	Impact	Mitigation Control Technology	Mitigation Technology Cost Estimate*
Advanced Lead-Acid Battery	Normal	Air, Land	Hydrogen, Arsine, Stibine	Catalytic recombination, special scrubbing system including instrumentation	18% plus land fill costs- $7-9/ton waste ($1979)
	Excursion (fire, acid spill)	Air, Land, Water	Noxious gases, particulates, acid water	Venting, scrubbing (CaO), dilution	
Underground Pumped Hydro-Electric	Normal	Land, Water, Biosystems	Chemical and biological contamination, cyclic stress, entrainment	Design and site selection criteria, fracture detection instrumentation	Not available
	Excursion (structural failure)	Land, Water	Leakage, flooding, collapse	Stabilization, diversion, design and site selection criteria	
Superconducting Magnet	Normal	Land, Biosystems	Magnetic field, cyclic stress	Structure reinforcement, distance, shield coil	4.3% of main coil or land < $10,000/ acre for 10 G spec. - $2,000/ acre for 0.3 G spec.
	Excursion (structural failure, explosion)	Land	Fracture - Wall collapse, emergency shutdown	Stabilization	
Compressed Air	Normal	Air, Land, Water, Biosystems	Chemical - Biological contamination, wall degradation, subsidence	Cavern design and site selection criteria development, cyclic stress studies, detection instrumentation	Aquifer - not available Water injection equip.-$5-50/kW, Wells and instrumentation - 5%
	Excursion (structural failure)	Air, Land, Water	Blowout, cave-in, subsidence, seismicity	Stabilization, sealants, design and site selection criteria	
Flywheel	Normal	Land	Safety	Design of flywheel and containment system, Sensors	5% of installed system-utility and residential Moving base-5-10%

TABLE 1. SUMMARY OF PRIMARY ENVIRONMENTAL IMPACTS, CONTROL TECHNOLOGY AND COST ESTIMATES (cont)

Storage Technology	Operation	Affected Environment	Impact	Mitigation Control Technology	Mitigation Technology Cost Estimate*
Thermal	Excursion (structural failure)	Land	Rotor burst, vacuum bearing failure, secondary-particulates	Design	
	Normal				
a. Sensible		Air, Land, Biosystems, Water	Leakage	Leakage design, dilution, biodegradation	3%
b. Sensible-Aquifer		Land, Biosystems, Water	Contamination, Chemical, Biological, Subsidence (See list of potential impacts in text)	Site selection, geohydrological studies	Not Available
c. Latent Heat		Air, Land, Biosystems, Water	Leakage	Leakage design, detection instrumentation	8% plus land fill costs - $7-9/ton waste ($1979)
d. Reversible Chemical Reaction		Air, Land, Biosystems, Water	Chemical and solution leakage	Leakage design, detection instrumentation	8% plus land fill costs - $7-9/ton waste ($1979)
	Excursion (major leakage, fire, explosion)	Air, Land, Water	Chemical and solution leakage	Scrubbing, land fill disposal	
Fuel Cell	Normal	Air, Land	Electrolyte disposal, thermal (2nd gen.)	Neutralization, land fill disposal	10% plus land fill costs - $7-9/ton waste ($1979)
	Excursion, (fire, leakage)	Air, Land	High temperature chemical release	Neutralization, land fill disposal, safety shield (2nd gen.)	

*Percentage costs are related to total capital cost of the energy storage system.

Storage Systems (STOR). STOR is divided into several activities with battery
development programs under the guidance of the Electrochemical Systems Activity. The goals of this branch include the development of batteries for utility and vehicle applications, industrial processes, and solar/wind energy
utilization. To meet these goals, part of the program is concentrating on
upgrading the performance characteristics (energy density, power density, and
cycle life) of the state-of-the-art lead-acid battery. A lead-acid battery
that meets the upgrading performance characteristics is referred to as an
"Advanced Lead-Acid Battery" Detailed specifications for such a battery
include:[4]

A. The battery* shall be capable of operating for its rated life under
    indoor ambient temperatures ranging from 0° to 40°C with a
    maximum relative humidity of 99%. The battery shall be capable of
    delivering full output at an ambient temperature of 25°C.

B. The battery shall be capable of withstanding without damage exposure
    to accidental acid spills and a water fog operation of 0.25
    gal/min./ft².

C. The capability of the battery measured at the direct current (dc)
    bus shall be as follows:

    • Discharge - The battery shall be capable of being discharged
      routinely between its 1 and 10 hr rates with infrequent dis-
      charges at a 15-min rate.

Condition	Time hr	Power kW	Current A
Constant Current	1/4	--	10,000
Constant Power	10	180	--

    • Charge - The battery shall be capable of being charged to 70% of
      its 10 hr rated capacity in ampere-hours within 2 hr followed by
      not more than 5 hr of additional charging to achieve 100% ampere-
      hour capacity.

    • The end of charge voltage and/or equalization voltage for all
      four strings in series shall be 1,000 (+0, -2) volts, and the
      maximum end of charge voltage for all strings in parallel shall
      be 250 (+0, -2) volts.

    • The allowable voltage and current ranges shall be

dc Current Range for any Voltage Within the Voltage Range	dc Voltage Range for any Current Within the Current Range
Amperes	Volts
60 to 2,500	
60 to 10,000	500 to 1,000
	125 to 250

---

*The battery may be composed of any number of cells.

D. The battery shall be capable of 1,750 cycles over a minimum of 10 yr.

E. The electrolyte shall be within 1.205 to 1.215 specific gravity at 25°C and filled to the high-level mark.

F. The battery shall not evolve more than 140 mg/min of stibine and 10 mg/min of arsine during any charge or discharge regime.

G. The amount of hydrogen gas emitted by the battery on charge or equalization charge shall be less than 600 ft^3 during any half-hour period at an ambient temperature of 35°C.

H. The amount of acid emitted by the battery at any time shall be less than 280 mg/min and/or 50 g in any 8-hr period.

I. All materials should be noncombustible or fire-retardant. If combustible, provision for placing 1 in. of noncombustible material between modules shall be provided.

J. The cells shall be capable of withstanding, without rupture of the fluid-containing cell cases, an explosion of the hydrogen mixture within the cell. The explosion shall not cause the failure or rupture of adjacent cells or modules.

K. The maximum ripple current that may be applied to the battery is:

Condition	Ripple Current as Percent of dc Current Peak-to-Peak
Charge or Discharge	20%
Equalization Charge	100%

Lead-acid batteries will be installed at the National Battery Energy Storage test (BEST) facility initially for shakedown purposes. This facility is funded jointly by the DOE, the Electric Power Research Institute (EPRI), and Public Service Electric and Gas Company of New Jersey (PSE&G). The objectives of the BEST facility are 1) to serve as a testing site for evaluating and assessing the performance of advanced battery systems for load-leveling and peak-shaving applications by electric utilities and 2) to evaluate new power conversion equipment associated with new batteries. The BEST facility will be completed in 1981 with the test program beginning in FY'82.

The DOE also plans a Storage Battery Electric Energy Demonstration (SBEED).[5] The objective of SBEED is to demonstrate the technical and operational characteristics of dispersed lead-acid battery energy storage on a commercial scale. The contractor consortium for SBEED is being selected at the time of this writing. Construction of the SBEED plant is to begin in late FY'81 and plant operation is scheduled to begin in mid FY'84.

As noted in Table 1, the primary impacts from large scale battery usage involves the production of hydrogen, stibine and arsine gases. The hydrogen evolved can have serious safety impacts if its concentration in air is not kept below its explosive limit of 4%.

The arsine and stibine evolved can also have potential adverse occupational and environmental impacts. These two compounds are extremely toxic, as exemplified by their low threshold limit values (TLV) of 0.2 mg/m^3 (arsine) and 0.5 mg/m^3 (stibine). The advanced lead-acid battery specifications limit production to 10 mg/min of arsine and 140 mg/min of stibine. This generation rate occurs for a short period during the recharging of the battery. These compounds would normally be released to the facility environment and should react <u>readily</u> with the oxygen in the environment to produce antimony oxide and arsenic oxide--both solids at the facility operating conditions. The TLV for arsenic oxide (0.5 mg/m^3) is higher than that of arsine. The oxidation of arsine and stibine can be accomplished with various oxidants, e.g., air, manganese dioxide.

Calculations conducted in this study based upon battery specifications and assuming a constant stibine and arsine generation rate (worst case)[3], suggest that the TLV's will be exceeded and therefore these gases or oxidation products should not be permitted to escape into the facility environment. Work done at ANL suggests a lower generation rate for stibine and arsine.[6]

The evolution of hydrogen during the charge mode of a lead-acid battery can have a potentially serious impact on safety. The concentration of hydrogen in the battery facility can be kept below the explosive limit by diluting the cell vent gas and installing a catalytic recombiner.

The processing of the cell offgas to recombine the hydrogen and oxygen to form water offers the opportunity for further processing to control the emissions of compounds of arsenic and antimony and sulfuric acid mist. Figure 2 is a diagram of a conceptual process for this purpose. The vent gas stream is diluted with oxidizer (air), before it enters any manifolding, to assure that the concentration of hydrogen does not exceed its explosive limit of 4% and to promote the oxidation of arsine and stibine to their particulate oxides. The gas is then scrubbed to remove these particulates. The scrubber is equipped with a deentrainment pad to separate the entrained water/sulfuric acid mist from the gaseous stream. After passing through the scrubber, the gases enter a hydrogen-oxygen recombiner. This device catalytically produces water from the hydrogen and oxygen in the gas stream. These recombiners are currently and extensively used in the nuclear reactor industry.

This process illustrates that control technology is currently available to control the possible environmental impacts associated with the normal operation of an advanced lead-acid battery storage system.

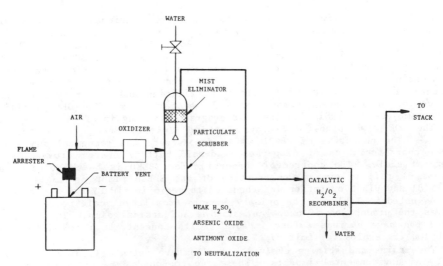

Fig. 2.   Conceptual control process for advanced lead-acid
battery facilities.

The mitigating control technology is comparable to pollution control technology and processes used in the chemical process industry. The cost of such pollution control processes is not expected to exceed 10% of the capital costs of the facility.

The cost of electrolyte neutralization and subsequent disposal of the neutralization products, for the case of potential impacts from spillage, leakage, or disposal upon decommissioning, are estimated to be 8% of the capital cost of the facility. The major cost is attributed to the neutralization process as land-fill costs for the disposal of the neutralization products. This cost escalated to July 1979 is from $7 to $9 per ton.[7]

2.2   Underground Pumped Hydroelectric

Although pumped hydroelectric storage, using above-ground reservoirs, has been used by utilities for almost 50 years, underground pumped hydroelectric storage (UPHS) has been considered only recently. UPHS plants have not yet been constructed, but at least two electric utilities, Commonwealth Edison Co. in Chicago and Potomac Electric Power Co. (PEPCO) in Washington, DC are conducting feasibility studies. The U.S. Department of Energy (DOE) and the Electric Power Research Institute (EPRI) are also involved in the PEPCO study.

The DOE goals include development of the necessary technology for the completion of a demonstration plant in the late 1980's.[8] Argonne National Laboratory (ANL), the lead laboratory, is currently assessing markets and potential for UPHS commercialization and sponsoring development of reversible pump-turbines that can operate at very high heads.

Renewable, but intermittent, energy resources such as solar or wind require efficient energy storage systems to make them viable. Small scale low-head UPHS, as well as flywheels and compressed air, are under consideration as options. Sandia Laboratory (Albuquerque, NM) is currently the lead laboratory in this development program.[9]

Basic elements of a UPHS system have been described previously.[2] These elements are generally similar to those in above-ground plants with the main difference being in the powerhouse and lower reservoir which are located below ground (2000-5000 ft.). The greater the distance between the upper and lower reservoirs (head), the less water that will be required to be pumped for a given energy output. This distance is limited by present reversible turbine design (maximum of about 2000 ft of head), necessitating the use of an intermediate reservoir and powerhouse for greater depths. New research is directed towards reversible turbines that can operate at heads considerably in excess of 2000 feet.

Many potential impacts can be identified for UPHS systems but the primary ones relate to chemical and biological water contamination and cyclic stressing of the walls of the underground reservoir.[3] Reduction in water quality could result from mineralization, entrainment of fish and other biota and turbidity. Local aquifers could be contaminated during construction. A permanent pathway for contamination could result. Cyclic stressing of reservoir walls due to daily water level fluctuations could result in significant erosion and fracturing leading to ultimate wall collapse.

Siting and design considerations are perhaps the most important tools available for mitigating environmental impacts. In addition, the development of fracture detection instrumentation would be helpful.

Because the environmental impacts for UPHS are so closely linked with siting and design considerations, it is difficult, if not impossible, to assign costs to their controls at this time. The costs for environmental controls can be determined for a project once a specific site and equipment design is chosen, but cost estimates for an "average" UPHS plant are not likely to be meaningful.

2.3   Superconducting Magnet

Details of superconducting magnetic energy storage (SMES) devices have been given previously.[2]   At the present time, devices have been planned or constructed for diurnal energy storage (large amounts of energy), system stabilization and reactive power control (VAR).[3]

Of the various proposed designs and applications of SMES, only the large units to be used for diurnal storage (1000-10,000 MWh) in electric utility grids pose significant problems for the environment. Smaller magnets might, in some instances, involve shielding of personnel against the effects of magnetic fields; but this concern should be easily manageable by conventional methods.

The primary environmental impacts associated with large SMES systems involve the effects of high magnetic fields upon animals and man, cyclic stressing of rock walls induced by high magnetic forces and potential problems arising from emergency shutdown of the SMES system.[3]   As yet no standards have been decided upon for exposure tolerances of animals and man to magnetic fields.   Hassenzahl et al have considered this problem and selected three field levels (200 G, 20 G and 0.3 G) for consideration.[10]   In general, tradeoffs exist between additional land area (the field falls off roughly as $1/r^{-3}$ where r is the distance from the solenoid axis) and the introduction of superconducting shield coils. Iron shielding is very expensive in all cases.   Proper site selection, design criteria, and further attention to microfracture and fatigue strengths of rock should assist in mitigating cyclic stress problems.   Magnet quenching (irreversible transition from the super-conducting to the normal state) is the most probable cause for an emergency shutdown.   The following items could lead to such a shutdown:

- Loss of Dewar vacuum;
- Loss of refrigeration
- Major rock fracture;
- Earthquake or other natural disaster;
- External power system failure.

Technological controls now available, as well as improvements that will be developed can be effective in protecting a SMES system against those events which could lead to a quench.   The general philosophy guiding the shutdown planning to date has been: first, to limit the extent of the quench; second, to retard the growth of the quench; and third, to transfer from the cryogenic environment the energy development in a quench.   More specifically, the schemes involve: the control of the normal zone propagation velocity by means of heat sinks; electrical and thermal isolation; and the transfer of excess energy from the cryogenic environment. Cost estimates relating to the magnet-ic field control technology are shown in Table 1.   Costs are not available concerning mitigation of cyclic stress and emergency shutdown effects.

2.4   Compressed Air

Compressed Air Energy Storage (CAES) is perhaps the nearest to commer-cialization of the new energy storage schemes under consideration for use by utilities for meeting peaking power needs.   One CAES plant has been built in West Germany, and at least three other plants are in the planning stage in the United States.   The DOE in conjunction with the Electric Power Research

Institute (EPRI) is actively pursuing commercialization of the CAES technology for utilities through the preliminary system design and site selection of three CAES plants, one in each of the technically feasible geologic storage media--hard rock, salt, and aquifer. Battelle Pacific Northwest Laboratory (PNL) has been designated by DOE as the lead laboratory for CAES research and development. Near term goals are for the commercialization of isothermal CAES plants which require some petroleum-based fuels for operation. Long-term goals for 1990 and beyond are for the development and commercialization of CAES cycles which do not use petroleum-based fuels (no-oil). Consideration has also been given to the small scale use of compressed air for energy storage (<100 kWh) for renewable, but intermittent energy sources such as solar and wind. Near term goals for this program are the completion of feasibility studies in FY'79, and the technical development of the concept by FY'82.

The two major components of a CAES system include: a) the above-ground turbomachinery and energy conversion equipment, and b) the underground geologic storage media. An extensive discussion of both the above-ground turbomachinery and the three generic underground geologic storage formations has been given previously.[2]

As was the case with UPHS, many potential and similar impacts can be identified for CAES systems. In addition, storage emphasis in aquifers and in salt generate additional potential environmental impacts. The primary impacts nevertheless are similar to those of UPHS and include chemical and biological aquifer (water) contamination, cyclic stressing of cavern walls, perturbations of ground-water flow and possible subsidence. NO_x emissions from CAES turbines may not be a serious problem. Initial design studies and operation of the Huntorf plant (West Germany) indicate that NO_x emissions can be significantly lower than those from a standard gas turbine peaking plant burning the same amount of fuel depending upon the equipment configuration. The high pressure cavern air (200 - 1000 psi) used with different fuel ratios and reheat between the high and low pressure power turbines allows lower flame temperatures to be used (1000° - 1200°F) and therefore less NO_x is formed. Low NO_x emissions from a CAES plant would eliminate the need for water or steam injection systems used for NO_x control.

The Huntorf West Germany CAES plant built by Brown Boveri Corporation for NWK (a German utility) has been in operation since 1978 and should provide an excellent resource base for determining environmental impacts associated with salt cavern operation as well as certain types of turbomachinery. Operating experience of the plant to date has shown the following:[12]

- There has been no significant leakage of air from the cavern. The leakage rate was estimated at 1/1000 (0.001) to 1/10,000 (0.0001) of a percent per day.

- There has been no significant carryover of salt with the compressed air.

- The only significant environmental control that was necessary for the plant was that concerned with the rate at which solution-mining brine could be disposed of in the ocean.

Siting and design criteria are important considerations in mitigating many of the environmental impacts. Accurate geohydrologic knowledge is required. Fracture detection instrumentation should be developed and a well monitoring system similar to that used in the natural gas industry will also be required.

Many of the potential environmental impacts associated with CAES plants are either very site-specific or are so poorly defined at this time that cost estimates on their controls are very difficult, if not impossible to determine

with any sort of confidence. Since many of the environmental impacts can be
mitigated by proper site selection and/or equipment and cavern design there
will be no costs directly attributable to these control techniques. Environmental controls such as grouting to control air leakage from a cavern could
potentially be so expensive as to make a CAES plant uneconomical at that site.
Water injection equipment for NO_x control (if it is needed) can add anywhere
from \$5 to \$50/kW to the construction costs depending upon the availability
of water and the water treatment needed (\$10/kW might be considered an average
value). Assuming that three observation wells (including instrumentation)
are needed to determine the integrity of the operation of a hard rock or
aquifer storage cavern, this equipment would likely add less than 5 percent
to the cost of developing the cavern.

2.5   Flywheel

Energy storage through the kinetic energy stored in rotating masses
(flywheels) is a technology that has been used by man for hundreds of years.
Flywheel energy storage systems (FESS) can thus make a contribution to our
future needs for energy storage in both large and small scale applications.
Near term objectives of DOE for flywheel energy storage research are
the development of regenerative braking systems for vehicles (both electric
and heat engine) and energy storage for solar-photovoltaic and wind energy
systems.[11]
Most of the flywheel energy storage devices of importance will have the
following components:

- Flywheel Rotors - using different materials and geometries.

- Bearings - that use roller, fluid film or advanced magnetic designs.

- Vacuum Systems - using a vacuum pump or a sealed and periodically
  maintained container.

- Seals - magnetic couplings or integration of the electric motor
  inside the vacuum contained so power can be transmitted outside the
  container without the loss of vacuum.

Of these components, the flywheel rotor has the greatest potential to
produce environmental effects. Current flywheel research and development is
directed towards rotor materials that have high strength to weight ratios and
geometries that make the best use of these materials. Metal flywheels have
been used extensively, have well-known design strengths and manufacturing
techniques, but unfortunately, have poor energy densities, have catastrophic
failure modes into chunks of shrapnel which requires a heavy containment ring,
and are costly. Composite materials such as E-glass, kevlar, or graphite in
an epoxy matrix have much higher energy densities and have potential for lower
costs. They also have the potential for a much more benign failure mode in
which the outer layers of material delaminate and slow the wheel down without
catastrophic failure. Cellulose, in the form of plywoods, fiberboards or
paper rolls, is also being seriously considered as a flywheel material because
of its low cost, good strength to weight ratio and the possibility of a benign
failure mode like the composites.
During normal operation of a flywheel energy storage system, either in
a fixed or moving base application, there are no known significant environmental impacts. For a large utility type FESS peaking plant, some concern
has been voiced about noise problems from the motor/generators and related
machinery, but these are problems associated with any large scale power

generating operation and should be considered site-and design-specific and
relatively minor. Aesthetic and land use considerations are site-specific
potential problems, but the land requirements for a FESS are relatively small
and can be placed on existing utility-owned sites. Residential and industrial
FESS land requirements are similarly small and can be easily built underground.

The only major environmental impacts from flywheel energy storage systems appear to be health and/or safety problems related to flywheel rotor
failures and accidents, which are excursions from normal operation. The
primary method of controlling these impacts will be to ensure that excursions
are extremely uncommon occurrences. Thus, research into material properties
and stress failure modes are required. Proper design of containment systems
is necessary. Possibly sensor development indicating a forthcoming rotor
failure is also required.

Mitigation technology costs related to containment and sensor systems
for utility and industrial applications are estimated to be on the order of
5% of the installed system costs. Slightly higher costs (5%-10%) are estimated for vehicular applications.

2.6   Thermal

The DOE has recognized the importance of the development of thermal
energy storage (TES) by establishing several programs under its Division of
Energy Storage Systems (STOR).[13] Table 2 shows the division of responsibilities at the national laboratories and NASA.

Discussion of these systems is best considered by type, i.e., sensible
heat, latent heat and reversible chemical reaction. For purposes of this
paper, sensible heat systems are further subdivided into standard liquid or
solid materials and aquifers.

Sensible heat systems use the internal energy of the material for heat
storage. The materials composing the system experience no phase change. The
quantity of heat that can be stored is dependent upon the mass and heat
capacity of the material as well as the differential temperature during the
heat transfer process. Some materials that are currently in use are commercial heat transfer oils, rock, and building materials such as concrete and
water. These materials are confined by containers, or by their own structural
strength, insulated, and exposed to a source of thermal energy. Heat is
transferred to the material by conduction, convection, radiation, or by a
combination of these heat transfer modes. Some typical heat transfer fluids
include: Caloria HT43, Exxon (petroleum distillate); SF-96-(50), General
Electric (silicone oil); Dowtherm SR-1, Dow Chemical (glycol type); Therminol
66, Monsanto Chemical (high aromatic hydrocarbon).

Storage of a liquid in an aquifer is also a form of sensible heat
storage. In this process, water is heated above ground (by solar energy,
waste heat, etc.) then injected into a confined aquifer. The over- and
underburden act as efficient insulation to prevent heat transfer out of the
aquifer. The water is stored until it is needed for utilization above-
ground. Such utilization is easily adaptable to seasonal heating and cooling
requirements of large facilities such as airports, shopping centers, office
buildings, etc.

Latent heat energy storage systems utilize the heat absorbed or released
by a material undergoing a change in phase, i.e., liquid to solid or liquid
to gas, change in composition (loss or gain of waters of hydration). The
heat can be extracted from the system by effecting heat transfer from the
system to a utilization medium. Some typical materials under investigation
at the present time include: Glauber's salt, sodium pyrophosphate decahydrate, calcium chloride hexahydrate and copper sulfate pentahydrate.

TABLE 2.   SUBPROGRAM RESPONSIBILITIES

National Laboratory	Thermal Energy Storage Subprogram
Oak Ridge National Laboratory	Low Temperature Thermal Energy Storage (LTTES) (Sensible and latent heat, < 250°C)
NASA Lewis Research Center	High-Temperature Thermal Energy Storage (HTTES) (Sensible and latent heat, > 250°C)
Sandia Laboratories, Livermore	Thermochemical Energy Storage and Transport (TEST) (Reversible chemical reaction)
Argonne National Laboratory	Thermal Energy storage for electric load-leveling behind the meter and thermal energy storage for solar applications.
Battelle Pacific Northwest Laboratories	Thermal energy storage in aquifers.

Storage utilizing the heat of reaction of reversible chemical reactions is considered as another form of TES. Chemical heat pipes or pumps are one form of this type of system. Included are such typical reaction systems as sulfuric acid-water, sulfur dioxide-oxygen, calcium oxide-carbon dioxide, ethylene-hydrogen, etc.

The environmental impacts associated with thermal energy storage systems relate primarily to the loss of the energy storage material from the system. While this is true of all the systems discussed, additional potential impacts can be associated with the aquifer storage system. Toxicity, corrosion and in many cases, fire hazard problems also exist with many of the materials in use.

Impacts associated with aquifers are similar to and typical of those previously discussed under UPHS and CAES systems. Site selection criteria and adequate geohydrologic knowledge should help mitigate many potential concerns.

In general, practical control technologies for sensible, latent heat and reactive systems would include: a) engineering design for chemical solution leakage, material biodegradation, scrubbing systems, development of leakage detection instrumentation and land fill disposal. Mitigation costs associated with these activities should not exceed 3%-8% of the system plus some additional land fill costs. Costs relating to aquifers are unknown.

2.7   Fuel Cell

A fuel cell is an electrochemical device in which the chemical energy of a fuel is converted directly into low voltage direct current (dc) electrical energy. This process theoretically occurs isothermally and is therefore not limited by the Carnot efficiency; that is, the fuel cell makes it possible to eliminate the high temperature combustion and mechanical-to-electrical processes associated with conventional power producing schemes.

Although the production of electricity is isothermal, the process releases the heat of combustion (or reaction) and this heat can be removed and utilized. Since fuel cells produce dc, an inverter is an integral part of a fuel cell system. The inverter transforms the dc to alternating current (ac).

In the strict sense fuel cells are not energy storage systems but rather are highly efficient power producing devices. Although they have often been described as primary batteries with their fuel and oxidant stored externally, only one reference in the literature described the development of a fuel cell as an energy storage device. This "water battery" operates in the discharge mode as an H_2/O_2 fuel cell and in the charge mode as a water electrolyzer.[14] Another fuel cell system that approaches the concept of energy storage systems is the regenerative fuel cell system in which the products of the fuel cell process are recycled to a fuel producing process.[15]

The DOE is developing fuel cells in an effort to conserve fossil fuels. The DOE program objectives are to develop fuel cell power plants leading to commercialization for electric power generation (peaking, intermediate and base load), cogeneration (electrical power and waste heat), Onsite Integrated Energy Systems (OS/IES), and waste conversion-fueled systems.

Fuel cells utilizing phosphoric acid as an electrolyte are generally referred to as first generation fuel cell systems. The 4.8 MW Utility Demonstration Program utilizing phosphoric acid cells is intended to demonstrate the viability of fuel cells on a utility grid. The delivery of the power plant to the demonstration facility is to be complete in October 1979. To date the stack performance is meeting design goals, and testing of the facility is to begin in February 1980. The fuel cells have been designed by United Technologies Power Systems Division.[16]

The fuel cells commonly referred to as second generation are those that utilize a molten carbonate electrolyte. The DOE development objectives are to advance the state-of-the-art of molten carbonate electrolyte fuel cells, thus providing for early commercialization. The schedule for development of the molten carbonate fuel cell is to have a full scale cell/stack by 1982.

Fuel cells in themselves (both phosphoric acid and molten carbonate) are virtually emission free.[17] Significant impacts do result however from the processing and refining of the fuel for the fuel cell systems. The latter presumably are controlled by USEPA regulations. Emissions from fuel cell operation are considerably below those for power-producing facilities. Electrolyte disposal could be another primary impact but it appears that lime (CaO) neutralization for phosphoric acid cells would be satisfactory. Molten carbonate cells could result a more serious solid waste disposal problem. The size and gravity of this problem is unknown at the present time.

Mitigation costs are not expected to be too high being related to available land fill disposal sites.

3.   APPLICATIONS

Many previous studies have considered the energy storage system to be an integral portion of a utility electric power generation system. The investigation into potential applications of the various storage and fuel cell technologies had a dual objective of identifying new applications and subsequently determining any environmental impacts associated with the new application. Table 3 summarizes the applications for the various technologies. Further details may be found elsewhere.[3]

Most of the applications shown have been chosen based on a number of economic market penetration studies made by selected organizations. Not all are shown in the table. In general, most environmental impacts appear to be similar to those previously discussed although they may differ in degree. Some impacts noted previously may not even be applicable.

TABLE 3.   POTENTIAL NEW APPLICATIONS FOR ENERGY
STORAGE AND FUEL CELL TECHNOLOGIES*

Advanced Lead-Acid Battery	1. Electric Vehicle (includes hybrids)	18
	2. Load-leveling by user- (Power Management Battery Storage System)	19
Underground Pumped Hydroelectric	1. Small Scale Low-Head (in combination with PV or wind turbine)	20
Superconducting Magnet	1. System stabilization	3
	2. Reactive power control	21
	3. Regenerative braking	22
Compressed Air	1. Intermittent energy storage (solar and wind)	23
Flywheel	1. Vehicular-energy storage and regenerative braking	24
	2. Residential and small substation storage (also for industrial-commercial)	25
	3. Stabilization, reactive power control	26
Thermal	1. Heating and cooling (daily and seasonal)	27,28
	2. Cogeneration-district heating	29
	3. Vehicular	30
Fuel Cell	1. Water battery	31
	2. Transportation	32
	3. Cogeneration	33

*Does not include standard load-leveling and peak-shaving applications for centralized electric power generation.

4.   CONCLUSIONS

In virtually all cases it appears that some additional studies and control technology research and development are needed for most of the technologies reviewed. A brief discussion follows for each.

1. ADVANCED LEAD-ACID BATTERY: The use of the conceptual process shown
    in Fig. 2, if engineered properly, should prevent hydrogen, arsine
    and stibine from escaping from the facility and thus pose no prob-
    lem beyond the site boundary. Stibine retention (or elimination)
    is a more serious problem since its generation is greater by at
    least an order of magnitude over that of arsine. Instrumentation
    should be available for detection and monitoring so as to assist in
    the overall processing of the gas mixtures. The BEST facility will
    be helpful in determining appropriate needs and design engineering
    parameters for large scale battery installations.

2. UNDERGROUND PUMPED HYDROELECTRIC: Design and site selection cri-
    teria are most important for this system (see also 4, CAES).
    Attention to fish and other biological contamination should be made.
    As with the CAES system, cyclic stress studies (in-situ) should be
    made and fracture detection instrumentation developed.

3. SUPERCONDUCTING MAGNET: It is readily apparent that additional
    research is required on the effects of various magnetic field in-
    tensities on both biosystems and electrical and electronic systems.
    In addition, further studies are required on rock support problems
    related to structural design. Research is required on the effects
    of microfractures and cyclic stressing of rock.

4. COMPRESSED AIR: The potential fracture development within suppor-
    tive structures and subsidence induced by daily cyclic stress pat-
    terns have been determined to be an important impact. This has
    been identified previously.[2] In-situ stress analyses on candidate
    rock types should be made. Such studies should reveal the potential
    for mechanical failure. Development of appropriate instrumentation
    is implied. The use of aquifers for CAES involves a host of poten-
    tial chemical and biological impacts, the degree of importance of
    which is difficult to assess. Many studies upon individual problems
    should probably be done but it may be necessary to take some risk
    and actually construct a practical system for total study in order
    to observe and measure system interactions. Finally, it is recom-
    mended that the Huntorf plant be critically observed with respect
    to salt cavern operation.

5. FLYWHEEL: Environmental control technology, in a strict sense, is
    not required for flywheel systems. Major effort, however, is
    required for the overall characterization of failure modes for these
    various flywheel configurations. Included, of course, is the char-
    acterization of a number of materials.

6. THERMAL ENERGY

    6.1 Sensible Heat: Research is required whereby biodegradable
         liquids are identified and modified so that their physical and
         chemical properties are useful for these systems.

6.2 <u>Sensible Heat-Aquifer</u>: Considerable geohydrologic characterization is required prior to assurance that many of these potential environmental impacts will or will not actually be a problem. Because of site specificity for this particular type of storage, this is a difficult system upon which to generalize.

6.3 <u>Latent Heat and Reversible Chemical Reaction</u>: Designing to minimize leakage and the development of leakage detection instrumentation are necessary for these systems. The wide range of temperatures for the materials in these systems compound these problems.

7. FUEL CELL TECHNOLOGY: No immediate development work is required for first generation cells. Most of the major materials used for second generation cells appear to be non-toxic, at least those being used at present. Considerable research and development is still in progress on the second generation cell. Thermal and safety problems may be more important especially within the facility for the second generation cells.

REFERENCES

1. "Energy Systems Environmental Control Technology Planning Survey," Aerospace Corp. report No. ATR-76 (7518-1) (August 1975).

2. E. L. Kaufman, Ed., "Environmental Control Technology R & D Requirements for Energy Storage Systems," Los Alamos Scientific Laboratory report No. LA-6979-MS (September 1977).

3. M. C. Krupka, Ed., "Environmental Issues of Energy Storage Systems - An Update", Los Alamos Scientific Laboratory report No. LA-MS (to be published).

4. Public Service Electric and Gas Company Engineering and Construction Department Request for Quotation No. 77220, BEST Facility-Station Shakedown Battery (August 1, 1977).

5. Environmental Development Plan (EDP), Electric Energy Systems, U.S. Department of Energy report No. DOE/EDP-0038 (August 1979).

6. R. Varma and N. P. Yao, "Stibine and Arsine Generation from a Lead-Acid Cell During Charging Modes Under a Utility Load-Leveling Duty Cycle," Argonne National Laboratory report, ANL/OEPM-77-5 (March 1978).

7. W. D. Bassel, <u>Preliminary Chemical Engineering Plant Design</u>, (American Elsevier Publishing Company, New York, 1976), p. 268.

8. "Solar, Geothermal, Electric and Storage Systems Program Summary Document, FY 1979," U.S. DOE report DOE/ET-0041 (78) (March 1978).

9. "Mechanical Energy Storage for Photovoltaic/Wind Project, Technical Progress report," Sandia Laboratory (NM) SAND 79-0874 (May 1979).

10. W. V. Hassenzahl, M. W. Mahaffy and W. Weihofen, "Evaluation of Environmental Control Technologies for Magnetic Fields," U.S. Department of Energy report, DOE/EV-0029 (August 1978).

11. "Multi-Year Program Plan for Thermal and Mechanical Energy Storage Program," Div. of Energy Storage Systems, U.S. DOE, Review draft (November 1978).

12. Z. S. Stys, and H. Hans-Christoph, "Huntorf 290 MW – The Worlds First Air Storage System Energy Transfer (ASSET) Plant: Construction and Commissioning," paper at the American Power Conference (April 24-26, 1978).

13. U.S. Department of Energy, "Project Summary Data: Thermal and Mechanical Energy Storage Program," FY 1979 (March 1979).

14. J. E. Clifford, E. W. Brooman, V. T. Sulzberger and Y. Z. El-Bradry, "Evaluation of a Water Battery Energy Storage Concept for an Electric Utility System," A Battelle Energy Program report, Battelle-Columbus Laboratories (June 1975).

15. R. F. Gould, Ed., Regenerative EMF Cells, American Chemical Society (1967).

16. United Technologies Power Systems Division, UTC, Demonstration Model Spec. FCS-0493, Cont. No. EX-76-C-01-2102 (March 15, 1978).

17. G. E. Voelker, "DOE Fuel Cell Program Overview," National Fuel Cell Seminar, Bethesda, Maryland (June 26, 1979).

18. E. P. Marfisi, C. W. Upton and C. E. Agnew, "The Impact of Electric Passenger Automobiles on Utility System Loads, 1985-2000," prepared by Math-tech, Inc., for EPRI, EPRI-EA-623 (July 1978).

19. O. P. Hall, "Power Management Battery Storage System" Paper 79-0982, AIAA Terrestrial Energy Systems Conference, Orlando, Florida (June 4, 1979).

20. Investigation of the Technical and Economic Feasibility of Using Pumped Well-Water Energy Storage," BDM Corp., BDM/TAC-79-321-TR (June 1979).

21. H. Boenig, Los Alamos Scientific Laboratory, informal report, patent applied for.

22. L. O. Hoppie, General Motors Corp., private communication (1979).

23. H. M. Dodd, et. al., "An Assessment of Mechanical Energy Storage for Solar Systems," 12th Intersociety Energy Conversion Engineering Conference Proceedings, Vol. 2, Washington, DC (1977).

24. "Study of Flywheel Energy Storage," AiResearch Mfg. Co. report for U.S. Department of Commerce, UMTA-CA-06-0106-77 (September 1977).

25. "The Application of Flywheel Energy Storage Technology to Solar Photovoltaic Power Systems," by A. Millner, MIT/LL, in "Proceedings of the 1978 Mechanical and Magnetic Energy Storage Contractors' Review Meeting," CONF-781046 (October 1978).

26. "Economic and Technical Feasibility Study for Energy Storage Flywheels," Rockwell Int. report prepared for U.S. Department of Energy, HCP/M1066-01 (May 1978).

27.   D. Balcomb, et al, "Solar Heating Handbook for Los Alamos," Los Alamos
      Scientific Laboratory report No. LA-5967 (May 1975).

28.   C. Wyman, "Thermal Energy Storage for Solar Applications: An Overview,"
      Solar Energy Research Institute report, SERI/TR-34-089 (March 1979).

29.   "Evaluation of Thermal Energy Storage for the Proposed Twin Cities
      District Heating System," Third Annual Proc. of Thermal Energy Storage
      Contractors' Info. Mtg., Report No. CONF-78123, December 1978).

30.   "A Road Vehicle Thermal Energy Storage Concept and Evaluation - Phase
      1," Final Report, Period January-October 1978, Sigma Research, Inc.,
      prepared for DOE, Report No. ANL-K-78-3983-1, (December 1978).

31.   "J. E. Clifford, et al., "An Off-Peak Energy Storage Concept for Electric
      Utilities: Part II - The Water Battery Concept," Applied Energy 3
      (1977).

32.   "Application Scenario for Fuel Cells in Transportation," Los Alamos
      Scientific Laboratory report, LA-7634-MS (February 1979).

33.   "Advanced Technology Fuel Cell Program," Annual report prepared by
      United Technologies Corporation for EPRI, EPRI EM-956 (December 1978).

Thermal Energy Storage in Aquifers for a Solar Power Plant

W.J. SCHAETZLE and C.E. BREET
The University of Alabama
University, Alabama 35486, USA

J.M. ANSARI
University of Petroleum and Minerals
Dhahran, Saudi Arabia

ABSTRACT

This project develops a theory for thermal energy storage systems in confined aquifers which can be utilized at over 100C (212F). The proposed operating temperature range is 100C to 600C (212F - 1112F). The storage system capital cost is estimated to be approximately 0.4 percent when applied to a solar power plant system. The transfer of thermal energy to the confined aquifer is made with pressurized water injection and withdrawal through a unique system. Thermal energy injection and withdrawal cycles operate on:

1. A daily basis for night time energy requirements,
2. A few days basis for poor weather energy requirements, and
3. An annual basis for lower winter insolation requirements.

Energy injection and withdrawal is accomplished by reversing flow through a system of wells connecting the aquifer to the surface. Each well contains a submerged pump and controls.

The thermal energy storage system is analyzed in conjunction with a solar power plant system. The storage system provides energy during night time, bad weather and the winter low insolation periods. The power plant system consists of a Rankine steam cycle, a two dimensional parabolic concentrating solar collector system, and the aquifer thermal energy storage system.

The per kilowatt plant capacity capital cost is approximately $10,000/kw. The storage system is only $40/kw or 0.4 percent of capital cost. Assuming an interest rate of 10 percent, a sinking fund depreciation rate for 20 years with 10 percent interest, and 90 percent utilization, the capital cost per kw hr is 14.5 cents.

1. INTRODUCTION

Thermal energy storage has been studied by a large number of investigators. Numerous storage media have been proposed such as storage tanks and underground aquifers.

The use of confined aquifers as natural containers for fluid such as water has been considered seriously for the past decade by Esmail and Kimbler (1) and Smith and Hamor (2). More recently, however, such attention has been focused on the use of aquifers for high temperature thermal storage.

The temperature range is 60C - 204C (140F - 400F). This study was conducted
by Meyers and Todd (3), Meyers (4), and Maltz and Bell (5). Fred J. Maltz
and James C. Warman (6) conducted experimental testing of the heat storage
well concept to provide data which would be utilized for calibration of
mathematical models. The tests show that slow natural flow of water in the
aquifer is essential for efficient storage. The injection of water into the
aquifer will not disrupt the layers within the aquifer.

Schaetzle, et al. (7,8) investigated the use of aquifers to store thermal
energy that can be utilized in community heating and cooling. Paired wells
are drilled in the same aquifer where water is injected into one and withdrawn
from the other. During the injection and recovery process, a water front and
temperature front move between the wells. Flow inside aquifers has been
determined to be approximately potential flow. The energy loss in thermal
storage is a function of surface area per unit volume and the conductivity and
temperature of the adjacent volume. The storage system must be designed so
that both the surface area and conductivity of the adjacent volume are small.
Collins et al. (9) investigated the storage of high pressured hot water in
aquifers. The water temperature ranged to 343C (650F) under 2700 psi. Tsang,
Buscheck, Mangold, and Lippman (10) investigated the hydrodynamic and thermal
behavior of an aquifer and its application for thermal energy storage. They
developed a three-dimensional numerical model of the fluid and heat flows in
aquifers used for hot water storage. The result shows a recovery factor of
more than 80 percent. In estimating the economic feasibility and the optimum
aquifer storage system efficiency they emphasized:

1. Thermal behavior and heat losses during successive cycles,
2. Pressure changes throughout the aquifer, and
3. Chemical reactions and its effect on aquifer permeability and
    porosity.

The mesh in this model is very fine near the well in order to reduce any
potential error in studying the temperature variations within the aquifer.
Mathy and Mejoz (11) investigated the different techniques for underground
storage of heat. Underground heat storage is based on adjusting production
and consumption of thermal heat. Their study indicates that heat transfer
within the aquifer by natural convection during injection of hot water is very
difficult to control and account for in numerical modeling.

W. S. Duff and W. W. Shamer (12) looked into the economical aspect of
solar power generation. The study focused on temperatures of 250C (482F) and
below. Two concepts of solar energy collection systems were utilized: the
tower/heliostat system and the distributed collector system using point focus
Fresnal reflectors. The estimated minimum cost of electricity delivered to
the power grid is about 2.4 cents per kwh excluding storage and maintenance
costs. The optimization used in their study is based on numerical analysis of
the different parameters involved to minimize costs.

W. W. Shamer and W. S. Duff (13), N. T. Pierce (14), and B. P. Gupta (15)
studied various types of solar collectors in reference to cost and maximum
production of electricity and minimum dollars per kw hr. They evaluated the
performance of three types of solar collectors: parabolic trough, fixed stats
with movable abscriber, and movable stats with fixed absorber. The results
indicate the advantages of using a parabolic collector for high temperature.
This result, based on cost and performance, is very significant when applied
to power plants of a 40 MW production capability.

2. SYSTEM DESCRIPTION

This thermal energy storage system is designed for high temperature sotrage. A solar power plant is analyzed as a potential source of energy for the storage system. The corresponding solar collectors are selected on the basis of temperature range, amnufacturing cost, and efficiency. Thermal energy storage with the solar power plant application is shown in Figure 1.

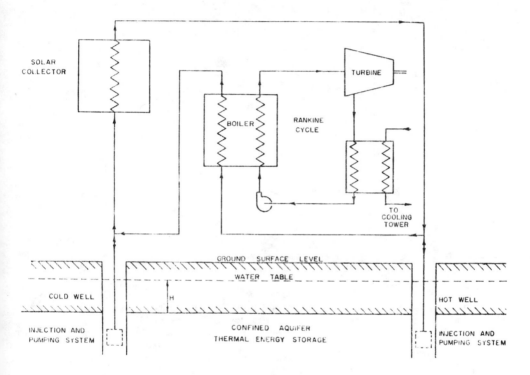

FIGURE 1.   SOLAR POWER PLANT SYSTEM – SCHEMATIC

2.1.   The Thermal Energy Storage Module, or Aquifer

Aquifers are water-bearing rocks found near or at nominal depths below the soil zone of the earth's surface. Such rocks have physical properties of porosity and permeability, i.e., the presence of intergranular voids, fractures, solution channels, or other interconnected openings capable of receiving, storing, and transmitting water. Thermal energy storage in aquifers is a function of porosity and rock type. Most rocks have a specific gravity of approximately 2.6 and a specific heat of about 0.2 But lbm F. Porosity can range from less than 5 percent to more than 30 percent with 10 percent to 20 percent as general averages for most aquifers. The thermal storage capacity per degree temperature change is just over 30 BTU/ft^3, varying with porosity. At this value, a rock volume of 100 ft. by 200 ft. by 30 ft. has a storage capacity of 4 X 10^9 BTU with 200F (93C) temperature change. General values for the storage capacity of aquifers as a function of size, porosity, and temperature are shown in Figure 2.

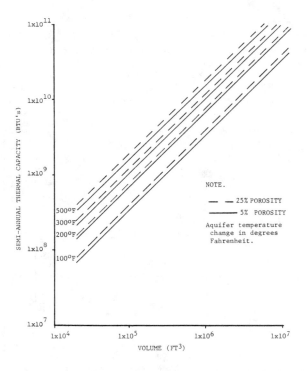

FIGURE 2. AQUIFER THERMAL ENERGY STORAGE CAPACITY

For the proposed system, a confined aquifer, one with a confining layer above and below the aquifer at a depth of over 175 meters (575 ft.), is required. The top of the aquifer must be a minimum distance h below the water table (Figure 3). This minimum distance is required to provide a water head (pressure) sufficient to prevent boiling in the aquifer at the proposed temperature. Thermal energy transfer to the aquifer thermal energy storage is accomplished by transferring hot and warm water to and from the aquifer by a paired system of wells.

2.2. Two-Well System

A system of wells is required as a conduit for water between the aquifer and ground surface. Well diameter is expected to be fifteen inches or larger. Well size and number of wells will be a function of water demand and aquifer characteristics, primarily those of permeability, porosity, and thickness. Water flow increases with diameter (logarithmic function). Well design for a particular aquifer will be established by a trade-off between well size and number of wells. Energy transfer per unit cost is the controlling factor.

Each well will have a pumping system (Figure 3) and controls to provide specific characteristics in the pipe line. Figure 3 is a schematic of the general concept of the injector system in the well.

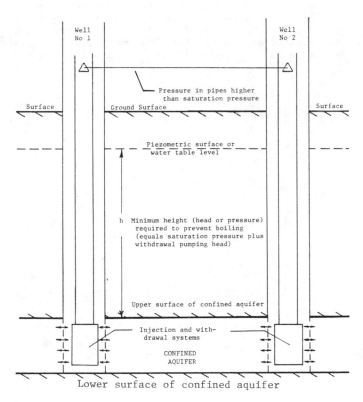

FIGURE 3.   SCHEMATIC OF HIGH TEMPERATURE STORAGE SYSTEM

2.3. Injector System

To prevent boiling, the liquids in the system are compressed or pressurized before heating. This raises the boiling point temperature and therefore allows the substance to be heated to higher temperatures without boiling. The proposed injector system allows the placement of high temperature liquid into an aquifer at normal aquifer pressure for thermal energy storage.

If water at a pressure above atmospheric pressure at the surface is injected into a confined aquifer, the aquifer injection pressure will equal the above ground pressure plus the pressure resulting from the surface to the aquifer head. As a result, the pressure is normally greater than aquifer pressure. By maintaining a very high flow rate, the aquifer can be pressurized to the new pressure. In this case the aquifer can be blown with a loss of pressure or water losses in many directions can result. If the flow rate is not maintained, the lower pressure will transfer to the surface and cause boiling.

In the proposed injector system the pressurized water is injected at aquifer
pressure.

To prevent boiling at the higher temperatures, aquifers are required at
sufficient depths so the normal aquifer pressure is above the saturation
pressure at the temperature of the injected liquid. Figure 3 shows a schematic diagram of the pressure system. Note the pressure head "h" to the water
table. The head (pressure) must exceed the fluid temperature pressure plus
the pressure drop created by pumping fluid from the aquifer. The variation
of required aquifer depth from the water table is shown in Figure 4.

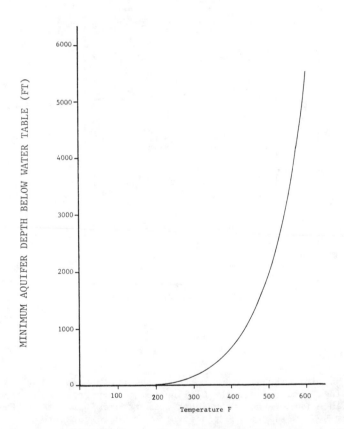

FIGURE 4. MINIMUM AQUIFER DEPTH BELOW WATER TABLE VS. TEMPERATURE

The fluid pressure in the pipe extending into the aquifer, at aquifer
level will equal the fluid pressure in the pipe at the surface plus the
pressure due to the head (pressure) resulting from the fluid column height
from the surface to the well. This pressure must be reduced to aquifer
pressure level. Two methods are proposed:

1. Place a throttling valve (pressure regulator) at the end of the

pipe to reduce the pressure.
2. Place a turbine at the end of the pipe to remove work from the injected liquid. A generator turbine combination is a potential
system.

Both of these systems will reduce the well pipe pressure to aquifer pressure.
The second system is preferable based on energy requirements.

To remove the thermal energy, the flow is reversed and a pump (pumps)
in the aquifer must pressurize the fluid to the surface fluid pressure plus
aquifer head. The surface pressure must be greater than temperature saturation pressure in order to prevent boiling. This pressure is not necessary
in some cases where flashing at the surface is desired.

The work required for the injection-recovery system is calculated with
regard to the two different types of injection-recovery systems:

1. A throttling valve (pressure regulator) to reduce pressure for
injection.
2. A motor-pump generator-turbine combination that recovers part of
the work lost in injection and utilize this work in recovery.

The work involved in the first system is twice the work required to pump the
water to the required pressure, once for storage and once for recovery.

The work required is given by

$$w = 2 \, \Delta p \, v \, \frac{1}{\eta_p}$$

where

w = Work in BTU/lbm

Δp = The pressure drop between aquifer and the surface, and pressure
safety factor to assure zero boiling

v = The specific volume of fluid at given temperature

η_p = The motor-pump efficiency or turbine-generator efficiency with
subscript 't'

With the turbine-generator injection, part of the ideal work is
recovered. In this case

$$w = 2 \, \Delta p \, v \, (\frac{1}{\eta} - \eta_t)$$

Calculations are made to give work required per unit energy storage. For
these calculations, temperature differences utilized are 38C (100F) and
93C (200F). The generator-turbine injection system and motor-pump efficiencies are assumed equal and are 40 percent, 60 percent, and 80 percent.
The results are shown in Figures 5 - 8.

The injection-recovery system allows storage of thermal energy at
temperatures well above the normal atmospheric boiling temperatures. Storing
thermal energy at temperatures over 100C (212F) is practical with this system.

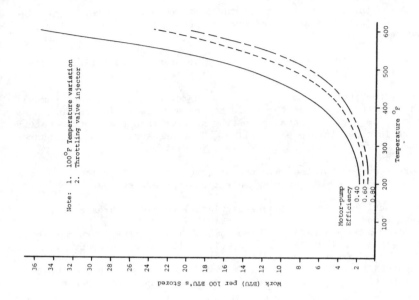

FIGURE 6. WORK REQUIRED PER UNIT THERMAL
ENERGY STORAGE

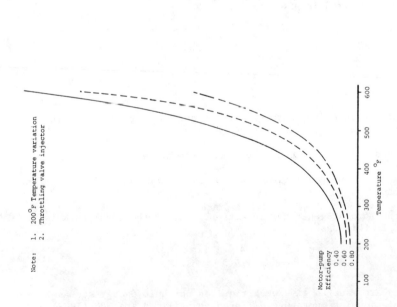

FIGURE 5. WORK REQUIRED PER UNIT THERMAL
ENERGY STORAGE

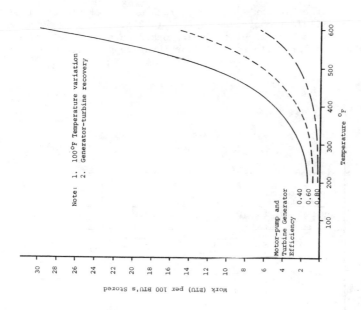

FIGURE 7.   WORK REQUIRED PER UNIT THERMAL
ENERGY STORAGE

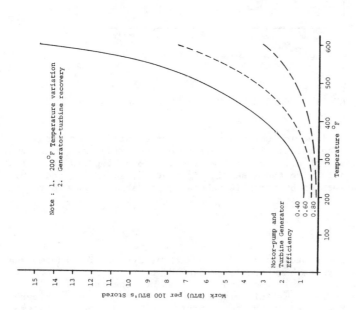

FIGURE 8.   WORK REQUIRED PER UNIT THERMAL
ENERGY STORAGE

2.4. Solar Collectors

A parabolic solar collector system and tracking system are utilized to achieve the required fluid temperature of 200C (392F). The axis is aligned on a north-south basis. The calculations are based on a Honeywell Incorporated (15) parabolic solar collector. The performance data are shown in Figure 9.

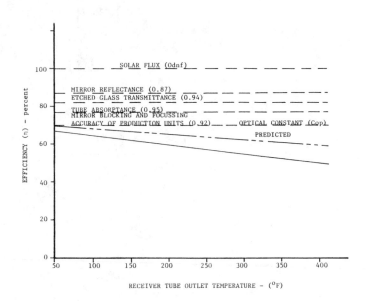

FIGURE 9.   PERFORMANCE OF THE CONCENTRATING COLLECTOR

Solar energy collection calculations are made using solar data from Tucson, Arizona. The calculation F chart technique using only incident insolation of Beckman, et al., is utilized (16). The collectors are oriented on a north-south axis with the axses inclined at 0°, 10°, 20°, and 30°. The average daily collection by month is tabulated in Table 1 and plotted in Figure 10.

2.5. Power Plant System

The power plant system is based on a Rankine Cycle. The thermodynamic efficiency of the system has been investigated for two types of working fluids. Preliminary calculations indicate that superheated steam is most efficient for the cycle. Efficiencies of up to 23 percent are achieved for the system with 492F as the upper temperature and 90F as the lower temperature.

TABLE 1: Daily Insolation Collected per Day per Square Meter* (Tucson, Latitude = 32.1°N.)

Month	Jan	Feb	March	April	May	June	July	Aug	Sept	Oct	Nov	Dec	
$H(MJ/m^2)$	13.09	16.81	22.83	27.85	30.95	29.65	26.26	24.67	24.30	18.69	14.85	12.42	
k_t	.66	.68	.75	.77	.78	.72	.65	.66	.75	.71	.71	.67	
T °F	10	11	19	18	22	27	30	28	26	20	14	10	
H_d^a/H	.25	.24	.18	.17	.16	.20	.26	.25	.18	.20	.20	.24	

Solar Insolation Collected mJ/m^2 Day

Inclination													Average
0°	4.92	7.50	12.99	18.48	23.20	21.64	17.53	16.06	15.48	10.29	6.35	4.47	13.24
10°	7.21	10.17	15.83	21.25	24.31	21.59	18.21	17.41	18.07	12.57	9.04	6.74	15.20
20°	9.53	12.40	17.96	22.06	24.06	21.29	17.78	17.76	19.84	14.98	11.69	9.13	16.54
30°	11.51	14.23	19.09	21.76	22.35	19.30	16.13	17.06	20.41	16.66	13.89	11.22	16.96

Table 2. Estimated Cost of the System*

Solar Collector Inclination Angle	Area of Solar Collector	Installed Collector Cost	Piping Cost	Cost of Power Plant	Cost of Wells	Total Cost	Capital Cost	Cost*
degrees	m^2	$ millions	$ millions	$ millions	$ millions	$ millions	$/KW	$/KW hr
0	1,300,000	300	75	16	1.6	392.6	10,000	0.145
10	1,140,000	257	64	16	1.6	333.6	8,500	0.123
20	1,045,000	236	59	16	1.6	312.6	7,800	0.113
30	1,096,000	230	57	16	1.6	304.6	7,600	0.110

*Assumes an interest rate of ten percent, a sinking fund depreciation rate for twenty years with ten percent interest, and 90 percent utilization.

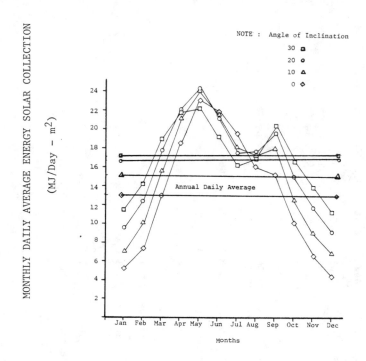

FIGURE 10.   SOLAR ENERGY COLLECTION FOR TUCSON, ARIZONA

3. CONDITIONING THE AQUIFER

The aquifer system is conditioned by passing hot water into the aquifer
for a period of time prior to aquifer utilization as an energy storage system.
Injecting energy till the thermal front reaches the second well is antici-
pated. The loss of this energy is not accounted for in recovery efficiency
calculations. This allows the aquifer and aquifer boundaries to reach higher
temperatures and allows higher thermal energy recoveries. Analysis indicates that 99 percent of the energy injected can be recovered in a daily
cycle, 95 percent plus of the energy injected can be recovered in a few days
cycle, and over 80 percent of the energy injected can be recovered in an
annual cycle. Recovery is primarily a function of aquifer thickness.

4. CO-GENERATION CAPABILITIES

By using small distributed solar power plants, the systems can also be
utilized for district heating. The basic storage system for thermal energy
has the storage capability without additional expansion.

The water is stored and retrieved at 200C (392F). After supplying the
energy to the power cycle, the water is at 145C (292F). This temperature is
more than sufficient for district heating. By withdrawing thermal energy for
another 55C (100F), the temperature decreases to 90C (192F). This energy is
of sufficient quality for heating. The storage system for the power plant
has the capability to handle storage for district heating. No storage system
capacity or other changes are required, only the operating sequences need to
be changed. Overnight, few days, and annual storage are available. However,
additional collectors, collecting on an annual basis, will be required.

5. SYSTEM ECONOMICS

A 40 MW power plant system is utilized for economic analysis. The
system consists of the Rankine steam cycle, a two-dimensional parabolic concentrating collector system, and the aquifer thermal energy storage system.
The Rankine steam thermodynamic cycle efficiency is 20 percent.

The required solar collector area is calculated according to the annual
daily average solar energy collected for a given inclination angle. The installed solar collector cost is based on vendor's quotation of $20 - $22/ft^2.
The cost of piping is based on 25 percent collector cost as estimated by the
collector vendor. The cost of boilers, steam cycle, and generating equipment
is estimated as $400/KW. The cost of drilling 16 wells required for the
system based on $100,000 per well is $1,600,000. The results of the individual calculations and the overall costs are tabulated in Table 2. The
storage system is less than 1% of capital costs in all cases. The KW hr
cost of electricity is based only on capital cost. It includes a ten percent
interest rate and a sinking fund depreciation over twenty years with a ten
percent interest rate. A drop in collector costs would show an almost
proportional decrease in electric cost.

6. CONCLUSION

Thermal energy storage in aquifers is a potential solution which is
technically and economically feasible for annual energy storage. The system
has been investigated for direct heating and cooling of community buildings.
The application of aquifers for high temperature thermal energy storage has
promise for electric power generations and process heat. The economic
feasibility for this solar power plant system may occur in the near future
as fuel costs continue to rise. The cost of collectors is expected to decrease with additional solar collector research and production. Again, the
plant storage system capital cost is in the range of 0.4 to 0.5 percent of
the total plant capital cost. This removes a major barrier to solar electric
power generation.

REFERENCES

1. Esmail, O. J., and Kimbler, O. K., "Investigation of the Technical
    Feasibility of Storing Fresh Water in Saline Aquifers", Water Resources,
    1967.

2. Smith, G. G., and Hamor, J. S., "Under Ground Storage of Treated Water:
    A Field Test", Ground Water, 1975.

3. Meyer, C. F., and Todd, D. K., "Heat Storage Wells", Water Well Journal,
    1973.

4. Meyer, C. F., "Status Report On Heat Storage Wells", Water Resources Bull,
    1976.

5. Molts, F. J., and Bell, L. C., "Head Gradient Central In Aquifers Used
    For Fluid Storage", Water Resources, 1977.

6. Molts, F. J., and Warman, J. C., "Confined Aquifer Experiment-Heat
    Storage", proceeding of thermal energy storage in aquifers workshop.
    (Conf-7805140). U.S. Department of Energy. Lawrence Berkeley Laboratory,
    University of California, 1978.

7. Schaetzle, W. J., Brett, C. E., and Grubbs, D. M., "Annual Energy Storage
    In Aquifers For Community Energy System," International Energy Symposia
    Montreux, Switzerland, 1979.

8. Schaetzle, W. J., and Brett, C. E., "Heat Pump Centered Integrated
    Community Energy System Interim Report", NTIS No. ANL/ICES-TM-30,
    March 1979.

9. Collins, R. E., Fanchi, J. R., Morrell, G. O., Davis, K. E., Guha, T. K.,
    and Henderson, R. L., "High Temperature Underground Thermal Energy Storage"
    Presented in Berkeley, California, 1978.

10. Tsang, C. F., Buscheck, T., Mangold, D., and Lippman, M., "Mathematical
    Modeling of Thermal Energy In Aquifiers", Lawrence Berkeley Laboratory,
    Berkeley, California, 1978.

11. Mathey, B., and Manjoz, A. "Underground Heat Storage: Dimensions, Choise
    of a Geometry, and Efficiency", Center d' Hydrologeotogie de L' Universite,
    CH-2000 Neuchatel Swiss, 1977.

12. Duff, W. S., and Shaner, W. W., "Solar Thermal Electric Power Systems:
    Manufacturing Cost Estimation and System Optimization," Energy Conversion,
    1978.

13. Shaner, W. W. and Duff, W. S., "Solar Thermal Electric Power Systems:
    Comparison of Lime-focus Collectors," Solar Energy, 1979.

14. Pierce, T., "Efficient Low Cost Concentrating Solar Collectors," Solar
    Energy, 1977.

15. Gupta, B.P., "Development and Evaluation of a Medium Temperature Concentrat-
    ing Collector", Energy Resource Center, Honeywell, Inc., Minneapolis, Min-
    nesota, 1977.

16. Bechamn, W.A., Klein, S.A. and Duffie, J.A., Solar Heat Design, John Wiley
    and Sons, Inc., New York, 1977.

Seas: A System for Undersea Storage of Thermal Energy

J.D. POWELL
Scripps Institution of Oceanography
La Jolla, California 92093, USA

J.R. POWELL
Brookhaven National Laboratory
Upton, New York 11973, USA

ABSTRACT

Methods of storing medium grade (200 to 275° C) thermal energy are of great economic interest for electric generation and process heat. Storage of water heated by land based energy sources and pumped to ocean or lake depths great enough that the ambient hydrostatic pressure equals the saturation pressure appears to be a low cost, safe, and technically feasible method for storing large volumes of hot water.

INTRODUCTION

Low cost methods of storing medium grade thermal energy are of great interest for load leveling in proposed solar energy central power plants, increasing the baseline electrical output of nuclear power plants, and supplying industrial process heat. Thermal energy obtained from hot water or steam is very attractive. Water costs little, is nontoxic, nondegradable, relatively noncorrosive and has high specific heat. Unfortunately, the high pressure level, large volume requirements, and the associated hazards of land based storage require very expensive thick-walled storage vessels.

Underwater storage appears to be a low cost, safe, and technically feasible method for storing large volumes of hot water. In the SEAS (Storage of Energy at Sea) concept, hot water is stored offshore or in deep lakes at a sufficient depth that the ambient hydrostatic pressure equals saturation pressure. Storage containers then need only be thick enough to counteract small pressure differentials due to density differences between hot and cold water, and to keep thermal losses to an acceptable level. As an example, hydrostatic pressure at 300 m (990 feet) permits storage at a temperature of 232° C (451° F) and a pressure of 2.95 MPa (428 psi). Such depths can be reached at a substantial number of locations by a transmission line and economic characteristics of storage systems, transmission lines and power cycle use are examined.

1. STORAGE SYSTEMS

The containers that incorporate the most feasible design ideas are shown in Figures 1-4. Figure 1 is a flexible insulated bag retained by steel nets. The nets are anchored by driven piles, clump anchors, or embedment anchors.

Figure 2 is an excavated flattened hemisphere covered with an anchored
diaphragm. The excavated volume is lined with an impermeable membrane and
fastened to the diaphragm support ring which is then anchored with skirt
pilings. Backfill is then placed over the membrane for ballast and thermal
insulation.

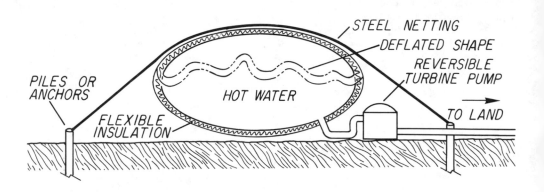

FIGURE 1. FLEXIBLE BAG CONTAINER

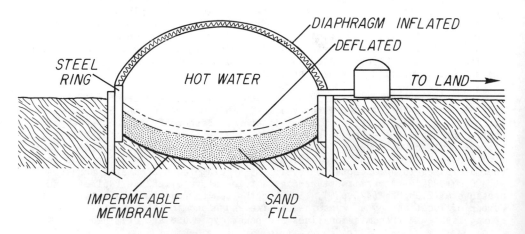

FIGURE 2.   FLEXIBLE DIAPHRAGM CONTAINER

Figure 3 is a tank formed from a large rigid ring with a movable insulated
lid sealed with a rolling diaphragm. The fourth type, Figure 4, is a constant
volume rigid tank with a movable inner disc separating the two fluids, the only
seal being a minimum radial clearance. In each case hot water is withdrawn
from the container at constant temperature. After being used for electrical
generation, the cool water product is stored on land.

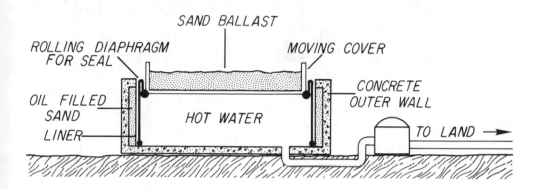

FIGURE 3. VARIABLE VOLUME RIGID CONTAINER

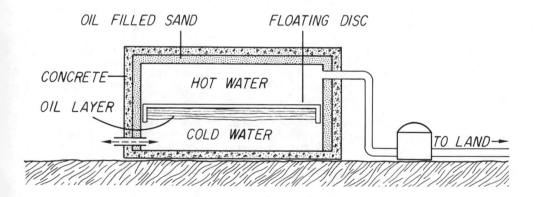

FIGURE 4. CONSTANT VOLUME RIGID CONTAINER

The fluctuation volume of hot water is accommodated in the first two designs by changes in shape and volune of the flexible bag or diaphragm, in the third type by vertical movement of the insulated disc, and in the fourth by admission and/or expulsion of cold ambient water to or from the bottom of the constant volume container.

1.1.   Structural Considerations

The first two types of containers depend on the structural integrity of the flexible membrane. The only type of fabric presently feasible is a teflon impregnated glass cloth. This particular fabric was developed by Owens-Corning Fiberglass for space suits. There are numbers of very large inflatable structures that utilize this fabric.   For example, the roof of the 80,000 seat stadium in Pontiac, Michigan is 46,500 m^2 (10 acres). The largest bag in the SEAS system for a 2,000 MW (thermal) system (1.24×10^5 m^3) would typically have a 40 m diameter, length of 99 m and a surface area of 15,000 m^2. The membrane for type Figure 1 would have a surface area of only 7000 m^2. The glass cloth now used (Beta fiber) would eventually deteriorate due to the moisture and temperature. However, there are experimental fabrics using glass with a greater silica content which show great promise for temperatures up to 250°C. For the design of a storage container such as Figure 2, there is considerable promise in the use of metal diaphragms. A 5,000 m^2 stainless steel membrane roof was developed by D. A. Sinoski and International Nickel Company for a sports center at Dalhousie University, Nova Scotia. Supported by air pressure in normal operation, it can revert to a concave suspended-type roof upon air pressure loss. Special contraction joints are welded to 1.5 mm thick nickel stainless steel membrane segments.

Storage container (Figure 3) would be double walled construction with the inner wall corrosion-protected steel or concrete. The outer wall would be concrete. The space between the walls would be filled with an insulating substance such as oil saturated sand or dredge spoil. The rolling diaphragm would be made out of teflon-glass fiber. The floating top of the tank is a concrete dish filled with dredge spoil, which would serve the dual function of insulation and ballast. The bottom could be concrete or similar to the design of Figure 2. Buoyancy forces would be resisted by the lid and bottom weight, and wall weight if the bottom is a diaphragm. Structural stresses due to density differences are minimal. For a 2,000 MW (thermal) tank (68 m dia., 34 m high, .75 m wall thickness) hoop stresses in the wall would be 1.9 MPa (275 psi) compressive at the bottom if the ambient pressure was 3.45 MPa (500 psi) and there was a 3.45×10^4 Pa (5 psi) internal over-pressure at the top. Stress design of the tank would be dictated more by the strength required for construction, towing and placement.

Tank (Figure 4) would have wall and bottom construction similar to Figure 3.   The top would be an arched reinforced prestressed concrete diaphragm for resisting the slight positive internal pressure. The floating disc would be corrosion protected concrete with an oil layer. Tank (Figure 4) could store water at greater temperatures than previous designs because fabrics are not required.

The offshore construction industry is quite capable of building the concrete storage containers which are small compared to massive structures such as Ekofisk or gravity platforms such as the Condeep series. The primary feature of the designs presented that will require creative development will be the flexible components, if further studies show that they are the most economical means of storage.

1.2. Insulation

Heat losses through rigid walls or sand bottoms are not a significant problem because of the large quantities of water stored with the consequent favorable surface to volume ratio and large thermal time constant. For example, tank (Figure 3) with a storage capacity of 1.25×10^5 m^3 and an insulation thickness of 30 cm of oil filled sand would lose heat at the rate of 0.5°C/day. For flexible membrane containers, insulation is required to decrease heat loss and is desirable for increasing structural life. Exterior insulation might consist of oil filled fluoro-elastomer foam or tubes filled with oil saturated mineral wool to reduce convection. An inorganic wool could be used to line the interior to reduce membrane temperature.

1.3. Pressure Control

The saturation pressure of the hot water at the supply source on land plus the elevation difference to the storage site would exceed the ambient hydrostatic pressure for the proposed working depths. After passage through the transmission line, hot water pressure will be reduced to slightly above ambient hydrostatic by a turbine before entering the storage container. Any net pressure increase gained by elevation drop can be converted into electrical energy and transmitted back to land. The turbine will be reversible and used to pump the stored water back to land.

2. TRANSMISSION COSTS

The transmission line cost per unit length is in $/km year is given by (MKS units)

$$C_T = C_{pp} + C_{HL} + C_{cc} \tag{1}$$

$$= 10^{-9} \left(\frac{f}{2}\right) \left(\frac{4}{\pi}\right)^2 \left(\frac{\dot{V}_F}{D^5}\right)^3 \rho HW \, t_o \, c_{pp}$$

$$+ 10^{-9} \, U(Tw - Ta) 2\pi D t_o \, C_{HW} + 0.15 \, C_c$$

where the first term represents pumping power cost in $/km year; the second, cost of heat loss in $/km year; and the third, fixed charges on the capital cost C_{cc}, in $/km year (15% fixed charges per year are representative of the utility industry). The cost of pumping power is taken as $5/GJ (1 GH = 10^9J) and the value of the thermal energy of the transmitted hot water as $1/GJ; time period, t_o, is 3.15×10^7 sec (1 year); friction factor, f, is 0.01; density of hot water, ρHW, is 800 kg/m^3; and the difference of temperature between hot water and ambient outer water (Tw - ta) is 230°C. The hot water flow rate, $\dot{V}_f$, is determined by total thermal energy storage requirements with the hot water pumped to the tank half of the cycle time, and pumped from it for half. The pipe is insulated with a middle layer of oil-saturated sand. For an insulation thickness of 10cm, U is 2w/m$^{2\circ}$K.

TABLE 1: Thermal Energy Transmission Costs

Transmitted Thermal Power MW(th)	500			1000			2000		
Water Flow Rate, m³/sec	0.3	0.72		1.42			2.86		
Pipe ID, m	0.98	0.5	0.75	0.5	0.75	1.0	0.75	1.0	1.25
Pumping Power, MW(e)/,m	0.076	0.010	0.61	0.080	0.019	0.64	0.152	0.050	
Heat Loss, MW(th)/km	0.44	0.72	0.72	1.09	1.45	1.09	1.45	1.81	
Capital Cost, $/km	1.5×10^5	2.5×10^5	4×10^5	2.5×10^5	4×10^5	5.5×10^5	4×10^5	5.5×10^5	7×10^5
Pumping Cost, $/km yr.	1.6×10^5	1.2×10^4	1.6×10^3	9.6×10^4	1.26×10^4	3×10^4	1.01×10^5	2.4×10^4	7.9×10^3
Heat Loss Cost, $/km yr.	1.4×10^4	2.3×10^4	3.4×10^4	2.3×10^4	3.4×10^4	4.6×10^4	3.4×10^4	4.6×10^4	5.7×10^4
Capital Charges, $/km yr.	2.25×10^4	3.75×10^4	6.10^4	3.75×10^4	6×10^4	8.25×10^4	6×10^4	8.25×10^4	1.05×10^5
Total Line Cost, $/km yr	1.9×10^5	7.2×10^4	9.6×10^4	1.6×10^5	1.1×10^5	1.6×10^5	1.98×10^5	1.5×10^5	1.7×10^5
Energy Revenues, $/yr	0.79×10^7	0.79×10^7	0.79×10^7	1.58×10^7	1.58×10^7	1.58×10^7	3.15×10^7	3.15×10^7	3.15×10^7
Line Capital Cost, $/KW(th) km	0.6	1.0	1.6	0.5	0.8	1.1	0.4	0.55	0.7
Critical Length, km	42	110	82	99	144	99	159	210	185

NOTE: Product value (hot water at 230 C=$1/GJ), Pumping power cost=$5/GJ, Capital charges=15%/year

Table 1 shows transmission costs for pumping hot water at $230°$ C. to an underwater storage tank. Three thermal power ratings are examined (500, 1,000, and 2,000 MW (th)) with pipe inner diameters ranging from 0.3 to 1.25 m. Capital costs of the transmission line are taken as 2.5 times those of oil and gas pipelines (1). This cost increase is for thermal insulation and laying the pipe underwater. Capital charges and heat loss increase linearly with pipe diameter, while pumping cost decreases as D^{-5}. As a result, there is a pipe diameter corresponding to minimum total cost at each power rating. The minima are at D ~ 0.5 m (500 MW (th)), ~0.75 m (1,000 MW (th)), and ~1.0 m (2,000 MW (th)). The minima are relatively broad, so that a pipe diameter greater than that corresponding to the minima does not cost much more. Assuming that more than one underwater storage tank and transmission line are in place, it may be advisable to choose a pipe diameter somewhat larger than the minimum cost value in case a line should fail.

Two economic figures of merit are shown in the capital cost in $/kw (th) km, and the critical length. The latter is the value of the thermal energy carried per year by the transmission line divided by the total line cost in $/km year. Using either figure of merit, line lengths of 20 to 30 km are practical, even for relatively low thermal power ratings, e.g., 500 (MW (th). At higher power ratings, e.g. 2,000 (MW (th) line lengths of 40 to 50 km in length appear practical, and leave ample margin for the other costs for a thermal storage system (underwater storage tank, supply cost of thermal energy, etc).

3. POWER CYCLES

In this paper we assume that thermal energy is stored as sensible heat of hot water. Table 2 lists storage media of land-based systems and their energy densities. Thermal energy stored in the form of steam has significant advantages from the power cycle standpoint. However, the density difference between stored steam and the ambient outside water is several times greater than the density difference between hot and cold water creating more of a storage problem. Steam storage has attractive benefits and will be investigated in future studies.

TABLE 2: Thermal Energy Storage Densities

Concept	Thermal Energy Storage Density (Btu/ft^3)	Temperature Swing $(°F)$
Steam Storage	10,600	420-250
Feedwater Storage	15,850	422-168
Liquid Sodium	11,690	1,000-300
Fused Salt	26,030	950-750
Fuel Oils	10,700	530-250
Metal Hydride	1,820	200
Thermal Well	2,450	340-180

There are some possible improvements that could be made to the water used for thermal storage. If an inexpensive substance such as an inorganic salt was added, this would increase density and decrease vapor pressure. Storage containers could be a very simple design, perhaps just a pool of water with an insulated lid if the storage water was denser than the surrounding water. Also the decrease in vapor pressure would allow storage at a water depth considerably less than if pure water was used for a given amount of thermal energy. However, we suspect there is a problem involved in finding an inexpensive and not too corrosive an additive.

Table 3 lists the principal supply and use modes for thermal energy of hot water, together with their advantages and disadvantages. An important limitation to all supply/use modes is the requirement that the storage/supply and/or consumption points be within a few tens of kilometers of each other. This restricts the number of available storage/supply/use points. If solar collector cost projections are achieved, solar supply will probably be the principal mode of supply. Storage and transmission costs are then low enough that all power generation would be by solar heated hot water. If solar collector costs turn out to be much higher than the projected costs, then extraction steam from central station coal fired or nuclear power plants would be the principal supply mode, with the stored thermal energy used to generate peaking and intermediate load power. In effect, base loaded central station plants would then supply all electrical energy, with time variations of demand met using the stored hot water for power generation.

TABLE 3:   Supply and Use Modes for Thermal Energy in Hot Water

SUPPLY MODES	ADVANTAGES	DISADVANTAGES
Solar collectors	Unlimited supply	Possible high solar-collector cost.
Extraction steam	Relatively low cost	Supply limited to large central station power plants

USE MODES		
Power generation - feedwater heating	High marginal usage efficiency-cheap eqpt.	Must be tied to central station power plant
Power generation - steam flashed peaking turbine	Not tied to central station plant	More expensive equipment - lower efficiency
Power generation - organic vapor turbine	Not tied to central station plant	More expensive equipment - lower efficiency
Process heat	High value for thermal energy and directly displaces oil and gas	None except limitations on site availability

Of the power generation modes in Table 3, feedwater heating is the most attractive. In studies of Light Water Nuclear Reactors (LWR's), for example, plant output can be increased by up to 40% if feedwater heating and steam retreat is carried out with an outside source of thermal energy rather than the customary mode of using primary steam from the steam generator and extraction steam from the turbine. Figure 5 shows the flow sheet for this feedwater heating mode, for both supply and delivery phases to the underwater storage tank.

For a 1200 mw(e) LWR, output can be increased by 440 MW(e) by using 1600 MW(th) of auxiliary thermal input to supply steam reheat and feedwater heating (3). The hot water supply temperature is ~300°C. with the sensible heat of the supply water being utilized at successively lower temperature levels in the various reheaters and feedwater heaters, down to a discharge temperature of ~50°C. The cold discharge water can then be stored in an insulated pond (similar to the solar pond designs developed for temperatures up to ~90°C.) until the need for power augmentation disappears. The cool water is then reheated and pumped back to the underwater storage tank.

The marginal efficiency for electric generation from the hot water thermal addition is on the order of 28%. This is relatively high, considering that the average temperature of the energy addition is only about 175°C. The other power generation modes, peaking turbine with steam flashing, and a Rankine-organic vapor turbine, will have substantially lower power cycle efficiencies on the order of 20% for hot water heat to electricity.

These two power generation modes will also require more expensive capital equipment. Their principal advantage is that they are not tied to large central station coal or nuclear plants, and can be used for distributed load generation for whatever demand pattern is desired.

Low and intermediate temperature process heat applications would have the highest economic payoff for the SEAS system, since the hot water would directly displace the oil and gas fuel presently used for this purpose. Hot water would be worth on the order of $3/GJ (oil at 12 barrels, 70% efficiency) for our purpose as compared to its value of ~1/GJ for power generation.

4. GEOGRAPHICAL LOCATIONS

There are a considerable number of locations in the world which would be suitable for the SEAS concept. Figure 6 shows coastline locations where water depths are sufficient. One economic advantage of the SEAS system is due to population concentration in coastal areas with the consequent need for nuclear, solar and future fusion plants. In addition, there are numerous lakes in North America such as Lake Superior, 1300 ft.; Lake Michigan, 1000 ft.; Lake Powell, 560 ft., and a choice solar location. World-wide there are locations such as Loch Ness, 850 ft.; Lake Baykal, 5300 ft.; Lake Tanganyika, 1800 ft.; Boden See, Germany, 800 ft., and many others.

5. SYSTEM APPLICATIONS AND ECONOMICS

Storage is important whenever energy sources, e.g., solar collectors or energy demands, e.g., electric loads on a grid, are available. Storage can also serve as a spinning reserve to meet demand if a supply source should

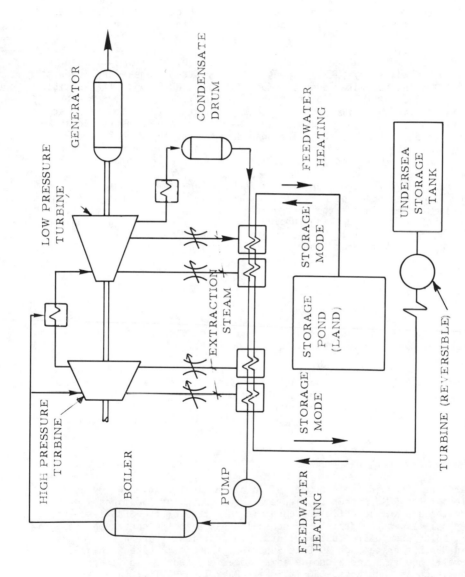

GENERATOR

CONDENSATE
DRUM

LOW PRESSURE
TURBINE

FEEDWATER
HEATING

STORAGE
MODE

EXTRACTION
STEAM

STORAGE
POND
(LAND)

UNDERSEA
STORAGE
TANK

HIGH PRESSURE
TURBINE

STORAGE
MODE

BOILER

PUMP

FEEDWATER
HEATING

TURBINE (REVERSIBLE)

FIGURE 5. SEAS SYSTEM USED FOR FEEDWATER HEATING

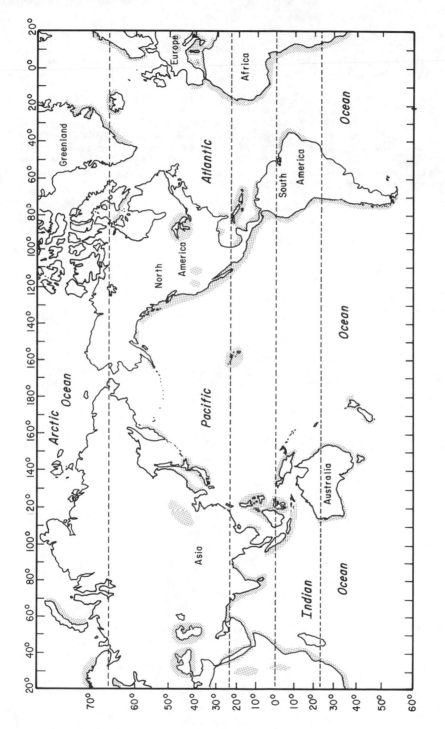

FIGURE 6. LOCATIONS WITH NEARSHORE DEPTHS SUFFICIENT FOR ECONOMIC STORAGE

suddenly fail. Three timescales are of primary interest for storage: daily,
weekly and seasonal. Storage costs using previously proposed methods (hot
water in above ground steel tanks, hot oil, fused salts, etc.), are too
expensive for most applications, including daily storage. (2)

Table 4 shows a representative range of energy storage costs for
the underwater SEAS system. A 20 km long, 2000 MW(th) transmission line is
assumed. The range of capital costs for the underwater storage tank is taken
as $3, $10, and $30 per m^3 of storage capacity ($300,000 to $3,000,000 for a 10^5
m^3 tank). $30 per m^3 seems realistic relative to present material and construc-
tion technology. Reductions significantly below this figure would depend on major
technological advances. For comparison, the cost of storage of hot water in
above ground steel tanks has been estimated at $1000/m^3. (2)

TABLE 4: Effect of Storage Period on Energy Storage Costs

	Storage Tank Cost ($/m^3)		
	3	10	30
Transmission Cost, $/GJ (20 km line, 2000 MW(th))	0.10	0.10	0.10
Capital Charges on Storage Tank, $/GJ			
Daily Storage	0.002	0.0065	0.019
Weekly Storage	0.014	0.046	0.138
Seasonal Storage	0.71	2.38	7.08
Total Cost for Energy Storage, $/GJ			
Daily Storage	0.10	0.11	0.12
Weekly Storage	0.11	0.15	0.24
Seasonal Storage	0.81	2.48	7.18
Total Cost for Energy, $GJ (Solar Collector at $50/m^2)			
Daily Storage	1.05	1.06	1.07
Weekly Storage	1.06	1.10	1.19
Seasonal Storage	1.76	3.43	8.13

Assumptions: 50% charge time
50% discharge
15% annual fixed charges on capital investment

The costs for underwater storage are very low for daily and weekly storage,
on the order of 5-10% of the value of the stored thermal energy which can be
valued at $1/GJ for electric power generation and $3/GJ for process heat
applications. The cost of the storage tank becomes important for seasonal
storage, i.e., energy is collected for approximately 50%. Seasonal storage
would be economical only if storage tank cost is on the order of $3/m^3 of
capacity.

The last part of Table 4 shows the total cost of energy assuming a solar
collector system with an average insolation of 250 watts/m^2 (averaged

diurnally and over the year) and a capital cost of $50/m^2. Studies of line focus collector system generating temperatures of 400°C project cost on this order. [4] Considering that the average collector temperature of the hot water stored in underwater tanks is only about 175°C, it should be possible to achieve collector costs considerably under those quoted above (or conversely, if the above estimates are too low, true costs for lower tempera- ture collection should be closer to the $50/m^2 value than those for higher temperatures.

The total energy costs for a solar collector/underwater energy storage system are then on the order of $1/GJ. This has important implications for a solar energy economy. One of the principal barriers to the use of solar energy for power generation and/or process heat has been the high cost of storage, requiring either an expensive backup system or a solar/ conventional power (e.g., fossil) hybrid approach. With cheap storage, solar power could supply all power generation/process heat applications if the solar collectors are economic.

The low cost of storage, also implies that utilities can run only base- loaded nuclear or coal-fired plants and meet all daily and weekly load variations without resorting to high fuel cost peaking and intermediate load gas turbines. Besides reducing power generation costs, this will save considerable amounts of scarce oil and gas fuels.

Finally, the low cost of storage for extended periods means that utilities can use long-term, cheap storage to reduce excess capacity generation requirements at present because of unexpected plant outages, transmission line failures, scheduled maintenance, etc. The excess capacity of a grid system must be approximately 30 to 40% above its peak load. The use of cheap long term underwater storage with low cost steam turbines would considerably reduce both excess capacity requirements and capital investments associated with them.

REFERENCES

1. Karkhack, John, Long Distance Transmission of Hot Water for District Heating. Proc. of IECEC Conf., 2, pp. 1022-1028, San Diego, California 1978.

2. An Assortment of Energy Storage Systems Suitable for Use by Electric Utilities. Public Service Electric & Gas Company, New Jersey, EPRI EM-264, Vol. 2, 1976.

3. Benenti, R.R. and J.R. Powell, SOAR (Solar Assisted Reactor) Power System. Trans. Amer. Nuclear Soc., 32pp. 10-12, 1979.

4. Conceptual Design Analysis of 100 MW Line Focus Solar Central Power Plant, EDM/TAC-79-051-BR, Large Solar Thermal Power Systems Meeting, Reston, Va., March 20, 1979.

Thermal Storage Cell for High Temperature Solar Systems

LEE FELLOWS
U.S. National Park Service
Titusville, Florida 32780, USA

ABSTRACT

Large scale solar thermal electric power generation is practical at present if higher temperatures can be developed. This has been proven in the United States and other countries with the use of large concentrating collector systems which attain tremendous temperatures and high thermal flux. The concept of solar power as a sole source of power, rather than a supplementary supply, is feasible also, and though further research is needed in this area, a simple, and relatively inexpensive storage method is currently being examined, utilizing high temperature steel alloys.

1.0 INTRODUCTION

Solar-generated electricity as a sole source of utility power is not only technically feasible, it can also be economically attractive to both the consumer and the utility company. A low-cost, efficient means of energy storage is the vital element in full time generation of solar-derived electricity, and the simplest means may be the most practical. The following comparisons outline such a means from both the technical and economic standpoints.

We convert fossil fuels into electrical current at average efficiencies of 16 - 30 per cent, at a mean cost to the consumer of ten cents per kilowatt hour (kWh). For example, in a given area, the consumer may be paying four cents per kWh to the utility company; but this does not reflect the portion of his or her personal income tax which ultimately helps to defray the cost of fuel. Strangely enough, opponents of solar energy usually compare the installation costs of fossil or nuclear plants with those of a solar power plant, and ignore costs of long term operation. In actuality, the only type of conventional power generation which is comparable on the basis of economic similarity is hydroelectric; usually considered the cheapest form of electric generation. Like solar, the initial investment is high, but operating costs are low, since the fuel is supplied by nature. Also, since rains and reservoir levels are as variable as solar radiation, the analogy is more than apt. Both require large land areas for collection of energy: acres of collectors for solar; acres of reservoir for hydroelectric. Both are soft energy path sources and, from the standpoint of by-products, environmentally benign.

In nuclear and fossil fueled systems, installation is only the tip of the iceberg where costs are concerned. Fuel deposits must be searched out, the fuel refined for use and the waste products disposed of. Not only are the

costs for such activities enormous, these chemical and radioactive by-products
will continue to cost everyone inconceivable sums in the form of environmental
damage, health care, life and property insurance, legislation and legal con-
tests. As stated before, only a fraction of these costs are evident in the
consumer's monthly power bill. For example, a gallon of refined petroleum
fuel will generate an average of ten kWh* of electricity. This figure will
vary only slightly, and can be considered a fixed figure for present generating equipment efficiencies. Since ten kWh per gallon is all the energy we can
extract, it is quite easy to see the direct cost of the energy we buy. If
fuel costs sixty cents per gallon, then the cost per kWh is six cents. This
price disregards the initial cost of equipment, the cost of continuing maintenance, and the indirect costs mentioned above. A solar system, on the other
hand, is immune to most of these complications. Though it generally will cost
more to install than the fossil or nuclear plant, the fuel is cost-free.
Therefore, in terms of the useful life of the plant and its impact on society
and the environment, the solar system is less expensive to operate.

    In a simple comparison of direct operational costs, and disregarding
other factors, let us say that a fossil plant costs $1 million to install,
while a solar plant of equal output costs $3 million. Both plants over an
arbitrary period of twenty years, generate 100 million kWh of current. If
the cost of refined fuel remains at six cents per kWh over that period of
time, then the operation of the fossil plant, ignoring maintenance, will be
the installation cost plus the cost of fuel; or $7 million. The cost of the
solar plant, again ignoring maintenance, will also be the installed cost plus
the cost of fuel; or $3 million. The cost per kWh for the fossil plant opera-
tion is more than twice that of the solar plant over the life of the system.
If the solar plant can provide power on a full time basis, as does the fossil
fueled plant, then the solar system is far superior in all respects. Contrary to popular belief, solar energy as a sole source of electric power is
not only feasible, it is easily obtainable. Common sense tells us that the
amount of energy available in a hydroelectric system is equal to the amount
of water stored in the reservoir, plus the new supply of water flowing in
each day. In direct analogy, the amount of energy available in a solar thermal electric system is equal to the amount of heat energy we can store, plus
the heat collected each day. In a concentrating collector system, an average
of 800 watt-hours of heat can be collected over each square meter of collector
surface, for each hour of direct sunlight, between the 45 degree latitudes.
This figure includes the efficiency factor for a typical mirrored collector
surface. Collection of energy does not pose any particularly difficult problems at present, but an economical energy storage medium has been elusive.
In order for the solar system to compete on a true economic basis with other
forms of power generating systems, it must produce all the power needed, not
merely supplement other systems. Though hybrid generating systems are certainly practical, they are more costly to install and they do not eliminate
the need for fossil or nuclear fuel for utility electric generation. While
total solar dependency may not be practical in some instances, it is practical for the major portion of the populated world if used on a large scale.
Just as conventional utility power plants buffer each other by being connected to a common grid or distribution network, and by buying and selling
power to one another; so can solar generating plants. It might be cloudy
and raining over one city for example, while a city fifty miles away is ex-
periencing sunny weather. Peak storage capacity for individual solar plants
is reduced in a regional grid, thereby reducing installation costs.

* 11,600 Btu of fuel consumed at power plant per kWh delivered to consumer.
(Assume 10,536 Btu/kWh station beat rate for all stations, 9% line loss as
reported for 1971 by Edison Electric Institute.)

2.0 THE STORAGE CONCEPT

The storage medium, in order to be economically practical, and to suit the requirements of varied climates and loads, must meet certain criteria:

-- It must be relatively inexpensive with regard to cost per Btu stored.

-- It must have a very rapid charge-discharge capability to take full advantage of all available solar radiation.

-- It must have long life.

-- It must exhibit high energy density.

The medium considered here is highly alloyed steel shot. In comparison with other storage media, it is low in cost, exhibits a high rate of thermal conductivity and does not deteriorate significantly over long periods of time if operating conditions are compatible. At elevated temperatures, it can have a specific heat one-fourth that of water. In the high temperature system envisioned, alloys with scaling temperatures above 1500°F are best for long life, and this oxidation threshold increases almost proportionally with the percentage of chromium in the metal.

The choice of alloy used depends, of course, primarily on its ability to store heat, and due to the unwritten law of compromise, the common alloys which exhibit the highest heat capacity at elevated temperatures are generally the least refractory and most susceptible to corrosion. At 1800°F, Type 430 stainless, for example, with a chromium content of 14 - 18 per cent and a scaling temperature of 1500°F, has almost twice the heat capacity of Type 310 stainless, which has a chromium content of 26 per cent and a scaling temperature of 2000°F. Conductivity is 50 per cent greater also for the Type 430. Type 446 stainless exhibits good parameters also, with higher scaling temperatures than the Type 430. Figure One illustrates the specific heat versus temperature of these alloys.

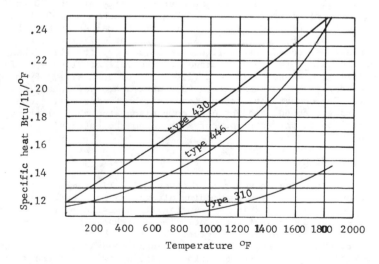

Fig. 1 - Heat Capacity of Stainless Steels

Since a constant annealing process is occurring within the cell, the formation of oxides and sediments will constitute the greatest threat to system life; therefore low sulphur-high silicon content alloys are preferred for the sake of minimizing oxide formation in the metal and the resulting sedimentary blockages, loss of conductivity and reduced cell life. The addition of small amounts of yttrium to the alloy mix also help to reduce oxide formation in chromium steels. The nickel content of austenitic stainless steels can be detrimental in the sodium environment since the nickel will leach out of the steel and deposit in the cold sections of the system, causing blockages. Therefore, the ferromagnetic stainless steels are preferred also for their low nickel content.

For cell temperatures below 1600°F the 430 and 446 stainless steels appear to be good compromises, exhibiting high specific heat and thermal conductivity, and are suitable for use with metallic sodium, the proposed heat transfer fluid. By operating the cell at a pressure just below atmospheric, the boiling point of sodium can be reduced to 1600°F with improved operating characteristics and longer cell life. The liquid alkali metals transfer thermal energy at a minimum rate over fifty times that of water, and at very low vapor pressures. Sodium exhibits a vapor pressure of 760 mm Hg at 1621°F, making light weight containment and piping possible. Oxygen getters and cold traps are necessary to reduce sediment and blockages in the cell and oxidation of the transfer agent, even with the higher temperature alloys.

A large portion of the thermal energy is stored in the transfer agent; the amount of energy being determined by the temperature of the system and mass of liquid metal contained. Since approximately half the cell volume will be occupied by the transfer fluid, a 192 ft³ cell contains about 4500 pounds (2042 kg @ 870°C) of liquid metal, with a specific thermal storage capacity of over two million Btu. Add to this the 1800 Btu/lb latent heat of vaporization for an additional 8.1 million Btu and total energy potential of the transfer agent is 10.1 million Btu for this size cell. The cell also contains approximately 23 tons (20,865 kg) of steel shot, and at a temperature of 1600°F and a specific heat of .23 Btu/lb, 16.9 million Btu are stored within. Combining the thermal energy of the storage medium and the heat transfer agent yield 27 million Btu, for an energy density of 140,770 Btu/ft³. With a minimum cell temperature of 400°F and a maximum of 1600°F, a useful temperature range of 1200°F is available for thermal storage with a deliverable capacity of 116,196 Btu/ft³. At 33 per cent efficiency, an engine/generator set can convert this figure to 38,345 Btu/ft³ (11.23 kWh/ft³), for a total deliverable energy storage for the cell of 2156 kWh. In cube form this cell will measure six feet (1.88 meters) on a side. By mass, this is an energy density of approximately 42.7 watt hours per pound, deliverable energy. For comparison, todays lead-acid batteries will store about 18 watt hours per pound, and for cost comparison, the thermal storage cell will cost five cents per watt hour of storage to the battery's eleven cents per watt hour.

3.0 THERMAL STORAGE CELL

The cell design is based on several premises which fit the model storage criteria. The geometry of the storage medium allows the greatest surface area practically possible, with which a transfer fluid must come in contact. As shown in Figure Two, 23 tons of two centimeter shot has a surface area of 8800 square feet (817 m²). Uniform spherical shot must settle into a container in a uniform manner; therefore, the exposed surface area of the shot and the voids between the shot must be uniform also, allowing the transfer fluid to rise evenly throughout the storage medium. The result is the most

2000 kWh thermal storage cell
compared to surface area of
shot contained within

————————— 28.6 meters —————————

1.9 m

Fig. 2 – Conductive Surface Area of Medium

rapid energy exchange possible for the cell, with evenly distributed thermal
flow and minimal temperature gradients throughout the cell. The spherical
shape also allows the most rapid saturation of the medium possible, since
thermal energy from the surrounding transfer fluid must flow and equalize
across a distance of half the diameter of the shot. (See Figure Three.) For
example, if a steel sphere of two centimeters diameter were unfolded, so to
speak, for all practical purposes it would equal a flat plate one-third centimeter thick and 12.56 cm^2 in area. Thus the path of energy flow within the
cell is extremely short.

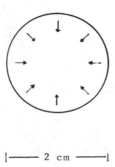

|——— 2 cm ———|

Fig. 3 – Thermal Penetration of Medium

The specific heats of the alloys being examined vary from about .10 to
.12 Btu/lb/°F at room temperature, increasing to approximately .14 to .25
Btu/lb/°F at 1800°F. An input of 500 kW_{th} will charge the cell in 12.6 hours.
A typical energy extraction and conversion system will convert this stored
energy to 2080-2160 kWh_e of delivered energy. This is sufficient energy to
supply ten average homes for eight days. The conductivity of the cell is
such that it can potentially accept or discharge energy at the prodigious
rate of 140.7 X 10^6 Btu/hr, or 41,200 kW_{th} instantaneous flow. The engine-
generator sets used to extract the thermal energy can be Stirling free-piston
linear alternators, steam turbines, other expander-type machines or thermionic
generators. In fact, the stored thermal energy can be used for a multitude of
purposes, including desalinization, space heating and air conditioning, as
well as electrical generation; and system size can be scaled upward almost
indefinitely.

Waste heat storage is another possibility for this cell configuration,
and by substituting common steel peening shot for the storage medium and a
gas such as helium for the thermal transfer agent, cost per kWh stored will
remain about the same. The cell temperatures will be far lower, however;
pressures far higher and end-use and conversion more limited. About 24,000
Btu/ft^3 gross storage is obtainable (about 114 kWh/m^3 recoverable). A cell
of this type would require 75 tons of shot and be 2.6 meters on a side to
have the same storage capacity as the 2000 kWh sodium cell.

In practice, a reinforced concrete vault is constructed and ureathane
foam poured around the outside (Figure Four). Dry sand is poured into the
floor of the vault, and the steel storage cell set in place. After the cell
has been connected, evacuated, sealed and tested, sand fills the remaining

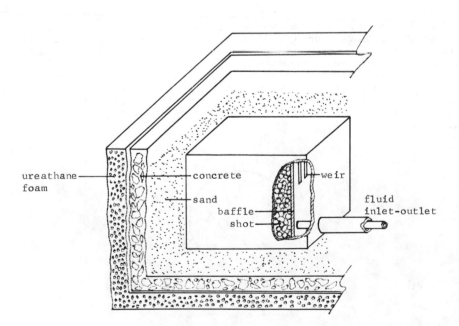

ureathane
foam

concrete   weir

sand
baffle   fluid
shot   inlet-outlet

Fig. 4 - Thermal Storage Cell

space around the cell and the vault is capped with concrete and foam. This
procedure insures low heat loss and provides safe containment. For small
installations, independent parabolas may be used, while a large utility project would likely utilize a central receiver and field of heliostats to collect radiant energy for the cell. In the latter case, the receiver tower
would ideally be located directly over the storage vault to keep the heat
transfer path as short as possible.

The cell casing and piping can be Type 316 or 347 stainless steel, or
high-temperature glass such as Corning Vycor. The spherical shape of the
shot allows automatic relief of stresses which would occur in a more rigid
medium due to thermal expansion; another benefit of cell geometry. As a
result, loading within the cell is never excessive. In the six-foot cube,
for example, the floor of the cell is 36 ft^2 in area, and the 26 tons of
shot and liquid metal resting on it produce a dead load of less than 10 p.s.i.
The inner surfaces of the cell walls are embossed diamond plate to prevent a
uniform plane forming in the layers of shot next to the cell walls, which
would cause expansion stresses at the corners of the cell. Expansion space
within the cell should be sufficient to accommodate only the thermal expansion of the shot, and the external piping sized to provide for expansion of
the transfer agent. When the temperature within the cell falls, a vacuum
will be created due to volume reduction of the transfer agent, and the shot
will prevent buckling of the cell walls.

4.0 UTILITY APPLICATIONS

Though redundant, it is emphasized again, that if a fossil-fueled plant
burns 50,000 gallons of fuel per day, and one week of storage capacity is
desired between fuel shipments, then in effect we must store 350,000 gallons
of fuel and add 50,000 gallons per day to maintain one week of storage
capacity. While the weather has historically been less dependable than the
fuel truck, this may no longer be the case; and in any event, we can size a
solar system to compensate for the worst vagaries of the weather. For example, if it is necessary to generate 1000 kWh per day, and one week of
storage capacity is desired; the storage system is sized to store 7000 kWh.
Collection is a distinctly separate consideration from that of storage. To
illustrate: one hypothetical area of the world where this particular 7000 kWh
per week is needed, receives only one hour of direct sunlight per week in the
worst season of the year. Therefore, the collector area must be sufficient
to collect 7000 kW$_{th}$ instantaneous input. In another locale, the desired
storage capacity and daily energy consumption are the same, however this area
receives six hours of direct sunlight per week under worst-case conditions,
therefore the collector area need be only one-sixth that of the former system, while the storage capacity remains uneffected. In a properly designed
system, collector area will vary according to worst-case weather conditions
for a given area, storage size will vary according to the heaviest usage
period of year, and generator capability will be determined by the maximum
peak demand. These three parameters should be considered separately for
each installation and designed to accommodate residential, industrial, or
any combination of such uses.

In order to serve a hypothetical city of 100,000 population, or 30,000
homes plus 600 attendant businesses, a solar thermal electric utility might
develop as follows:

If the average energy demand per customer is 900 kWh per month during
the peak season, and one weeks' storage capacity is desired, then the thermal

storage cell must have a <u>deliverable</u> capacity of 6.4 X 10^6 kWh. If the conversion efficiency of the engine-generator sets are taken as 33 per cent, then the actual storage capacity of the cell must be 19.4 X 10^6 kWh, or 66.5 X 10^9 Btu. This will require 72,000 tons of steel shot at an estimated cost of $200 million, 6900 tons of sodium at $9 million, 600,000 cubic feet of storage vault at an installed cost of $1.5 million, and a central receiver tower at $278,000. A peak generating capacity of 160,000 kW will be required at a cost of $31 million. Now, if this city receives a minimum of 30 hours of direct sunlight per week, 186 acres of collector surface would be required at a cost of $18 million. This brings the capital investment to $260 million.

5.0   CONCLUSIONS

$260 million for a 160 Megawatt hydroelectric project would not be considered excessive, therefore sole-source solar power generation is competitive with the most economical form of conventional power generation in use today. At an operating life output of seven billion kWh of electricity, the installed cost per kWh is 3.7 cents. Fossil or nuclear generated power costs three times as much at present, and will cost much more in twenty years. Nuclear plants are currently costing approximately $1200 per installed kW of generating capacity, which would amount to $192 million for a comparable nuclear facility.

With development of alloys of higher specific heat and resistance to corrosion, and mass production of system components, the cost of generated current can be further reduced in the solar system, while fuel costs for conventional power generation continue to rise. By charging 4.7 cents per kWh, the solar utility can realize $70 million over a twenty-year period; $22 million for maintenance and personnel costs, and $48 million in gross profits for an 18.5 per cent investment return.

In the next twenty years, no present form of fossil or nuclear power generation will be able to compete economically with solar thermal power generation if the latter is put into wide-scale use in utility-sized generating plants.

ACKNOWLEDGEMENTS

Mine Safety Appliances Company
Evans City, Pennsylvania

Callery Chemical Company
Callery, Pennsylvania

Republic Steel Corporation
Independence, Ohio

Corning Glass Company
Corning, New York

USI Chemical
Charlotte, North Carolina

Advanced Solar Thermal Storage Medium Test Data and Analysis

HRISHIKESH SAHA*
Alabama A&M University
Normal, Alabama 35762, USA

ABSTRACT

The primary objective of this paper is to present a comparative study of the experimental data of heat transfer and heat storage characteristics of a solar thermal energy storage bed utilizing containerized water and/or phase change material (PCM) and rock/brick. This experimental investigation was initiated to find new usable heat intensive solar thermal storage device other than rock storage and water tank which have been the basic storage used thus far. To serve this need four different sizes of soup cans filled with water were tested. These cans were stacked in a chamber in three different arrangements-vertical, horizontal, and random. Air is used as transfer medium for charging and discharge modes at three different mass flow rates and inlet air temperatures respectively. These results were analyzed and compared, which show that a vertical stacking and medium size cans with Length/Diameter (L/D) ratio close to one have better average characteristics of heat transfer and pressure drop. A similar experimental study is underway with three different sizes of metal and high density plastic jars filled with phase change material (Dow-Calcium Chloride Hexahydrate). Due to the delay in acquisition of PCM, these results were not available for this paper which will be compared with the containerized water and rock bed test results and published at a later date. The containerization process can be made economically acceptable if it is produced commercially in large quantities. Due to the internal anti-rusting coating of metal cans and very small corrosivity of water and PCM, the containers will be usable for fifteen to twenty years. Outside moisture rusting can be prevented by dipping the containers in an appropriate paint vat. These types of containerized fluid and salt thermal storage medium have a lower pressure drop, lower volume requirements and higher heat transfer and heat content values than other usual types; also these do not need any special type of storage chamber or heat exchange device. This containerization

*The author is the principal investigator of a NASA/MSFC grant No. NSG 8041 "Parametric Study of Thermal Storage Containing Rocks or Fluid Filled Cans for Solar Heating and Cooling." This paper presents some test data produced for this grant.

allows the storage chamber to be horizontal or vertical with respect to the air flow. The test results and analysis thus far
show that this type of storage device will be well suited for use
with solar air systems for space and hot water heating in both active and passive systems. Some test results with bricks as thermal storage medium are also shown. Standard bricks with appropriate holes make excellent storage medium.

INTRODUCTION

     The literature survey (1) through (7) revealed that there was
a need for basic experimental study on heat transfer and content
characteristics, pressure loss, flow channeling, temperature stratification, and other properties of advanced solar thermal storage
mediums and storage beds. For this purpose of an extensive parametric study to investigate the efficiencies of different thermal
storage beds containing rocks, other solids, and containerized
liquid and PCM a multy flow cycles storage test facility was designed and a series of tests were conducted. The intent of the
test series is to find the influences of the various parameters on
the performance of the storage beds, such as size, type, and orientation of stacking of containerized medium; area and height of
the storage unit; air flow rate; pressure drop across the test
bed; inlet and outlet temperature of air; and temperature distribution in the test bed. Some of these results are discussed below.

DESCRIPTION OF THE TEST FACILITY

     The general design information of the test facility is given
in (Fig.1). A temperature controller regulates the four resistance heating elements of 5 KW each respectively. The electric
blower with variable speeds can reach an air mass flow rate of
800-1600 cfm (0.377-0.755 cms) for a 1.0 inch (2.54 cm) of water
pressure drop. The inlet temperature range of the test-section is
70° - 200°F (21.1°-93.3°C). The storage test section height can
be varied from 2 ft. to 8 ft. (0.6-2.4 m). Integration of the a-
bove using ducts, turning vanes, dampers, intake and outlet valves,
etc., forms an air tight and thermally insulated system. Temperature measurements of air, water, PCM, bricks, and surfaces at various points are taken by using copper-constantan Thermocouples
(Type T) with a tape driven multipoint data logging system and a
thermocouple reference junction. The reference (8) contains further detailed design and instrumentation information of the test
facility.

THERMAL STORAGE MEDIUM AND TEST SPECIFICATIONS

     The Tables 1. and 2. show the test specifications and storage
medium properties. The test were conducted in a storage bed of
2 ft. (0.61 m) height and 4 x 4 ft.^{2} (1.2 x 1.2 m^2) square. The
water filled metal soup can sizes tested were, L/D = 1.09, 1.45,
0.807, and 1.59. Cans were stacked in random, vertical, and horizontal arrangements (Fig. 2.). Standard ten hole bricks of 7.9 x
3.6 x 2.3 in.^{3} (20 x 9.2 x 5.7 cm^3) size were also tested in horizontal arrangements. Inlet air temperatures for charging mode

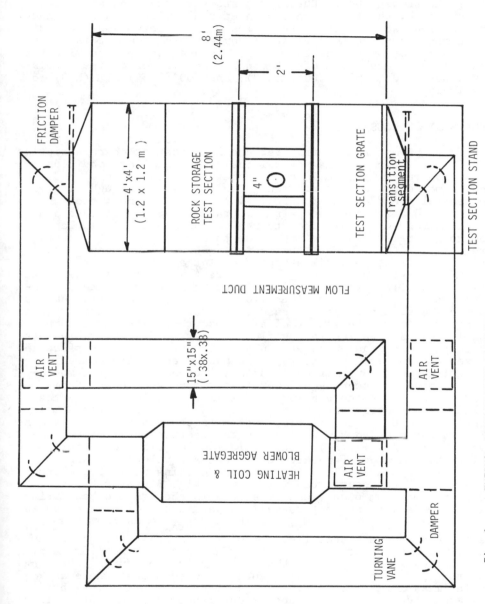

Fig. 1. THERMAL STORAGE TEST FACILITY

100° (37.7), 130° (54.4°), and 160°F (71°c) and inlet air velocities were 500 (2.54), 600 (3.048), 700 (3.556) FPM (MPS). During
the charging mode, the hot air is blown from the top to the bottom of the storage bed. During the storage mode, the storage bed
is isolated by shutting down the inlet and outlet dampers to measure the heat storage characteristics of the medium and the bed.
During the discharge mode constant temperature room air is blown
through the bottom to the top of the storage bed to determine the
heat releasing mechanism of the cans. The air velocities of discharge were 300 (1.52), 400 (2.032), and 500 (2.54) FPM (MPS) respectively.

RESULTS AND DISCUSSIONS

To compare the heat transfer and heat content characteristics
of the four sizes of cans several tests with similar conditions
(mass flow rate, maximum inlet temperature, total mass of H_2O, and
can arrangements) were performed. The figures 3 through 11 give
the general trend of the results. The figures 3 and 4 show the
charging and discharge mode air temperatures across storage bed
with time respectively. The storage bed was heated in a closed
loop to gradually heat the medium up to a present inlet temperature of 130°F (54.4°c) and then this temperature was kept constant
for a period of time until a preset outlet air temperature of
125°F (61.66°c) was reached. The figure 4 shows the air heat gain
from the cans in a reversed flow direction. The pressure drop a-
cross bed (Fig. 5.) for various flow velocities, can arrangements,
and can sizes. In all three cases, the vertical arrangement (Fig.
2.) seems to have the lowest of all pressure losses. Pressure
loss increases with increasing flow velocity. Fig. 6. represents
the temperature gradients between air flow and the can surface and
between can surface and water in a can located at the center of
the storage bed. Smaller mass/surface area have better temperature acceptance producing a lower temperature gradient. From
these measurements the can internal film coefficient h_i and the
outside film coefficient h_o are computed. An apparent thermal
transmittance U for the storage system is derived using experimental data for each configuration and the coefficients computed a-
bove, $U = 1/(1/h_o + 1/h_i)$. The variations of U-factor with time
during charging mode for three different air velocities are
showned in figure 7. This figure also contains water temperature
variations of the center can with time. Fig. 8. through Fig. 10.
record the maximum usable energy stored in the storage bed during
charging mode with respect to various parameters. This energy was
computed by $Q_{total} = \overline{C}_p \cdot M \cdot \Delta T$, where Q_{total} = maximum usable
energy, $\overline{C}_p$ = apparent specific heat of the thermal storage medium,
and ΔT = bed temperature increment. The cans with smaller mass/
surface area store more heat than the larger ones during charging
mode for a given time. These figures also show that where as,
smaller mass/surface area cans store more heat in vertical arrangement mode, the larger ones favour random arrangement. Figure
11 shows the charging and discharge mode total heat gain and heat
release with time for three air flow rates. A few cases were run
where the storage bed areas charged from the bottom through the
top and kept for several hours in storage mode to see the effect
of heat rise due to conduction and convection in the bed. The
temperature stratification upwards was found to be very insignifi-

cant. A similar parametric study was made with standard ten hole

Table 1.   Properties of the water filled cans used and test speci-
           fications (in SI units).

	TEST STORAGE SPECIFICATIONS AND PROPERTIES			
Can Type / Property	I	II	III	IV
Length (l) (cm)	10.16	17.78	8.89	7.62
Diameter (D) (cm)	6.985	11.176	11.176	6.985
L/D	1.45	1.59	0.80	1.09
Surface Area (m^2)	0.029	0.082	0.051	0.024
Water Wt./can (kg)	0.292	1.418	0.680	0.198
Empty Wt./can (kg)	0.056	0.170	0.113	0.043
Volume/can (m^3)	0.388 x 10^{-3}	0.173 x 10^{-2}	0.877 x 10^{-3}	0.292 x 10^{-3}
Water Wt./Surface Area (kg/m^2)	9.741	17.275	13.371	8.145
Total cans	1310	270	563	1928
Void Fraction	0.44	0.48	0.45	0.37
Apparent Specific Heat of Can & Water (j/kg °c)	3587.83	3784.60	3654.81	3529.22

Storage bed Height = 0.6096 m.
Storage bed and plenum volume = 1.132 m^3.
Air Mass Flow Rates: cu. ms/sec (M/S).
Charging Mode - 0.368 (2.54), 0.442 (3.048), 0.516 (3.556).
Discharge Mode - 0.221 (1.524), 0.295 (2.032), 0.369 (2.54).
Can Storage Orientation = Random, Vertical, and Horizontal.
Inlet Air Temperature range = 32.2 °c - 93.3 °c.
Specific Heat of Water = 4186.5 (J/kg °c).
Specific Heat of Can = 460.52 (J/kg °c).

bricks arranged horizontally for better air passage characteristics. Figure 12. shows the air temperature distribution across
bed during charging mode. Figure 13. shows the heat storage
characteristics of bricks in charging mode and also reveals the
thermal transmittance characteristics of brick storage. Pressure
drop across storage bed for different flow velocities is given in
figure 5.

Table 2.   Properties of Brick

Length	7.875 in.	20 cm.
Width	3.625 in.	9.2 cm.
Height	2.25 in.	5.7 cm.
Surface Area	1.117 ft.^{2}	.104 m^2
Weight	3.87 lb.	1.76 kg.
Volume	.027 ft.^{3}	.762 x 10^{-3} m^3
Weight/Surface	3.465 lb/ft.^{2}	16.92 kg/m^2
Volume of Each Hole	7.5 x 10^{-4}ft.^{3}	2.12 x 10^{-6} m^3
Number of Holes in One Brick	10	10
Volume Ratio Holes/ Brick	.22	.22
Number of Bricks	504	504
Void Fraction	.575	.575
Specific Heat	.193 BTU/lb^{o}F	807.99 J/kg. ^{o}c
Thermal Conductivity	.25 BTU/hr.ft.^{o}F	4.327 J/s.m.^{o}c

CONCLUSIONS

The test data analysis for a parametric study to determine the optimum size of cans and arrangement with respect to heat storage, heat transfer, and pressure drop reveals the following:

a.)   The size L/D = 0.80 with mass/surface area of 2.74 in a random stacking arrangement has better heat transfer characteristics, and
b.)   the vertical stacking has the least pressure drop across the test bed compared to random and horizon-stacking arrangement.

Since the internal and external film coefficient of containers packed with thermal storage medium can be computed from the test data, an apparent U-factor, representative of heat transfer characteristics, of different types of storage mediums can be easily evaluated. The containerization process can be made economically acceptable if it is produced commercially in large quantities. The problems of container leakage and rusting can be controlled by selecting metal cans with anti-rust coated inner lining (soup cans) and/or high density plastic containers. The external moisture rusting of metal cans can be prevented by dipping them in an appropriate paint solution. These types of containerized fluid and PCM have a lower pressure drop across storage bed, lower volume requirements; due to uniformity of containers thermal channelling does not occur; these do not need any special type of storage chamber or heat exchange device. The test results and analysis thus far show that this type of thermal storage device will be well suited for use with solar air systems for space and hot water heating in both active and passive systems.

Since bricks with different arrangements of voids are easily available, these make excellent thermal storage medium. Due to the holes the apparent diameter, the surface to volume ratio, and the void fraction of this type storage system can perform better than rock storage. Bricks can be used for both horizontal and

TEMPERATURE MEASUREMENTS IN AND AROUND

A CAN FILLED WITH LIQUID

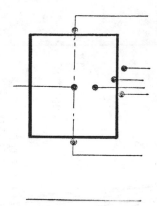

STORAGE TEST ARRANGEMENT

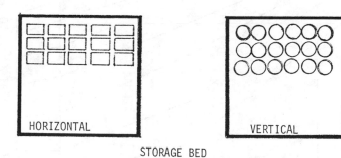

STORAGE BED

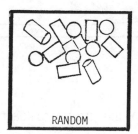

Fig. 2.   STORAGE CONFIGURATIONS AND TEMPERATURE
           MEASUREMENTS

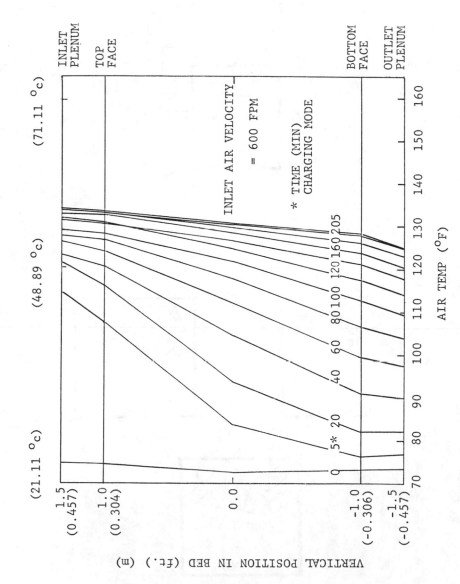

FIG. 3. AIR TEMPERATURE PROFILE IN BED DURING CHARGING MODE.

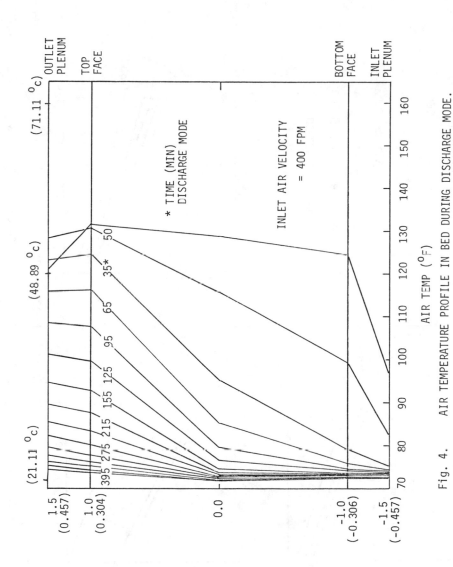

Fig. 4. AIR TEMPERATURE PROFILE IN BED DURING DISCHARGE MODE.

329

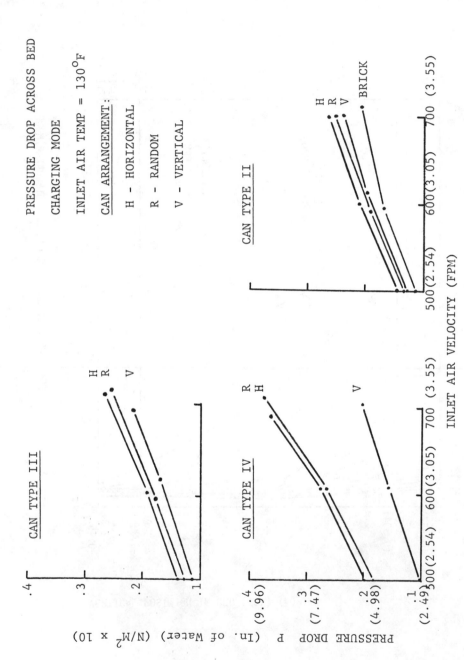

Fig. 5. PRESSURE LOSS CHARACTERISTICS ACROSS STORAGE BED DURING CHARGING MODE.

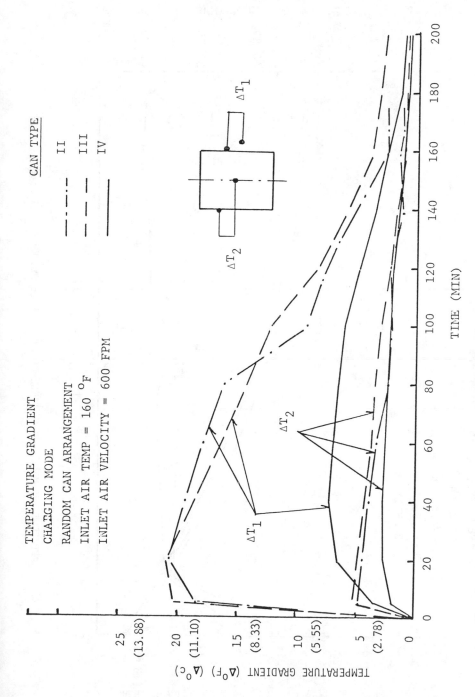

Fig. 6.   TEMPERATURE GRADIENTS DURING CHARGING MODE.

331

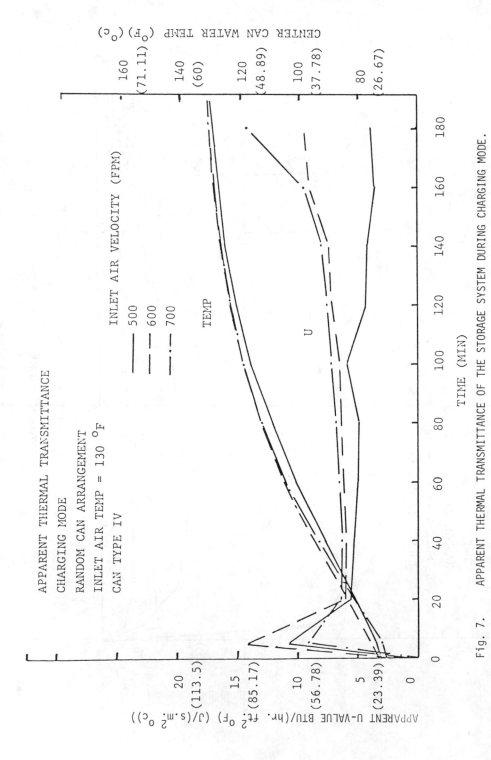

Fig. 7. APPARENT THERMAL TRANSMITTANCE OF THE STORAGE SYSTEM DURING CHARGING MODE.

332

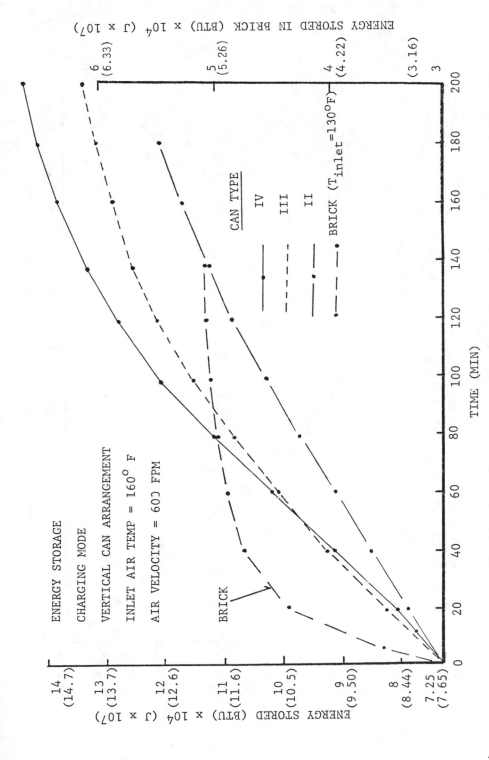

ENERGY STORAGE

CHARGING MODE

VERTICAL CAN ARRANGEMENT

INLET AIR TEMP = 160° F

AIR VELOCITY = 600 FPM

BRICK

CAN TYPE

IV

III

II

BRICK (T$_{inlet}$=130°F)

ENERGY STORED IN BRICK (BTU) x 10⁴ (J x 10⁷)

ENERGY STORED (BTU) x 10⁴ (J x 10⁷)

TIME (MIN)

FIG. 8.   ENERGY STORAGE DURING CHARGING MODE WITH VERTICAL CAN ARRANGEMENT.

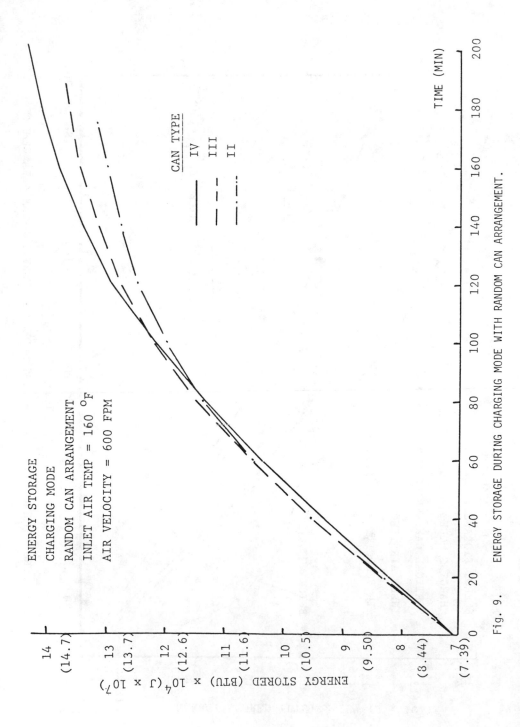

ENERGY STORAGE

CHARGING MODE

RANDOM CAN ARRANGEMENT

INLET AIR TEMP = 160 °F

AIR VELOCITY = 600 FPM

CAN TYPE

IV

III

II

Fig. 9.   ENERGY STORAGE DURING CHARGING MODE WITH RANDOM CAN ARRANGEMENT.

ENERGY STORED (BTU) x 10^4(J x 10^7)

TIME (MIN)

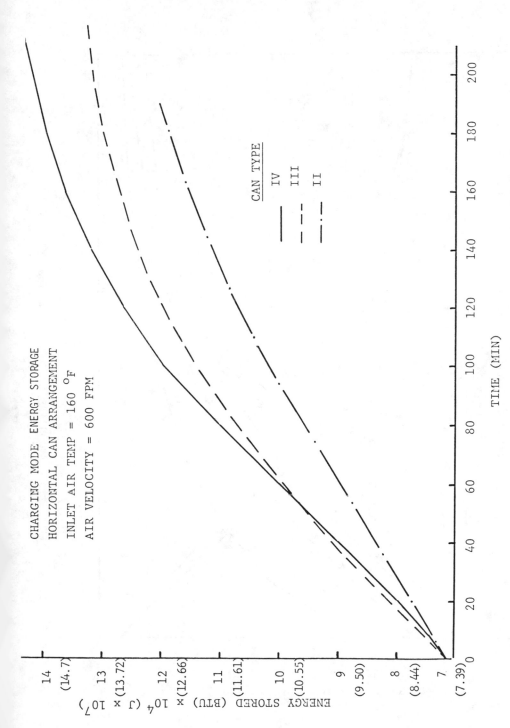

Fig. 10. ENERGY STORAGE DURING CHARGING MODE WITH HORIZONTAL CAN ARRANGEMENT.

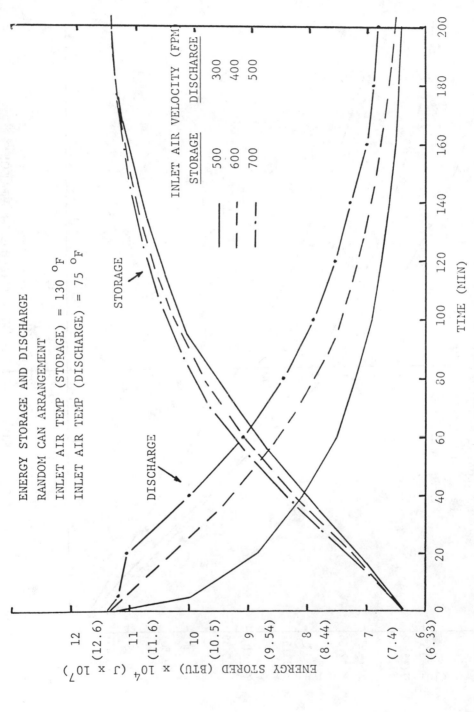

Fig. 11. ENERGY STORAGE AND RELEASE DURING CHARGING AND DISCHARGE MODE.

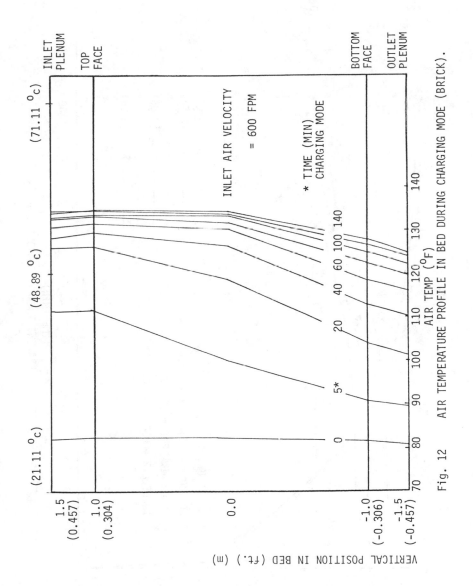

Fig. 12   AIR TEMPERATURE PROFILE IN BED DURING CHARGING MODE (BRICK).

337

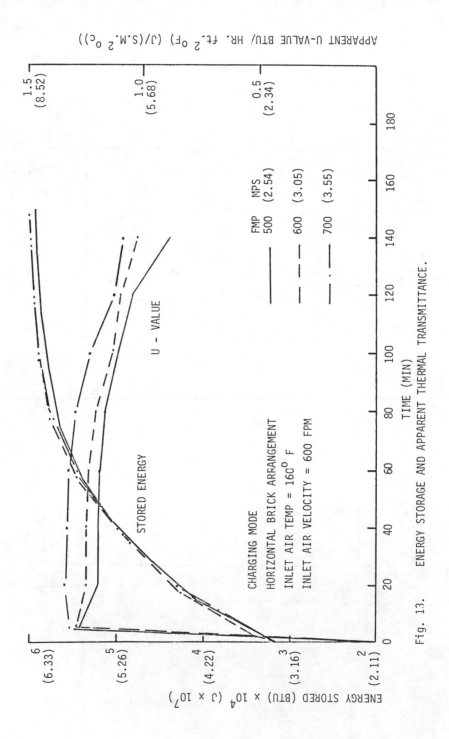

APPARENT U-VALUE BTU/ HR. ft.² °F) (J/(S.M.² °C))

1.5 (8.52)   1.0 (5.68)   0.5 (2.34)

	FMP	MPS
	500	(2.54)
	600	(3.05)
	700	(3.55)

U - VALUE

STORED ENERGY

CHARGING MODE
HORIZONTAL BRICK ARRANGEMENT
INLET AIR TEMP = 160° F
INLET AIR VELOCITY = 600 FPM

TIME (MIN)

ENERGY STORED (BTU) x 10⁴ (J x 10⁷)

6 (6.33)   5 (5.26)   4 (4.22)   3 (3.16)   2 (2.11)

Fig. 13.   ENERGY STORAGE AND APPARENT THERMAL TRANSMITTANCE.

338

vertical flow storage system and with appropriate stacking flow channelling can be minimized.

ACKNOWLEDGEMENT

The author wishes to acknowledge the helpful assistance and advice received from NASA/MSFC specialists. The author also appreciates the assistance of Alabama A&M University Graduate Assistants, Cephas Agola, Seyed Ahmad Alavi, John Y. Wu, and Althea Stirling (Secretary) in preparation of this report.

REFERENCES

(1)   D.J. Close, "Rock Pile Thermal Storage for Comfort Air Conditioning, "Mechanical & Chemical Engineering Transaction, May, 1965.

(2)   R.V. Dunkle, "Design Considerations and Performance Predictions for an Integrated Solar Air Heater and Gravel Bed Thermal Store in a Dwelling, "Victorian Technical Meeting on Applications of Solar Energy Research and Development in Australia, Melbourne, 2 July, 1975.

(3)   J.D. Balcomb, James C. Hedstrom, and B.T. Rogers, "Design Considerations of Air Cooled Collector/Rock-Bin Storage Solar Heating Systems, "1975 International Solar Energy Congress and Exposition, UCLA, July 28 - August 1, 1975.

(4)   S.A. Mumma, and W.C. Marvin, "A Method of Simulating the Performance of a Peble Bed Thermal Energy Storage and Recovery System, "Proceedings of the Conference on Improving Efficiency and Performance of HVAC Equipment and Systems for Commercial and Industrial Buildings, April 12-14, 1976, Vol. I.

(5)   G.O.G. Lof and R.W. Hawley, "Unsteady-State Heat Transfer Between Air and Loose Solids, "Industrial and Engineering Chemistry, June 1948, p. 1061-1070.

(6)   E. Alanis, L. Saravia, and L. Rovetta, Measurement of Rock Pile Heat Transfer Coefficients", Technical Note, Solar Energy, Vol. 19, No. 5, Pergamon Press 1977, pp. 571-572.

(7)   "LASL Solar Mobile/Modular Home Project", the Los Alamos Scientific Laboratory. 1976.

(8)   Hrishikesh Saha, "Parametric Study of Rock Pile Thermal Storage for Solar Heating and Cooling", Final Report, NASA/MSFC Grant No. NSG-8041, October 1977.

(9)   Hrishikesh Saha, "Solar Energy Storage Via Liquid Filled Cans, Test Data and Analysis, "Proceedings of the International Conference on Alternative Energy Sources, 5-7 December 1977, Miami Beach, Florida.

(10)   George A. Lane, "Macro-Encapsulation of Heat Storage Phase-Change Materials", Dow Chemical Report, Third Annual TES Contractors' Information Exchange Meeting, December 5-6, 1978, Springfield, Virginia.

Physical and Chemical Processes for Latent Heat Storage at Low Temperatures

F. REITER
EURATOM J.R.C.
Ispra, Italy

ABSTRACT

A survey on physical and chemical processes, which are suitable for latent heat storage between 20° C and 250° C, is given. As a high heat storage capacity per unit of volume is desirable the survey is limited to processes in or between condensed phases. A classification is made of processes with one, two, three and four components. The number of types of reactions suitable for latent heat storage increases very considerably in three and four components systems. Therefore, only those types of reactions have been presented in three and four components systems, which seem to be interesting for low temperature latent heat storage.

The types of reactions investigated in this report are:

1. One Component Systems
1.1. Solid–liquid transitions
1.2. Solid–solid transitions
1.3. Liquid–liquid transitions
2. Two Components Systems
2.1. The components are completely miscible in the liquid phase
  - The components form a eutectic mixture
  - The components form mixed crystals with an extremum of melting temperature
2.2. A gap of miscibility exists in the liquid phase
  - The components are partially immiscible
  - The components are completely immiscible
2.3. Polymorphic transitions
2.4. The components form a compound
  - The compound melts congruently
  - The compound melts incongruently
-- Decay in a solid and a liquid phase
-- Decay in two liquid phases
  - Formation and decay of a compound in the solid phase
3. Three Components Systems
3.1. Eutectic mixtures
3.2. The components form compounds
  - Binary compounds
  - Ternary compounds
4. Four Components Systems
4.1. Solid–liquid transitions
  - Eutectic mixtures
  - Systems with compounds

4.2. Reciprocal salt pairs

Transformation temperatures and enthalpies of reaction of special reactions
are presented for each of these types of reaction. Own experimental results
are reported for the double conversion of reciprocal salt pairs, which seem to
be a very promising type of reaction for latent heat storage at low temperatures.

The report will be concluded with a discussion and comparison of the
various types of reactions.

1.   INTRODUCTION

Low temperature heat (solar energy, geothermal energy, industrial waste
heat) is largely available for domestic and other applications. In order to
be able to use this low temperature heat, the energy storage is of fundamental
importance, as herewith the fluctuating energy at our disposition can be
adapted to the energetic requirements.

Storage materials can accumulate thermal energy as specific heat or as
heat of physical or chemical reactions, including their combined effects.
Experimental work has been favouring rocks, water, salt-hydrates, paraffins
and other organic compounds. In rocks and water only specific heat (sensible
heat storage) can be stored, whereas in paraffins and other organic compounds
latent heat of fusion and specific heat can be used for energy storage
(physical latent heat storage).

Other types of latent heat storage systems are those using chemical
reactions, which run completely in one direction above a certain transformation
temperature and run completely in the opposite direction below this temperature
(chemical latent heat storage).

Latent heat storage systems have the advantage of a high heat storage
capacity within a small temperature range and are the most prospective systems
for short-term storage.

The objective of this report is to give as complete a survey as possible
on physical and chemical processes running completely in one direction above a
transformation temperature and running completely in the opposite direction
below this temperature.

This survey shall therefore contain all types of physical and chemical
processes suitable for latent heat storage at low temperatures, that means
between 20 and 250° C. As a high heat storage capacity per unit of volume is
desirable the survey is limited to processes in or between condensed phases.

A classification is made of processes with one, two, three and four components. Processes with more than four components have not been considered.

2.   ONE COMPONENT SYSTEMS

Phase transitions in one component systems do not have very high enthalpies
of transition. However, they have the advantage that no thorough mixing is
necessary.

The equilibrium between two condensed phases in a one component system is
non variant at a given pressure and phase transitions therefore occur at a

fixed temperature. Transitions can take place between a solid and a liquid
phase, between two solid phases or between two liquid phases.

2.1. Solid-Liquid Transitions

The fusion and solidification process is widely investigated and used in
heat storage systems. Most promising materials are organic compounds with
latent heats of fusion between 100 and 300 kJ/dm^3. A literature survey on heat
of fusion-materials for low temperature heat storage is given by Lane et al.[1].
Some of the most promising heat of fusion-materials are presented in Table 1.

Table 1. Solid-liquid transitions

Substance	Formula	$T_T(°C)$	d (kg/dm^3)	ΔH_T (kJ/kg)	ΔH_T (kJ/dm^3)	Ref.
Diphenylmethane	$(C_6H_5)_2CH_2$	25.2	1.001	109	110	2
Capric acid	$CH_3(CH_2)_8COOH$	31.2	0.894	163	145	2
Lauric acid	$CH_3(CH_2)_{10}COOH$	43.8	0.873	183	160	2
Dibenzyl	$C_6H_5(CH_2)_2C_6H_5$	51.1	0.969	167	162	2
Mystiric acid	$CH_3(CH_2)_{12}COOH$	53.7	0.853	197	168	2
Palmitic acid	$CH_3(CH_2)_{14}COOH$	62.6	0.853	214	183	2
Diphenyl	$C_{12}H_{10}$	69.1	0.991	121	127	3
Stearic acid	$CH_3(CH_2)_{16}COOH$	69.4	0.847	199	169	4
Acetamide	CH_3CONH_2	71	1.013	240	243	2
Propionamide	$C_2H_5CONH_2$	79.5	0.960			2
Naphthalene	$C_{10}H_8$	80.3	0.981	148	145	2
Acetanilide	$C_6H_5NHCOCH_3$	115	1.211			8
Urea	NH_2CONH_2	133	1.323	242	320	2 , 8
Adipic acid	$HOOC(CH_2)_4COOH$	153				8
Anthracene	$C_{14}H_{10}$	218	1.252	162	202	2 , 8
Phthalimide	$C_6H_4(CO)_2NH$	234				8
Aluminumchloride	$AlCl_3$	192	s:2.44 1:1.33	259	345	2

A wide spectrum of melting points can be realised with just commercially
available mixtures of fatty acids, paraffin waxes or polyethylene glycols.

AlCl$_3$ is added in the last row of Table 1. This substance has a high heat
of fusion, is a good thermal conductor in the solid state and a moderate thermal
conductor in the liquid state, but has an extremely high change of density
during fusion and can therefore not be used as a latent heat storage medium.

2.2. Solid-Solid Transitions

Several substances can exist in more than one solid phase. Reversible
transitions between different solid phases can be used for latent heat storage.

In Table 2 some substances with reversible solid-solid transitions are presented.

Table 2. Solid-solid transitions.

Substance	T_T(°C)	d (kg/dm^3)	ΔH_T (kJ/kg)	(kJ/dm^3)	Ref
CBr_4	46.9	2.96	20	60	2
C_2Cl_6	45	2.09	11	23	2
	72	2.09	35	73	
NH_4NO_3	32.1	1.72	20	34	2
	84.2	1.71	17	29	
	125.2	1.70	53	90	
V_2O_4	72	4.65	52	240	2
KNO_3	127.9	2.09	50	105	2
NH_4Br	137.8	2.55	33	84	2
FeS	138	4.82	27	130	2
$KHSO_4$	164.2	2.24	15	34	2
Na_2SO_3	177	2.633	25	65	2
KHS	180	1.71	32	54	2
KHF_2	196.0	2.37	142	337	2
WCl_6	226.9	3.52	36	126	2
WCl_5	230	3.875	49	189	2
Na_2O_2	237	2.802	69	192	2
Na_2SO_4	241	2.698	49	131	2
KOH	249	2.044	113	232	2

2.3. Liquid-Liquid Transitions

Liquid-liquid transitions mentioned for completeness do not seem to be suitable for latent heat storage because of the costs of the products and the smallness of the heats of reaction. Temperatures and enthalpies of transition of two p,n-alkoxybenzoic acids are given in Table 3. The pentyl-derivate can exist in two liquid phases, while the heptyl-derivate exists in three liquid phases.

Table 3. Liquid-liquid transitions.

Substance	Transition	T_T(°C)	ΔH_T (kJ/kg)	Ref.
p,n-pentoxy-benzoic acid	solid-nematic	125	105	5
	nematic-isotropic	149	10	
p,n-heptoxy-benzoic acid	solid-nematic	94	82	5
	nematic-smectic	100	46	
	smectic-isotropic	147	11	

3.   TWO COMPONENTS SYSTEMS

3.1. The Components are completely miscible in the liquid phase

The components form a eutectic mixture. Eutectic mixtures melt at a fixed
temperature and are characterised by the crystallisation behaviour shown in
Fig. 1. Some eutectic mixtures of two components suitable for low temperature
latent heat storage are presented in Table 4.

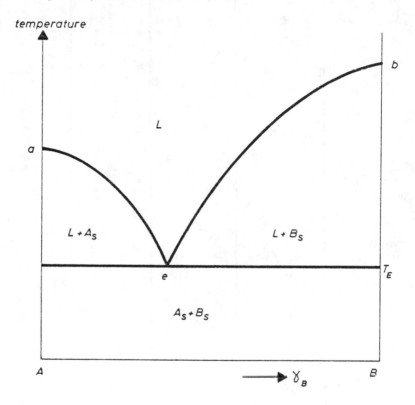

Fig. 1 MELTING DIAGRAM OF A TWO COMPONENTS SYSTEM WITH
        EUTECTIC CRYSTALLIZATION OF THE COMPONENTS A AND B

L = liquid mixture of A and B
A_S = solid A
B_S = solid B
a = melting point of A
b = melting point of B
e = eutectic point
T_E = eutectic temperature
γ_B = molar fraction of B

Table 4. Eutectic mixtures of two components.

Substances	$T_E(°C)$	d (kg/dm^3)	ΔH_T (kJ/kg)	(kJ/dm^3)	Ref
$LiNO_3$ H_2O	29				1
NH_4NO_3 CH_3CONH_2	36				1
NH_4NO_3 $CO(NH_2)_2$	45				1
$Mg(NO_3)_2$ H_2O	53				1
$CO(NH_2)_2$ CH_3CONH_2	53				1
$CH_3(CH_2)_{16}COOH$ $CH_3(CH_2)_{14}COOH$	60	1.0	200	200	
$C_{10}H_8$ C_6H_5COOH	67				1
$NaBr$ CH_3CONH_2	70				1
KNO_3 CH_3CONH_2	72				1
NH_4Br $CO(NH_2)_2$	76	1.8	150	270	
$C_6H_5CONH_2$ C_6H_5COOH	79				1
$NaNO_3$ $Mg(NO_3)_2$	130				2
$LiNO_3$ KNO_3	130–135	1.75	215	380	6
KNO_3 $Ca(NO_3)_2$	160				2
$LiNO_3$ $NaNO_3$	190–200	1.78	270	480	6
$NaNO_3$ $Ca(NO_3)_2$	230				2
$LiNO_3$ $Ca(NO_3)_2$	235				2

The components form mixed crystals with an extremum of melting temperature.
Two components systems which form a complete series of mixed crystals can have
an extremum of melting temperature. A system with maximum and a system with
minimum melting temperature is shown in Fig. 2. The curve of the beginning of
the melting process touches the curve of its conclusion in this extremum.
Therefore, the corresponding mixture crystallises at one temperature, contrary
to mixtures with other compositions which have a melting interval. The mixed
crystals cannot become heterogeneous, as they have the same composition as the
liquid from which they originate. Several mixed crystals with minimum melting
temperature are given in Table 5. Some metallic systems with maximum melting
temperatures are known but no system in the temperature range considered here.

Table 5. Mixed crystal systems with minimum melting temperature.

Substances	$T_T(°C)$	Ref
$SnJ_4 - SnBr_4$	19.4	2
$C_{16}H_{34}O - C_{18}H_{38}O$	48.5	2
$SbBr_3 - SbCl_3$	54	2
$SbJ_3 - SbBr_3$	84	2
$Se - S$	98	2
$SbJ_3 - ASJ_3$	128	2
$KNO_3 - NaNO_2$	140	2
$(C_6H_5)_2OTe - (C_6H_5)_2O_2Se$	146	2
$KNO_2 - NaNO_3$	150	2
$KOH - NaOH$	185	2
$KClO_3 - NaNO_3$	206	2
$KNO_3 - NaClO_3$	212	2
$NaNO_3 - NaClO_3$	218	2
$NaNO_3 - NaNO_2$	221.5	2
$KNO_3 - NaNO_3$	223	2
$KNO_2 - NaNO_2$	230.5	2
$KClO_3 - NaClO_3$	236	2

3.2. A Gap of Miscibility Exists in the Liquid Phase

The components are partially immiscible. Fig. 3 shows the melting
diagram of a two components system which has a gap of miscibility in the liquid
phase. The crystallisation of a homogeneous liquid mixture of the composition
l_2 starts in the point l_2. During further loss of heat, the system then moves
from l_2 to l_1 at a constant temperature T_T. The liquid L_2 of the composition
l_2 converts into a liquid L_1 of the composition l_1 with simultaneous precipitation
of B-crystals

$$L_2 = L_1 + B \qquad\qquad (1)$$

This latent transformation is finished at point l_1 where the initial liquid
L_2 is completely converted in the liquid L_1. Then, the temperature of the system

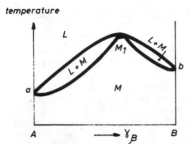

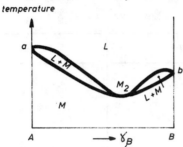

Fig.2 MELTING DIAGRAM OF A TWO COMPONENTS
SYSTEM WITH MAXIMUM M_1 OR MINIMUM M_2
MELTING TEMPERATURE OF THE MIXED CRYSTALS

L = LIQUID MIXTURE OF A AND B
M = MIXED CRYSTALS OF A AND B
a = MELTING POINT OF A
b =   "   "   " B
γ_B = MOLAR FRACTION OF B

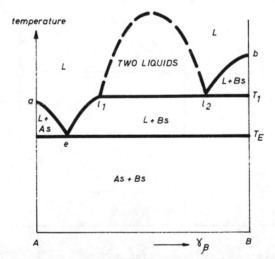

Fig.3 MELTING DIAGRAM OF A TWO COMPONENTS SYSTEM WITH A GAP OF MISCIBILITY
IN THE LIQUID PHASE
L = LIQUID MIXTURE OF A AND B
L_i =   "   "   " COMPOSITION l_i
As = SOLID A
Bs =   "   B
a = MELTING POINT OF A
b =   "   "   " B
e = EUTECTIC POINT
γ_B = MOLAR FRACTION OF B

decreases during further loss of heat until the eutectic point e is reached. Therefore, the solidification process terminates with the crystallisation of the eutectic mixture of A and B.

A two components system of the composition l_2 with a gap of miscibility in the liquid phase as in Fig. 3 has two points of solidification T_T and T_e and a solidification interval $l_1 e$. The system can be used for latent heat storage, when the liquid L_1 and the B-crystals remain in continuous contact during the conversion $l_2 l_1$ or when the density of L_1 is larger than that of L_2 and the density of B-crystals is larger than that of L_1.

Most of the systems of Fig. 3 are metallic or anorganic systems with transformation and eutectic temperatures above those of the temperature interval considered here. The only two systems of this temperature interval found in literature are toxic materials and too expensive for heat storage applications. They are given in Table 6.

Table 6. Two components systems with a gap of miscibility in the liquid phase.

Substances	$T_T(°C)$	$T_E(°C)$	Ref
UF_6/HF	61.2	− 85	2
$HgBr_2/AlCl_3$	132	126	2

The components are completely immiscible. The melting diagram for this case, that the two components are completely immiscible, is given in Fig. 4.

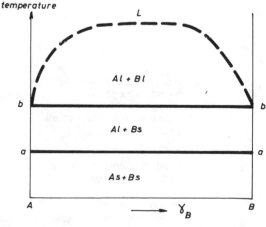

Fig.4 MELTING DIAGRAM OF A TWO COMPONENTS SYSTEM WHICH IS COMPLETELY IMMISCIBLE IN THE LIQUID PHASE

L = LIQUID MIXTURE OF A AND B
Al = LIQUID A
Bl = " B
As = SOLID A
Bs = " B
a = MELTING POINT OF A
b = " " " B
γ_B = MOLAR FRACTION OF B

This case is presented for completeness and will probably have no importance for latent heat storage. As there is no mutual solubility of the components in the liquid phase, both components melt and crystallise without reciprocal interference at their melting temperatures. Therefore, there are two complete melting equilibria: bb at the melting temperature of B with the phases liquid A + liquid B + solid B and aa at the melting temperature of A with the phases liquid A + solid A + solid B. When the solidification of the system is finished, the components of this system exist in two separate zones or in a tear-shaped distribution of one component in the crystallised phase of the other one.

3.3. Polymorphic Transitions

Transformations between different crystallic forms of one or both components of a two components system are also characterised by the existence of latent heats of transition. However, also this type of transformation will presumably not become important for latent heat storage. An interesting system with polymorphic transitions of both components but without importance for heat storage because of the price of one component is the system NH_4NO_3 – $AgNO_3$ (Fig. 5). This system has several polymorphic transitions of NH_4NO_3 (α/β at 32.1°C, β/γ at 84.2°C, γ/δ at 125.2°C) and $AgNO_3$ (158°C), the formation of a compound $NH_4NO_3 \cdot AgNO_3$ and two eutectic temperatures at 101.5°C and 109.6°C [7].

3.4. The Components form a Compound

The compound melts congruently. A compound melts congruently, when the solid and the liquid phase of the compound have the same composition at the melting point. The melting diagram of a two components system which forms a congruently melting compound is given in Fig. 6. The melting temperature of the system T remains virtually the same, when the composition of the system is not exactly that of the compound, because $\frac{dT}{d\gamma B} = 0$ at the melting point of the compound.

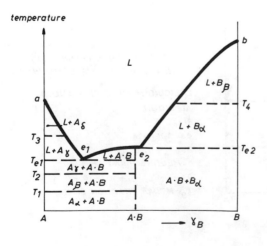

Fig. 5 MELTING DIAGRAM OF THE SYSTEM NH_4NO_3 (A)/ Ag NO_3 (B)

L= LIQUID MIXTURE OF A AND B
A_i = SOLID A IN THE STATE i
B_i = " B " "   "   "

A·B = SOLID COMPOUND OF A AND B
a = MELTING POINT OF A
b = "   "   "   " B
e_i = EUTECTIC POINTS
T_i = TRANSITION TEMPERATURE
T_{ei} = EUTECTIC TEMPERATURES
γ_B = MOLAR FRACTION OF B

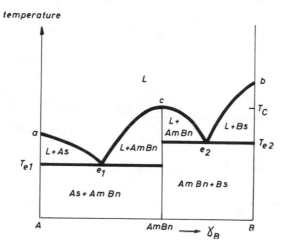

Fig. 6 MELTING DIAGRAM OF A TWO COMPONENTS SYSTEM WHICH
FORM A CONGRUENTLY MELTING COMPOUND

L= LIQUID MIXTURE OF A AND B   c = MELTING POINT OF THE COMPOUND
As = SOLID A e_i = EUTECTIC POINTS
Bs= " B T_C = MELTING TEMPERATURE OF THE COMPOUND
AmBn= SOLID COMPOUND OF A AND B T_{ei} =EUTECTIC TEMPERATURES
a = MELTING POINT OF A X_B = MOLAR FRACTION OF B
b= "   "   " B

Some congruently melting two components systems are presented in Table 7.

Table 7. Two components systems which form a congruently melting compound.

Substances	$T_C(°C)$	$T_{E1}(°C)$	$T_{E2}(°C)$	Ref
$NaCl \cdot AlCl_3$	152	108	152	2
$2KNO_3 \cdot Mg(NO_3)_2$	225	178	195	?
$KOH \cdot KNO_3$	236.5	217	223	2
$KCl \cdot AlCl_3$	257	128	257	2

The compound melts incongruently. An incongruently melting compound decays
during fusion in two phases, e.g. in a liquid phase and a higher melting solid
phase or in two liquid phases.

Decay in a solid and a liquid phase. The melting diagram of this very
frequent case is presented in Fig. 7. The compound melts with simultaneous
decay following the scheme

$$A_m B_n = pB + \left[(n - p)B + mA \right] \tag{2}$$

in p mole of B-crystals and in a liquid characterised by the term in square
brackets.

Crystallisation of this system during cooling proceeds as follows.

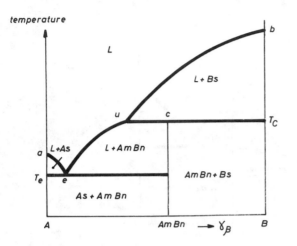

temperature

L = LIQUID MIXTURE OF A AND B
As = SOLID A
Bs = " B
Am Bn = SOLID COMPOUND OF A AND B
a = MELTING POINT OF A
b = " " " B
c = COMPOSITION OF THE COMPOUND
e = EUTECTIC POINT
u = TRANSITION POINT
T_C = MELTING TEMPERATURE OF THE
COMPOUND
T_e = EUTECTIC TEMPERATURE
γ_B = MOLAR FRACTION OF B

Fig.7 MELTING DIAGRAM OF A TWO COMPONENTS SYSTEM FORMING A COMPOUND
WHICH DECAYS DURING MELTING IN A LIQUID PHASE AND A SOLID PHASE

B-crystals precipitate in the concentration range of curve ub, the temperature decreases and the composition of the liquid changes until the point u is reached. Then the compound A_mB_n crystallises from the liquid with simultaneous consumption of B-crystals following the above reaction scheme, until the liquid or the B-crystals are consumed. The complete consumption of both phases takes place, when the initial composition equals c.

The compound A_mB_n crystallises from a liquid of the initial composition of the range of curve eu, the temperature decreases until the eutectic point is reached and the fusion process terminates with the crystallisation of the eutectic mixture.

The melting diagram of an incongruent melting compound goes over that of a congruent melting compound, when the concentrations of c and u are equal.

In Table 8 several two components systems are given which form an incongruent melting compound.

The most severe problems of these heat storage media are material segregation after melting, tendency to subcooling and low thermal conductivity in the solid state. Material segregation reduces the heat storage density from one melting/freezing cycle to another and can be prevented by embedding the heat storage medium in a binding agent, by intermixing or by additional water. Subcooling can be avoided by the addition of solid crystals which are isomorphous with the heat storage medium. Low thermal conductivity can be compensated by a large heat transfer surface to volume ratio.

Table 8. Two components systems which form an incongruently melting compound.

Substances	$T_C(°C)$	$T_E(°C)$	d (kg/dm^3)	H (kJ/kg)	(kJ/dm^3)	Ref
$CaCl_2 \cdot 6H_2O$	30	-49.8	1.68	174	290	9
$Na_2SO_4 \cdot 10H_2O$	32	- 1.2	1.464	214	310	8
$Na_2CO_3 \cdot 10H_2O$	32	- 2.1	1.46	240	350	9
$Na_2HPO_4 \cdot 12H_2O$	35	- 0.5	1.52	277	420	6
$Na_2S_2O_3 \cdot 5H_2O$	48	-10.6	1.685	209	350	9
$Ba(OH)_2 \cdot 8H_2O$	78	- 0.5	2.188			8
$MgCl_2 \cdot 6H_2O$	117		1.57	169	265	2
$NaNH_2 \cdot 2KNH_2$	120	92				2
$LiCl \cdot AlCl_3$	143.5	114				2
$4KNO_3 \cdot Ca(NO_3)_2$	174	145				2
$LiOH \cdot LiNO_3$	195	183				2

Decay in two liquids. The melting diagram of this case is presented in Fig. 8 and is characterised by a three phase equilibrium of the two liquid phases f_1 and f_2 and the solid phase of the compound. The formation and decay of the compound is given by the scheme

$$L_1 + L_2 = A_mB_n \tag{3}$$

Heat transformation at the transition temperature T_C is largest for a system of the composition c.

This case is represented by some intermetallic compounds and will probably have no importance for latent heat storage.

temperature

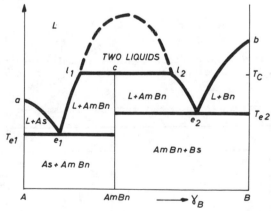

Fig. 8 MELTING DIAGRAM OF A TWO COMPONENTS
        SYSTEM FORMING A COMPOUND WHICH
        DECAYS DURING MELTING IN TWO LIQUID
        PHASES

L = LIQUID MIXTURE OF A AND B
As = SOLID A
Bs = "   " B
Am Bn = SOLID COMPOUND OF A AND B
a = MELTING POINT OF A
b =   "   "   "   " B
c = COMPOSITION OF THE COMPOUND
e_i = EUTECTIC POINTS
l_i = LIQUID OF COMPOSITION l_i
T_C = MELTING TEMPERATURE OF THE COMPOUND
T_{ei} = EUTECTIC TEMPERATURES
γ_B = MOLAR FRACTION OF B

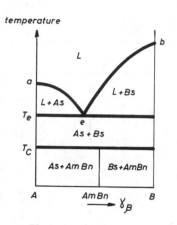 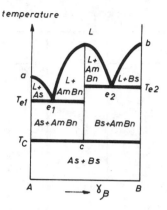

Fig. 9 MELTING DIAGRAM OF A TWO COMPONENTS SYSTEM WITH FORMATION
OR DECAY OF A COMPOUND IN THE SOLID STATE

L = LIQUID MIXTURE OF A AND B   c = COMPOSITION OF THE COMPOUND
As = SOLID A e_i = EUTECTIC POINTS
Bs = „   B T_C = TRANSITION TEMPERATURE OF THE COMPOUND
Am Bn = SOLID COMPOUND OF A AND B T_{ei} = EUTECTIC TEMPERATURES
a = MELTING POINT OF A γ_B = MOLAR FRACTION OF B
b =   „   „   „ B

Formation and decay of a compound in the solid phase. Two systems are
presented in Fig. 9, in the first system the compound is formed during cooling
in the solid state, in the second system at first the compound is formed
during crystallisation and then decays during further cooling in the components.
Both cases are connected with a latent heat of transition, but are observed only
in intermetallic systems and will therefore have no importance for latent heat
storage.

4.   THREE COMPONENTS SYSTEMS

4.1. Eutectic Mixtures

A three components system with three binary eutectic points e_i and a ternary
eutectic point E is presented in Fig. 10. Some promising systems are given in
Table 9.

Table 9.   Three components eutectic systems which do not form compounds

Substances	$T_E(°C)$	Ref
$KNO_3 - Ba(NO_3)_2 - CO(NH_2)_2$	80	1
$LiNO_3 - Ca(NO_3)_2 - CO(NH_2)_2$	82	1
$Ca(NO_3)_2 - KNO_3 - LiNO_3$	117	2
$NH_4NO_3 - KNO_3 - NaNO_3$	119	2
$KNO_3 - NaNO_3 - LiNO_3$	120	2

temperature

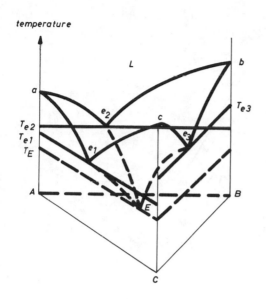

Fig. 10 MELTING DIAGRAM OF A THREE
        COMPONENTS SYSTEM WITH
        EUTECTIC POINTS OF THE
        COMPONENTS

L = LIQUID MIXTURE OF THE THREE COMPON.
a = MELTING POINT OF A
b =   "   "   " B
c =   "   "   " C
e_i = BINARY EUTECTIC POINTS
E = TERNARY EUTECTIC POINT
T_{ei} = BINARY EUTECTIC TEMPERATURES
T_E = TERNARY EUTECTIC TEMPERATURE

Table 9 (cont'd)

Substances	$T_E(°C)$	Ref
$NH_4H_2PO_4$ – NH_4NO_3 – NH_4Cl	131.5	2
$Ca(NO_3)_2$ – $NaNO_3$ – $LiNO_3$	170.3	2

4.2. The Components Form Compounds

The equilibrium conditions in a three components system become more com-
plicated, when the components can form compounds in the solid state, because
there are further equilibria between components and compounds and between the
compounds themselves. Binary and ternary compounds can occur in a three
components system.

Binary compounds. Several ternary eutectic points can occur in a three
components system with one or more binary compounds and that up to n+3 binary
and n+1 ternary eutectic points in a three components system with n binary
compounds. In Fig. 11 the melting diagram of the three components system
$ZnCl_2$ – KCl – NaCl is presented. This system has 4 binary compounds and 6
binary and 5 ternary eutectic points. It has only 6 binary eutectic points
instead of 7 possible ones, as the binary system KCl – NaCl forms mixed
crystals with a minimum melting temperature of 664°C.

Some three components systems which form one or several binary compounds
are given in Table 10.

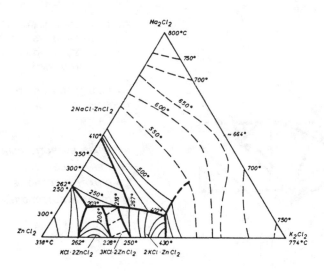

Fig.11 MELTING DIAGRAM OF THE THREE COMPONENTS
SYSTEM $ZnCl_2 - KCl - NaCl$

Table 10. Three components system which forms binary compounds.

Substances	Compounds	Ternary eutectic temperatures(°C)	Ref
$Mg(NO_3)_2$ – KNO_3 – $NaNO_3$	$2KNO_3 \cdot Mg(NO_3)_2$	130	2
KH_2PO_4 – KNO_3 – KCl	$KCl \cdot KNO_3$	235	2
		270	
$NaCl$ – KCl – $AlCl_3$	$KCl \cdot AlCl_3$	89	2
	$NaCl \cdot AlCl_3$	only one ternary eutectic temperature is known	
$ZnCl_2$ – $NaCl$ – KCl	$2NaCl \cdot ZnCl_2$	203	2
	$KCl \cdot 2ZnCl_2$	206	
	$3KCl \cdot 2ZnCl_2$	216	
	$2KCl \cdot ZnCl_2$	267	
		402	

The special case where two double salts are formed in a three components system permits a reaction of the type

$$A \cdot B + C \rightleftharpoons A + B \cdot C \tag{4}$$

This reaction is univariant, if only the solid phases of the two salts and the double salts which participate at reaction (4) are present. The equilibrium becomes nonvariant, if the liquid phase is present. Following the phase rule the four solid phases of (4) and the liquid phase can only exist at one temperature, the transition temperature or peritectic temperature of the system (4). This is the case of an incongruent melting ternary system in which the four solid phases and the liquid phase are in equilibrium only at the peritectic temperature, while at other temperatures only three solid phases and the liquid phase are in equilibrium. An example of this case is the ternary system $Na_2CO_3 - K_2CO_3 - CaCO_3$ with a peritectic transition at about 700°C. No system of this case with a transition temperature in the temperature range considered here is known.

Ternary compounds. Van't Hoff [10] and Roozeboom[11] have investigated the ternary system $Na_2SO_4 - MgSO_4 - H_2O$. They found a peritectic transition at 22°C:

$$Na_2SO_4 \cdot 10H_2O + MgSO_4 \cdot 7H_2O \rightleftharpoons Na_2Mg(SO_4)_2 \cdot 4H_2O + 13H_2O \tag{5}$$

The formation of the ternary compound astrakanite above 22°C is an endothermic process and the decay below 22°C is an exothermic process and reaction (5) can therefore be suitable for heat storage [12]. Reactions similar to (5) take place in systems where Na is replaced by K or NH_4 and Mg is replaced by Zn, Fe, Co, Ni, Mn, Cr, Vd, Cd or Cu. In the series $Na_2Me(SO_4)_2 \cdot 4H_2O$ the following peritectic temperatures were found:

Table 11. Ternary peritectic points in three components systems.

System	Peritectic ternary temperature T_p(°C)	Ref.
$Na_2SO_4 - MgSO_4 - H_2O$	22	10 , 11
$Na_2SO_4 - ZnSO_4 - H_2O$	8.7	13
$Na_2SO_4 - FeSO_4 - H_2O$	18.5	13
$Na_2SO_4 - CoSO_4 - H_2O$	17.4	13
$Na_2SO_4 - NiSO_4 - H_2O$	16.5	13
$Na_2SO_4 - CuSO_4 - H_2O$	16.7	13
$Na_2C_4O_6H_4 - K_2C_4O_6H_4 - H_2O$	55	14
$Na_2SO_4 - (NH_4)_2SO_4 - H_2O$	59.3	15

Sodium potassium tartrate melts incongruently above 55°C, the two single salts, d-sodium tartrate and d-potassium tartrate, are deposited and the water separated forms a saturated solution:

$$4NaKC_4O_6H_4 \cdot 4H_2O \rightleftharpoons 2Na_2C_4O_6H_4 \cdot 2H_2O + 2K_2C_4O_6H_4 \cdot \tfrac{1}{2}H_2O + 11H_2O \tag{6}$$

The ternary compound $Na_2SO_4 \cdot (NH_4)_2SO_4 \cdot 4H_2O$ also melts incongruently at 59.3°C and decays in solid Na_2SO_4, solid $(NH_4)_2SO_4$ and a saturated solution:

$$Na_2SO_4 \cdot (NH_4)_2SO_4 \cdot 4H_2O \rightleftharpoons Na_2SO_4 + (NH_4)_2SO_4 + 4H_2O \tag{7}$$

Eq. (5) or the transitions in the first six systems of Table 11 are represented by the general form

$$S_1 + S_2 = D + L \tag{8}$$

while Eq. (6) and (7) are given by

$$D = S_1 + S_2 + L \tag{9}$$

where D = double salt, S_1, S_2 = single salts and L = liquid.

5.   FOUR COMPONENTS SYSTEMS

5.1. Solid-Liquid Transitions

Eutectic mixtures. The melting diagram of a four components system which does not form compounds but can crystallise in eutectic mixtures has one quaternary eutectic point. This point represents the nonvariant five phase equilibrium

$$L = \alpha + \beta + \gamma + \delta \tag{8}$$

between the liquid phase L and the four solid phases α, β, γ and δ of the four components. No quaternary eutectic system is known which can be interesting for low temperature energy storage.

Systems with compounds. The number of binary, ternary and quaternary eutectic or peritectic transition points increases rapidly, if one or several compounds can be formed in a four components system, but up to now no system has been identified which can be suitable for low temperature latent heat storage.

5.2. Reciprocal Salt Pairs

A reciprocal salt pair is a system of two salts with different cations and different anions. In this system the reaction

$$AB + CD = AD + CB \tag{9}$$

can take place, called the double conversion of a reciprocal salt pair (A,C cations, B,D anions).

Reaction (9) has four substances, but the number of components is only three as the composition of system (9) can be expressed in terms of three components. Usually the double conversion of a reciprocal salt pair takes place in the aqueous saturated phase in order to get sufficiently high reaction rates. Then the system reciprocal salt pair + water is a four components system, as it can be described in terms of four components, three salts + water.

Process (9) has often been used for the production of a pure salt. Well known examples are the conversion of Chile saltpetre with potassium chloride:

$$NaNO_3 + KCl = KNO_3 + NaCl \tag{10}$$

and the ammonia-soda process:

$$NaCl + NH_4HCO_3 = NaHCO_3 + NH_4Cl \tag{11}$$

Six phases, four solid salts, saturated solution and vapour, can coexist in a four components system composed of two reciprocal salt pairs and water. Then the number of freedoms of this system is zero. The four solid phases in equilibrium with their saturated aqueous phase and their vapour phase can only exist at one temperature, the transition temperature. Below this temperature the reaction goes in one direction until at least one solid phase disappears completely. Above this temperature the reaction goes in the other direction until at least one solid phase of the other side of the equation of reaction disappears completely.

The approximate transition temperature can be determined by a method given by Van't Hoff [16]. He showed that that salt pair is the stable pair at any given temperature, which has the smaller product of solubility products. If we then plot the product of K_{AB} and K_{CD} (solubility products in gram equivalents per litre of water) and the corresponding product of the reciprocal pair AD and CB, both as $f(T)$, the intersection of the two curves should give, at least approximately, the transition point of pair stability. The Van't Hoff criterion becomes inexact at high concentrations and at incomplete dissociation of the dissolved salts. In both cases the solubilities must be multiplied by the corresponding activity coefficients, a property which is mostly unknown. However, the practical importance of the solubility criterion in these cases without knowledge of the activity coefficients is only slightly impaired, as it can still give an important indication of the existence and the approximate temperature of transition.

In many of these systems the transition point is not a point at which one salt pair passes into its reciprocal, but one at which a double salt is formed. Thus at 4°C, Glauber's salt and potassium chloride form glaserite and sodium chloride

$$2Na_2SO_4 \cdot 10H_2O + 3KCl \ \rightleftharpoons\ K_3Na(SO_4)_2 + 3NaCl + 20H_2O \tag{12a}$$

Above the transition point there could be $K_3Na(SO_4)_2$ and NaCl coexisting with KCl and solution; and it may be considered that at a higher temperature the double salt would interact with potassium chloride according to the equation

$$K_3Na(SO_4)_2 + KCl \ =\ 2K_2SO_4 + NaCl \tag{12b}$$

thus giving the reciprocal of the original salt pair. However, this point has not been experimentally realised.

The transition temperatures of more than 500 reciprocal salt pairs have been calculated by means of Van't Hoff's criterion and their enthalpies of reaction have been determined from the enthalpies of formation [17]. Following these calculations, about fifty of these reciprocal salt pairs with transition temperatures between 10°C and 120°C and enthalpies of reaction about 200 kJ/dm^3 seem to be suitable for low temperature latent heat storage, if no other criteria than temperature of transition and enthalpy of reaction are considered.

A research program has been started for the determination of

- temperature of transition T_T,

- enthalpy of reaction ΔH_R,

- specific heat and heat of solution above and below the transition point $(\frac{\Delta H}{T})_a$ and $(\frac{\Delta H}{T})_b$,

- heat storage capacity within a certain temperature interval $(\Delta H)_{T_1 \rightarrow T_2}$,

- heat storage capacity during a large number of heating and cooling cycles $(\Delta H)_{T_1 \rightarrow T_2}^{(n)}$

- solid and liquid density d_s and d_l at the transition point,

- compatibility with containment materials,

- costs of the material in technical purity and in large quantities.

The experimental studies have been started with two reciprocal salt pair systems:

$$2NaNO_3 + Ba(OH)_2 \cdot 8H_2O \rightleftharpoons 2NaOH + Ba(NO_3)_2 + 8H_2O \qquad (13)$$

$$2KNO_3 + Ba(OH)_2 \cdot 8H_2O \rightleftharpoons 2KOH + Ba(NO_3)_2 + 8H_2O \qquad (14)$$

with the following results:

Property	System (13)	System (14)
$T_T(°C)$	45	65
$\Delta H_R (J/g)$	187	204
$\Delta H_R (J/cm^3)$	414	440
$(\frac{\Delta H}{T})_a (J/cm^3 °C)$	5.5	7.0
$(\frac{\Delta H}{T})_b (J/cm^3 °C)$	2.9	2.5
$(\Delta H)_{30 \rightarrow 90}^{(o)} (J/cm^3)$	708	707
$(\Delta H)_{30 \rightarrow 90}^{(25)} (J/cm^3)$	637	709
$(\Delta H)_{30 \rightarrow 90}^{(50)} (J/cm^3)$	624	695
$d_s (g/cm^3)$	2.212	2.156
compatibility with Moplen	no attack during 3 months at 85°C	no attack during 3 months at 85°C
costs (US$/kg)	0.38	0.38

6.   COMPARISON AND DISCUSSION OF THE PRESENTED TYPES OF REACTION

A comparison of the various processes presented in the preceding chapters shall be given with a specification of their advantages and disadvantages and with remarks on the extent of investigations of these processes for low temperature latent heat storage. These data are generally referred to the systems given in the corresponding tables.

6.1. One Component Systems

Solid-liquid transitions.

Advantages: congruent melting, relatively high enthalpy of transition, non corrosive

Disadvantages: low thermal conductivity in the liquid and solid state, relatively high costs of the material, relatively high costs of the containment due to low heat transfer, flammable

Investigations: widely investigated and used in low temperature latent heat storage systems

Solid-solid transitions.

Advantages: no corrosion and containment problems, safe, reversible reaction

Disadvantages: low enthalpy of transition

Investigations: a few studies are going on

Liquid-liquid transitions.

Advantages: no corrosion and containment problems, good heat transfer properties

Disadvantages: very low enthalpy of transition, high costs

Investigations: no

6.2. Two Components Systems

The components are completely miscible in the liquid phase.

The components form a eutectic mixture.

Advantages: congruent melting, relatively high enthalpy of transition

Disadvantages: low or moderate thermal conductivity, relatively high costs if organic materials, corrosive if inorganic materials

Investigations: few studies have been made

The components form mixed crystals with an extremum of melting temperature.

Remarks: Composition and melting temperature of special systems are known from literature

Investigations: no

A gap of miscibility exists in the liquid phase.

The components are partially immiscible.

Remarks: No system which seems to be suitable for latent heat storage is known

Investigations: no

The components are completely immiscible

Remarks: No system which seems to be suitable for latent heat storage
                    is known

Investigations: no

Polymorphic transitions.

Remarks: No system which seems to be suitable for latent heat storage
                    is known

Investigations: no

The components form a compound.

The compound melts congruently

Remarks: There are some inorganic systems with relatively high
                    temperatures which could be of interest for latent heat
                    storage

Investigations: no

The compound melts incongruently.

Decay in a solid and a liquid phase

Advantages: high enthalpy of transition, low costs

Disadvantages: Incongruent melting and material segregation after melting,
                    tendency to subcooling, relatively low thermal conductivity in
                    the solid state, corrosive

Investigations: widely investigated and used in low temperature heat storage
                 systems

Decay in two liquids

Remarks: No system which seems to be suitable for latent heat storage
                    is known

Investigations: no

Formation and decay of a compound in the solid state

Remarks: No system which seems to be suitable for latent heat storage
                    is known

Investigations: no

6.3. Three components systems

Eutectic mixtures

Advantages: congruent melting, relatively high enthalpy of transition

Disadvantages: low or moderate thermal conductivity, relatively high costs
                   if organic materials, corrosive if inorganic materials

Investigations: no

The components form compounds.

Binary compounds

Remarks: composition and transition temperatures of special systems are
                   known from literature

Investigations: no

Ternary compounds

Advantages: relatively high enthalpy of reaction

Disadvantages: incongruent melting

Investigations: no

6.4. Four Components Systems

Solid-liquid transitions.

Eutectic mixtures

Remarks: No quaternary eutectic system which can be suitable for low
                   temperature latent heat storage is known

Investigations: no

Systems with compounds

Remarks: No system which can be suitable for low temperature latent
                   heat storage is known

Investigations: no

Reciprocal salt pairs

Advantages: There are systems with high or very high enthalpies of reactions
                   and with transitions in a temperature range suitable for house
                   or other applications

Disadvantages: costs of the materials, moderate thermal conductivity, corrosive

Investigations: A small experimental activity is going on at the J.R.C. Ispra,
                   Italy.

7. CONCLUSIONS

     Several of the presented types of reaction are not suitable for latent heat
storage. Other types of reaction are, or seem to be, suitable for low

temperature latent heat storage, namely:

- solid–liquid transitions of

-- one component systems,

-- two components eutectic systems,

-- two components mixed crystal systems with a minimum melting temperature,

-- congruently melting compounds of two components,

-- incongruently melting compounds of two components (e.g. salthydrates),

-- three components systems with binary eutectic points,

-- three components systems with ternary peritectic points,

- double conversion of reciprocal salt pairs.

Further investigations are necessary for all these types of reaction and for the identification of special systems. With this, the number of latent heat storage systems can be increased essentially and the interesting temperature range can be covered more completely.

REFERENCES

1. Lane G. A., et al. Heat of Fusion Systems for Solar Energy Storage. Proceedings of the Workshop on Solar Energy Storage Substances for the Heating and Cooling of Buildings, NSF-RA-N-75-041, ERDA, Washington D.C., (1975), p. 43.

2. Landolt – Börnstein: 1961. Zahlenwerte und Funktionen, 6 Auflage, Springer-Verlag, Berlin.

3. Friz, G. et al. 1968. Physical properties of diphenyl, o-, m- and p-terphenyl and their mixtures, Atomkernenergie, 13, 25.

4. 1972. Handbook of Chemistry and Physics, The Chemical Rubber Publishing Co., Cleveland Ohio.

5. Herbert, A. J. 1967. Transition Temperatures and Transition Energies of the p-n- Alkoxy Benzoic Acids, from n-Propyl to n-Octadecyl, Trans. Faraday Soc. 63, 555.

6. de Cachard, M. CEN Grenoble. 1972. 2. Symposium international sur l'energie d'origine radio-isotopique. Madrid.

7. van Zawidzki, J. 1904. Über Gleichgewichte im System NH_4NO_3 – $AgNO_3$, Z.f. Phys. Chemie 47, 721.

8. D'Ans – Lax. Taschenbuch für Chemiker und Physiker, Springerverlag, Berlin

9. Furbo, S. 1977. Heat storage in a solar heating system using salt hydrates. Thermal Insulation Laboratory, Technical University of Denmark, Report N° A/A/003/DK.

10. Van't Hoff, J. H. and van Deventer, C. M. 1889. Die Umwandlungstemperatur bei chemischer Zersetzung, Z.f. phys. Chem. $\underline{3}$, 165.

11. Roozeboom, H. W. B. 1889. Die Umwandlungstemperatur bei wasserhaltigen Doppelsalzen und ihre Löslichkeit, Z. f. phys. Chem. $\underline{3}$, 513.

12. Koppel, J. 1905. Die Bildungs- und Löslichkeits-Verhältnisse analoger Doppelsalze, Z. f. phys. Chem. $\underline{52}$, 385.

13. Reiter, F. 1978. Speicherung von Sonnenenergie bei niedrigen Temperaturen durch chemische Prozesse, 2. Internationales Sonnenforum, Hamburg, 12-14 July.

14. Findlay, A. 1951. The phase rule and its applications. 9th ed. Dover Publ. Inc., pp. 351-352.

15. Ricci, J. E. 1951. The phase rule and heterogeneous equilibrium. D. van Nostrand Company Inc., New York, pp. 309-310.

16. Van't Hoff, J. H. and Reicher, L.T. 1889. Die Umwandlungstemperatur bei der doppelten Zersetzung, Z. f. phys. Chemie $\underline{3}$, 482.

17. Reiter, F. 1978. Theoretical studies on the utilization of reciprocal salt pairs for solar heat storage, EUR 6044.

PCM Thermal Energy Storage in Cylindrical Containers of Various Configurations

**ARUN S. MUJUMDAR, F. ALI ASHRAF, ANIL S. MENON,
and M.E. WEBER**
McGill University
Sainte-Anne de Bellevue
Quebec, Canada

ABSTRACT

Experimental measurements are reported for the time variation of surface-averaged rate of heat storage during melting in single, thin-walled cylindrical containers of copper filled with a commercially available paraffin wax (Parowax, Esso). For the wax used the enthalpy-temperature curve was obtained using a differential scanning caloriemeter according to the ASTM method. Three lengths and three equivalent diameters of plain circular, plain square and internally partitioned cylinders were studied for their heat storage characteristics. The heat transfer measurements revealed the importance of natural convection during melting. The effects of cylinder geometry and temperature of the external fluid on instantaneous and integral heat storage rate were examined experimentally.

NOMENCLATURE

A area of cross-section for flow (m^2)

A_h area for heat transfer (m^2)

c_{p_l} specific heat of liquid wax $(kJ/kg\ K)$

c_{p_w} specific heat of solid wax $(kJ/kg\ K)$

c_p specific heat of water $(kJ/kg\ K)$

d diameter of the cylinder (m)

L length of the cylinder (m)

Q Volumetric flow rate of circulating fluid (m^3/sec)

Q_s Cumulative heat stored in tube wall and wax (kJ/m^2)

q_{total} instantaneous value of the amount of heat taken up or given out by the tube and wax contained in it (W/m^2)

q_{tube} instantaneous value of the amount of heat taken up or
             given out by tube only (W/m^2)

q_{net} instantaneous value of the amount of heat taken up or
             given out by wax only (W/m^2)

R   radius of the cylinder (m)

T   temperature of wax at any instant (K)

T_f temperature of the hot circulating fluid (K)

T_{cr} phase change temperature (K)

St   Stephan Number

Greek Letters

α thermal diffusivity (m^2/sec)

λ_s heat of solid-solid transition (kJ/kg)

λ_f heat of solid-liquid transition (kJ/kg)

λ total heat for transition, $(\lambda_f + \lambda_s)$ (kJ/kg)

ϕ fractional heat storage (dimensionless)

INTRODUCTION

     The development of practical thermal energy storage (TES)
devices is critically important for effective utilization of
solar energy for residential heating and cooling. The high
latent heats of fusion of materials, which melt (and freeze) in
the temperature range of interest in low temperature solar energy
applications, could be taken advantage of in the design of such
devices. One of the principal advantages of the concept of PCM
storage is that the storage function is performed essentially at
a constant temperature. A large number of potential candidate
materials for low (1, 2), intermediate (3), and high temperature
(4) applications have been identified.

     Design of PCM storage heat exchangers requires knowledge
about the effects of various governing parameters (e.g. geometry,
temperature, phase change material etc.) on the rate of heat
storage or release. The problem is extremely complex to handle
analytically because of its moving boundary characteristics
coupled with natural convection in the melt. This is particularly
true when the cylindrical container is other than plain circular.
The objective of this work was to examine experimentally various
simple configurations for the PCM container with a view to
identifying optimal configurations for further study.

     Most of the earlier analytical and experimental work on
melting and freezing has dealt with pure conduction. It is now
well known that natural convection indeed is the dominant heat
transfer mechanism during melting of a PCM in cylindrical

enclosures of practical interest (5, 6, 7). It is only recently
that computational models have been developed to predict the
rates of heat transfer and interface movement in horizontal and
vertical plain cylinders filled with PCM's (8). Results of the
simulation model compare favorably with the author's measurements
of heat transfer rate for paraffin wax filled cylinders of different diameters which are subjected to different overall
temperature differences. Extension of such models to noncircular
geometries is not a straight-forward task. Also, it is still
necessary to compare results of such models with further experimental measurements.

Recently Marshall (6) has reported experimental measurements
of heat transfer during the melting of paraffin wax in a
rectangular cavity. His objectives were to examine the effect
of aspect ratio of the cavity on the internal heat transfer
coefficient and to establish possible optimal geometries for
which the natural convection effects yield maximal heat transfer
rates. Values of the heat transfer coefficients obtained were
as high as 100 W/m^{2}K for gap spacings of 10 to 20 cm. Marshall
has provided empirical correlations in terms of the Nusselt and
Rayleigh numbers for various aspect ratios of the cavity. In
general, cavities with the aspect ratio of unity were found to
give optimal heat transfer rates for melting. It should be
noted, however, that the design of practical TES units are
probably controlled mainly by the thermal performance in freezing
rather than melting. Thus the optimal configurations for melting
may not be the optimal ones for real TES units.

Another experimental investigation relevant to the present
work was carried out by Rajagopalan et al. (9). They measured
the mean surface heat transfer coefficients for melting of
paraffin wax contained in a concentric annulus made of cylindrical
stainless steel tubes. Heat was transferred only from the outer
wall of the annulus, by circulating hot water through a concentric
shell. They observed that about 90 per cent of the total resistance to heat transfer resided in the molten layer of the paraffin
wax. Interestingly, they found favorable agreement of their
experimental results with the analytical solutions of Shamsundar
and Sparrow (10).

It may be noted that heat transfer during phase change is
also of interest in a wide range of technologies, e.g. geophysics,
freeze drying, crystal growth from solutions and melts, solidification in continuous casting, purification of materials etc.
Sparrow, Patankar and Ramadhyani (7), for example, cite a number
of references from the metallurgical literature which deal with
flow and heat transfer during melting and freezing including
effects of convection in the melt.

OBJECTIVES

The objectives of this work were to determine experimentally
the effects of the following parameters on the surface-averaged
heat transfer rate to vertical cylinders filled with a commercially
available paraffin wax: temperature of the external heat transfer
medium, shape of the cylinder and the height of the PCM column

in the container. The effect of external heat transfer coefficient
was minimized by utilizing the highest permissible flow rates of
the external fluid medium in order to minimize the external heat
transfer resistance. All containers were insulated at the bottom
and open at the top. Results are presented in the form of the
instantaneous surface-averaged heat transfer and the integral
rate of heat storage as a function of the time after immersion
in the external medium.

EXPERIMENTAL

A. Thermal Properties of Parowax

     The enthalpy-temperature relationship for the specific
paraffin wax used was determined using a Model DSC-1 Differential
Scanning Calorimeter (DSC) following the standard ASTM procedure.
Samples of the wax and standard reference materials were prepared
in flat aluminum pans (Perkin-Elmer No. 219-0041) according to
the DSC manual. A Cahn RG Electrobalance was used to determine
the sample weight accurate to 0.01 mg. Figure 1 shows a typical
DSC output for the Parowax sample used in this study. This
curve clearly displays the existence of two phase transitions
and a wide melting temperature range. Table I summarizes the
DSC measurements and compares them with literature data which
pertain to a different commercial wax. Reproducibility of the
DSC measurements was within 3%. About 5% lower values were
obtained for the heats of phase change when the DSC scanning
rates were 5 $^{\circ}$C/min and 2-5 $^{\circ}$C/min rather than the ASTM recommended rate of 10 $^{\circ}$C/min. Values reported in Table I were
obtained according to the ASTM standard.

Table I. Thermal properties of commercial paraffin wax

Property	Literature data	Measured for Parowax, Esso
1. Melting range, $^{\circ}$C	50-60	44-54
2. Heat of solid-solid transition (kJ/kg)	50	46
3. Heat of fusion solid-liquid kJ/kg	200	207
4. Heat capacity (kJ/kg K) (solid wax)	2.9	2.8
5. Heat capacity (kJ/kg K) (liquid wax)	2.1	2.1

     To check if repeated thermal cycling (melting and freezing)
caused demixing or any other chemical or physical change in the
wax, several DSC runs were carried out with Parowax samples
subjected to 20, 50, 100 and 200 cycles of melting and freezing.
No measureable difference in the enthalpy-temperature curve was
observed. Thus, it appears that no discernible degradation of

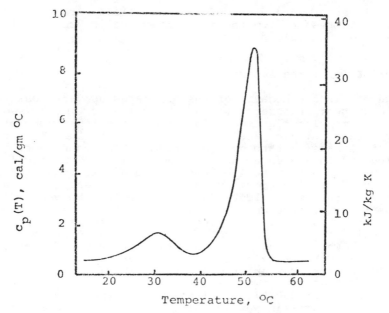

Fig. 1 Effective heat capacity of PCM used

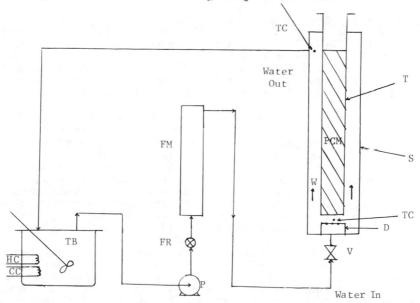

Fig. 2 Schematic diagram of the experimental set-up.

CC	cooling coil	HC	heating coil	TB	thermostalled bath
D	diffuser	P	pump	TC	thermocouple
FM	flow meter	S	shell	V	flow valve
FR	flow regulator	T	PCM container	W	water

the thermal performance of paraffin wax filled storage units may
be expected to result from the inherently cyclic process of solar
energy storage and release.

B. Description of the Apparatus

     Figure 2 is a schematic of the apparatus used to measure the
surface mean heat transfer rate as a function of time. Essentially
the PCM-filled cylindrical container forms the inner tube of a
concentric-tube single-pass heat exchanger with the heat transfer
fluid (water) passing through the annulus. The container
was sealed and insulated at the bottom and mounted perfectly
vertically. The outer shell of the apparatus was insulated with
1.8 cm thick fiberglas to minimize the heat losses. The transient
heat transfer rate was calculated by monitoring the water flow
rate, the inlet water temperature and the outlet water temperature.
For the melting runs water entered at the controlled temperature
of 70 ^{o}C while for the freezing runs water entered at 20 ^{o}C.   The
PCM container was initially at the room temperature and at 70 ^{o}C
for the melting and freezing runs respectively. (Results of only
the heating cycles are reported in this paper.)

     To obtain reproducible results it was found necessary to
ensure that the paraffin wax in the test containers did not
contain appreciable amount of entrapped air.

     The water flow rate was measured with an accurately cali-
brated rotameter. An HP model 2801 A quartz thermometer with
an HP model 2850 C, 111-25 temperature probe was used to measure
the water temperature accurately. The combined heat transfer
rate (to the tubular metal container and the wax) was computed
from the instantaneous drop in the temperature of water, the
water flow rate and allowance for heat losses to the surroundings.

     As the primary objective of this study was to study the heat
transfer in the PCM, attempts were made to minimize the external
heat transfer resistance by using the highest permissible water
flow rate in the annulus without jeopardizing the accuracy of
measurement of the temperature drop between the inlet and outlet
to the exchanger. To ensure that the external resistance was
indeed acceptably low, experiments were carried out at water flow
rates of 200, 300 and 400 cc/min. Figure 3 shows the results
obtained. On the basis of these results the water flow rate
selected for further tests was about 400 cc/min.

     To check if at the selected flow rate the external heat
transfer resistance was indeed minimized, runs were carried out
with external fins welded to the container wall and also with
additional turbulence promoters placed in the annulus. It was
found that enhancement of the external heat transfer coefficient
beyond that obtained in the smooth tube at a flow rate of 400 cc/min
did not influence the measured heat transfer curves within the
accuracy of measurement.

     Since the amount of heat stored as sensible heat in the
metal container itself can be of the order of 10 per cent of the
total heat stored in the PCM-filled containers before filling

each test container, "blank" runs were carried out in the same apparatus under identical operating conditions. The "net" heat stored in the PCM (sensible as well as heat of phase change) can then be estimated approximately by subtracting the "blank" tube heat transfer curve from the combined one. Since the phenomenon is not linear such decoupling is not strictly correct. However, for the purpose of this preliminary investigation heat transfer data are presented in terms of both the total (or combined) and net heat transfer rates.

RESULTS AND DISCUSSION

A. Role of Natural Convection

     Development of significant free convection patterns in the melt, during the melting of paraffin wax contained in transparent plexiglas cylinders, were observed in flow visualization studies using tracer particles embedded in the wax. The importance of natural convection in governing the heat transfer to PCM melts can be assessed by approximate calculation of the melting time for long cylindrical bars of solid paraffin wax using simple analytical expressions available in the literature.

     Table II lists the predicted melting times using the analytical results of Adams (11) and Solomon (12) for circular cylinders 0.019, 0.0254 and 0.03175 cm in diameter. Here it is assumed that the heat transfer takes place only by conduction through the melt layer. The predicted values are indeed under-estimates of the melting times since they are computed on the assumption that the solid is initially at the melting point. No allowance is made for superheating. For purely conductive transfer, the melt time are of the order of 1 and 3 hours for melting of 1.9 cm and 3.175 cm cylinders of paraffin wax respectively. The experimentally observed melting times are of the order of 0.5 hr in both cases. (Precise values of the melt times could not be determined in the present experimental set-up.) Thus, it may be concluded that conduction through the solid or the melt is a secondary mechanism for heat transfer in PCM's.

Table II. Melting times for pure conductive heat transfer

d, cm	Soloman (12) t_{melt} (hrs)	Adams (11) t_{melt} (hrs)
1.9	1.191	1.190
2.54	2.118	2.116
3.175	3.309	3.306

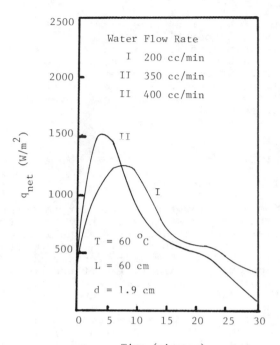

Fig. 3   Effect of water flow rate on q_{net}

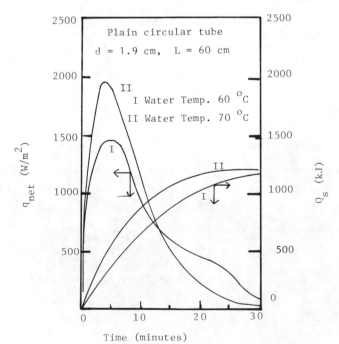

Fig. 4   Effect of water temperature

B.   Effect of Water Temperature

Figure 4 shows the effect of inlet water temperature on the
net heat stored in a 1.9 cm diameter plain cylinder (L = 60 cm)
as sensible heat and latent heat of fusion in the wax.   The
contribution of sensible heat stored in the metal container is
eliminated by utilizing data for a "blank" run with the same
container without the wax.   As may be expected the rate of heat
storage increases with the temperature of the heating medium
over most of the melting cycle.   However, after an immersion
time of about 13 minutes the instantaneous rate of heat transfer
to the PCM is actually lower for the higher water temperature
case.   The peak heat transfer rate at T_{water} = 70 $^{\circ}$C is about
30 per cent higher than the corresponding peak at T_{water} = 60 $^{\circ}$C
which occurs at about the same time of immersion.   Since the
initial transfer may be expected to be mainly by conduction,
the q_{net} curve rises more steeply for the higher water temperature
case.   Although the q_{net} curves cross-over, the cumulative heat
stored at any given time (t $\leq$ 30 minutes) is always higher at
the higher temperature.   Although the experiments were not
extended to longer times of immersion, it is clear that the
cumulative heat stored will reach an asymptotic value at sufficiently long times; the difference between the asymptotic
values at 70 $^{\circ}$C and 60 $^{\circ}$C should be equal to the enthalpy difference
for molten wax between these two temperatures.

Similar results may be expected for other PCM container
configurations.   However, in this study the inlet water temperature for all subsequent runs was held constant at 70 $^{\circ}$C.   In
future work it is proposed to study the effect of the heating
water temperature in detail.

C.   Effect of Cylinder Geometry

The cylinder geometry is specified in terms of the equivalent
diameter, d, shape of the cross-section and the length, L, (or
height) of the cylinder.   All cylinders were made of copper and
had the same wall thickness for a given d.   Figure 5 compares
the thermal storage characteristics of three types of circular
cylinders with d = 1.9 cm and L = 60 cm.   Because of the high
conductivity paths provided by the metal partitions, it is not
surprisingthat initially the transfer rates are significantly
higher for the partitioned cylinders than for the plain circular
cylinder of the same diameter.   The container with two partitions
shows a small but higher rate of heat storage until shortly after
the peak is reached.   The curve for the PCM container with a
single partition crosses over the curve for the former shortly
after the peak and then continues to store heat at a rate which
is consistently higher.   This may be attributed to deterioration
of the natural convection flow due to the partitioning.   The
data for the plain circular cylinder supports this hypothesis.
It may be noted that in the region where natural convection
dominates the transfer mechanism the transient rate of heat
storage is significantly higher for the plain (no partition)
cylinder than for the partitioned ones.

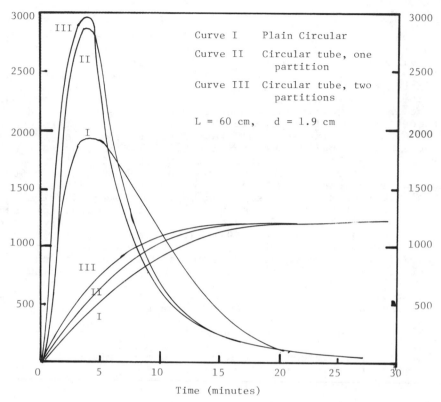

Fig. 5   Influence of container geometry on thermal storage

Figure 5 also shows the cumulative heat storage in the wax
as a function of time after immersion in the heat exchanger.
Here it may be observed that the cumulative rate of storage is
the highest for the double partitioned cylinder and the lowest
for the plain cylinder. However, the advantage of internal
partitioning is minimal. Effectively partitioning the cylinder
reduces the overall melting time only by about 5 minutes under
conditions of Figure 5.

Figures 6 through 8 display the influence of the cylinder
diameter on the net heat stored in the PCM. Each figure displays
measurements for a single height (L) of plain circular cylinders
with d as the parameter. It may be noted that the peak transfer
rate decreases with height of the cylinder for all diameters.
Also, the peaks in q_{net} spread out over longer times as the
cylinder height is increased. The diameter effect for a given L,
however, are not consistent. For example, the q_{net} peak occurs
at the following combinations of d and L: 1) d = 2.54 cm,
L = 30.48 cm, 2) d = 1.90 cm, L = 45.72 cm, and 3) d = 1.90 cm
and 2.54 cm and L = 60.96 cm. No generalization could be made

on the basis of the limited data obtained so far.   Further
investigation of the effect of d and L on thermal storage in
cylindrical containers of PCM is needed.

Figure 8 also includes the integral heat storage curves
normalized by the heat stored in each case at the end of 30 minutes.
Thus, the fractional heat stored, ϕ , is arbitrarily set equal
to unity at t = 30 minutes.   It can be seen that for L = 60.96 cm
the fractional heat storage, ϕ , (expressed as a fraction of total heat
stored in the PCM in the initial 30 minutes of immersion) decreases
with increase in the diameter of the container.   For L = 30.48 cm,
however, the intermediate size (d = 2.54 cm) yields the highest
ϕ while the largest diameter tube yields the lowest ϕ (Figure 6).
Once again the coupling between conductive and convective heat
transfer results in surface-average heat transfer characteristics
that cannot be interpolated or extrapolated empirically on the
basis of limited data.

Figures 9 through 13 display effects of the PCM container
geometry on the rate of integral heat storage in plain circular
and plain square cylinders of various lengths and equivalent
diameters.   From Figure 9 (curves I and II) it can be seen that
for d = 1.9 cm the square cylinder has a distinctly higher rate
of heat storage than the cylindrical cylinder with the same
cross-sectional area.   However, from Figure 11 it may be seen
that for larger cylinders (d = 3.175 cm) of the same length
the trend is actually reversed until t = 14 minutes after which
the integral rate of storage is again higher for the circular
cylinder.   Further, it can be seen by inspection of Figures 9
through 11 and 13 that the shorter the tube, the greater is the
integral rate of heat storage, regardless of the cylinder shape
and size.   For circular cylinders with d = 2.54 cm, Figure 10
illustrates this general trend very clearly.

Figure 13 is a plot of the time variation of Q_s/L i.e. the
amount of heat stored per unit length of the cylinder.   The length
effects seen in Figures 9 through 11 can be seen more clearly in
this figure.   For example, at t = 20 minutes, the amount of heat
stored per unit volume of the PCM in a plain circular cylinder
of 1.90 cm diameter increases by a factor of about three as the
length of the cylinder is reduced by a factor of two.   A similar
trend is exhibited by the surface cross-section cylinder.   It may
be expected from the trends seen in Figure 12 that the length
effect is possibly even more dramatic for shorter cylinders   i.e.
cylinders of small aspect ratios.   Marshall (6) has made similar
observations for the case of PCM's contained in rectangular
cavities.

CLOSURE

Preliminary experimental data are presented on the effect
of the container geometry, size and length on the thermal storage
performance of paraffin wax-filled vertical cylinders.   Over the
range of parameters studied the influence of geometry appears to
be of secondary importance in the design of practical PCM storage
units.   Cylindrical containers of small aspect ratios (i.e. small
L/d ratios) are expected to be of interest in practice and need

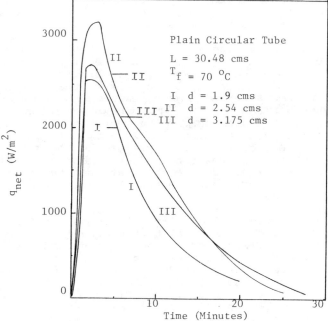

Fig. 6   Influence of diameter on thermal
           storage.

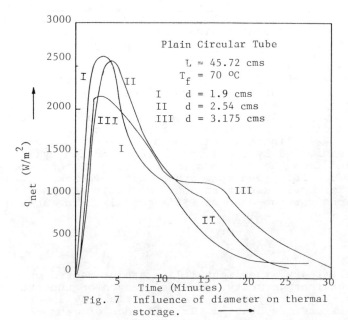

Fig. 7   Influence of diameter on thermal
           storage.

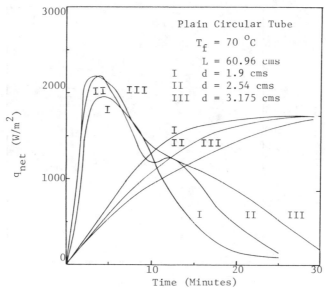

Fig. 8   Influence of container diameter
on thermal storage.

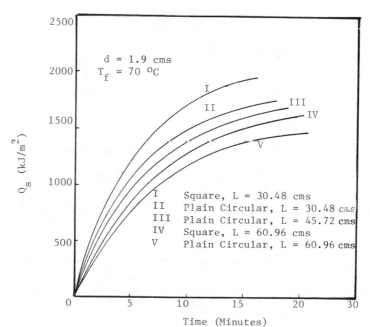

Fig. 9   Influence of container geometry and
length on thermal storage.

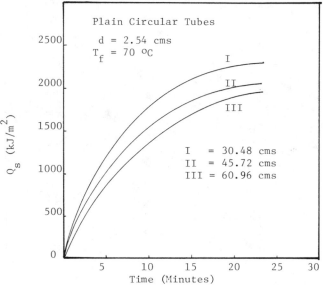

Fig. 10   Influence of container length on
          thermal storage.

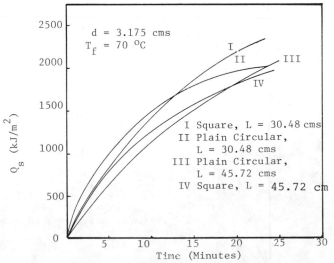

Fig. 11   Influence of container geometry
          and length on thermal storage.

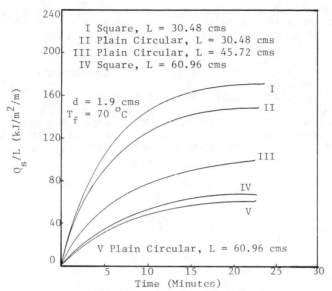

Fig. 12 Influence of container geometry
and length on thermal storage.

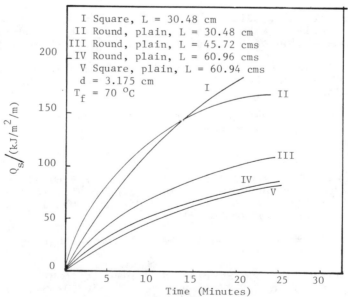

Fig. 13 Influence of container geometry and
length on thermal storage.

further investigations. It is observed that the time variation
of the surface-average rate of heat transfer to the PCM is
influenced by a complex interaction between conduction and
natural convection in the melt. The conductive transfer is
dominant only in the initial and final periods of immersion.
In view of the highly nonlinear and coupled nature of the heat
transfer process involved in this situation interpolation or
extrapolation of the experimental data must be made with caution.
Suitable mathematical models incorporating the natural convection
phenomenon are needed to supplement the experimental information
of the kind presented in this paper, to enable extrapolation of
the results over a wider range of parameters, including different
PCM's.

REFERENCES

1.   Lorsch, H.G., Kauffmann, K.W., and Denton, J.C., Energy
      Conversion, Vol. 15, No. 1/2, pp 1-9, 1975.

2.   Lane, G.A., Glew, D.N., Clarke, E.L., Rossow, H.E.,
      Quigley, S.W., Drake, S.S., and Best, J.S., Proceedings of
      the Workshop on Solar Energy Storage for Heating and Cooling
      of Buildings, ASHRAE, pp 43-55, 1975.

3.   Kauffman, K.W., and Lorsch, H.G., ASME Paper No. 76- WA/HT-
      34, 1976.

4.   Schroder, J., ASME Paper No. 74-WA/Oct, 1974.

5.   Barthelet, A.G., Viskanta, R., Leidenfrost, W., Experiments
      on the Role of Natural Convection and Heat Source Arrangement
      in the Melting of Solids, ASME Paper No. 78-HT-47, 1978.

6.   Marshall, R.H., Natural Convection Effects in Rectangular
      Enclosures Containing PCM, Thermal Storage and Heat Transfer
      in Solar Systems, Ed. Frank Kreith, ASME, New York, 1978.

7.   Sparrow, E.M., Patankar, S.V., and Ramadhayani, S., Analysis
      of Melting in the Presence of Natural Convection, Trans.
      ASME, Series C, J. of Heat Transfer, Vol. 98, No. 4, pp 520-
      526, 1977.

8.   Joglekar, G., Pannu, J., and Rice, P.A., Natural Convection
      Heat Transfer to Cylinders Containing PCM's, Paper No. 85C,
      To be presented at the AIChE 72nd Annual Meeting,
      San Francisco, November 1979.

9.   Rajagopalan, G., Personal Communication with Professor
      A.S. Mujumdar, 1979.

10.   Shamsundar, N., and Sparrow, E.M., Trans. ASME, Series C,
      J. of Heat Transfer, Vol. 98, No. 4, pp 550-557, 1976.

11.   Adams, C.M., Thermal Considerations in Freezing, Liquid Metals
      and Solidification, ASM, Cleveland, Ohio, 1958.

12.   Solomon, A.D., Melt Time and Heat Flux for a Simple PCM Body,
      Solar Energy, Vol. 22, pp 251-257, 1979.

A Cycle Life Tester for the Long-Term Stability of Phase Change Materials for Thermal Energy Storage

A. GRANDBOIS, J. SANGSTER, and J.R. PARIS
Ecole Polytechnique
Montreal, Canada

ABSTRACT

The choice of a phase change material (PCM) for a particular application of storage of low potential thermal solar energy depends upon several considerations, one of the more important being its long-term stability in repeated freeze-thaw cycles. An experimental apparatus has been constructed to simulate, on a greatly speeded-up scale, several years' use of macroscopic quantities of a PCM. The thermal behaviour of the substance is monitored by a microcomputer, which also controls the freeze-thaw cycles.

INTRODUCTION

Among the many alternative energy sources being considered today, solar energy is one of the more "benign" and immediately available. Like some of these sources, however, it is of intermittent nature and subject to the vagaries of daily weather changes. Some method of storage therefore is needed in order to take full advantage of its cheapness. Most of the houses, schools or commercial buildings which are partially or fully solar heated today store solar thermal energy as sensible heat in rock or water reservoirs [1]. Such materials are quite inexpensive and readily available; they in turn suffer from two inconveniences which, under certain circumstances, may become crucial. Relatively large volumes of material must be used and the temperature of delivery drops as the heat is extracted.

The use of the latent heat of fusion of phase change materials (PCM) greatly minimizes these inconveniences. The volume required is reduced by 3 to 10 times, depending upon the allowed temperature swing of the sensible heat substance [2, 3]. Since the temperature of a substance undergoing solid/liquid phase change remains very nearly the same, the PCM takes up and releases heat at essentially constant temperature.

Although PCMs represent an attractive way to store and deliver thermal energy, they have not yet been accepted for wide use because there are too many candidate materials on which the requisite data and experience have not been obtained. Denton [4a] has outlined a number of considerations for the choice of PCM, knowledge of which is lacking in very many instances (Table I).

Table I.   Considerations in choosing a PCM for thermal energy storage.

Safety

toxicity   oxidizing power
corrosivity   flammability

Market

availability
price

Physico-chemical properties

density   supercooling
volumetric expansion   nucleation
melting/freezing point   viscosity
heat of fusion   crystallization rate
congruence   thermal conductivity
specific heat   STABILITY

Engineering

containers and heat exchangers
scaling laws for heat and mass transfer
testing procedures
design guidelines

A number of surveys of candidate materials have been made, taking into consideration what is known of their safety and physico-chemical properties [4b, 5, 6], These properties have been or can be measured in a relatively straightforward manner. The present paper addresses itself to the long-term stability of the important properties of PCMs. Since a PCM's lifetime use would involve several hundred or more freeze-thaw cycles, its usefulness depends upon its long-term performance. In general, PCMs may not be simple chemical substances (salt hydrates and eutectics are promising candidates) and complications could intrude. It has been found, for example, that storage capacity and overall efficiency of systems using Glauber's salt $Na_2SO_4 \cdot 10H_2O$ [7] or $Na_2HPO_4 \cdot 12H_2O$ [8] decreased rather quickly with use. This was a result of incongruent melting and phase segregation in the melt.

THE CYCLE LIFE TESTER

It was desired to study the response of a PCM to a large number of freeze-thaw cycles on a greatly speeded-up scale. One way of doing this is by differential scanning calorimetry (DSC, ref. 9) or differential thermal analysis (DTA, ref. 4b). These techniques, because they use very small samples, tend to exaggerate the importance of supercooling. While this caution may be commendable for the evaluation of materials to be encapsulated in polymer or metal matrices, it is not realistic for those applications which will use the PCM in bulk.

We wished to test macroscopic samples of PCMs, and the apparatus here reported is an elaboration of a previous suggestion [4c]. A schematic diagram appears in Fig. 1.

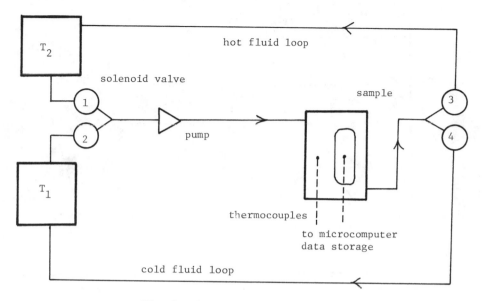

Fig. 1 - Apparatus Schematic

The temperature of the PCM sample (50-100 cc) in a small bottle can be
measured by up to seven thermocouples. The sample is successively melted
and frozen by hot and cold streams of a glycol-water mixture. The streams
are controlled by two-way valves operated simultaneously. A microcomputer
is programmed to operate the valves and to record the thermocouple data for
as many cycles as desired. The cooling (or heating) curve at any point in
the sequence as registered by a thermocouple may be retrieved from tape
cassette storage and displayed for inspection. A thaw-freeze cycle of a
PCM, as seen by a thermocouple in the sample, is shown in Fig. 2. Here T_1
and T_2 are respectively the temperatures of the cold and hot reservoirs of
heat transfer liquid. LRT and HRT are respectively the low and high
reference temperatures at which the microcomputer reverses the state of the
valves. T_f is the melting/freezing point of the PCM.

The thermal performance of the PCM may be judged from the heating or
cooling curves as a function of time. The melting point may be compared
to the freezing point, and the evolution of the following quantities could
be followed: degree of super-cooling, length and flatness of the temperature
plateau. Such information gives indications of development of chemical
decomposition, incongruency of melting or freezing, or other failure to
return to the same PCM state at the end of the cycle. The response curves
of multiple thermocouples in the PCM body may yield information on temperature
gradients (and hence on thermal conductivity and liquid/solid boundary
movement) and the rate of crystallization or fusion. If the PCM is in a
transparent container, it may be inspected visually.

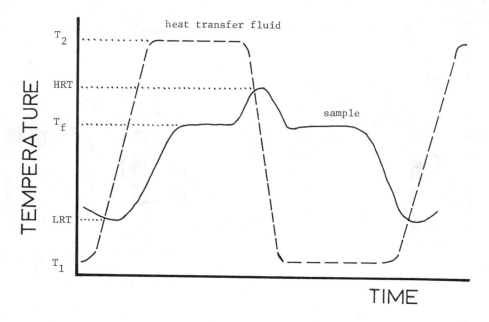

Fig. 2 - Thaw-Freeze Cycle

THE ELECTRONICS

Although it will not be of primary interest to all readers, a summary of the control and data acquisition system is given here in order to indicate, to those wishing it, how such a versatile system can be built around an inexpensive microcomputer. A simplified diagram of the microcomputer and peripheral components is given in Fig. 3.

The microcomputer is an Intel single-board SDK-85. It has a simple keyboard and display, and a standard video terminal has been added to facilitate programming and debugging. The SDK-85 also contains ROM, RAM and EPROM memory circuits for program debugging and storage and for temporary storage of temperature data. It includes a programmable timer (for real-time operations) and a microprocessor operating at a clock frequency for timing of data-bit acquisition.

The thermocouples are read through a Burr-Brown SDM-856 data acquisition subsystem, including an analog-to-digital converter and amplifier or gain 1000. The microcomputer transfers the data in blocks to the digital cassette (Phi-Deck, Triple-I Corp.).

The audio cassette is used for program storage during program development. This component, together with associated memory circuits on the SDK-85, is very convenient for initial program set-up and subsequent changes the operator may wish to make at a later stage.

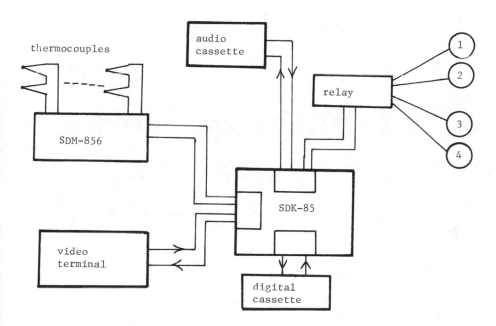

Fig. 3 - Simplified Electronics Diagram.

The solenoid valves are operated in pairs, through relay drivers and SPDT relays. This provides the necessary interface between the low voltage logic signal of the microcomputer and the 110VAC necessary to activate the valves. The valves are of the normally closed type, so that in the event of power (or system failure, flow to the sample chamberwould be cut.

The total cost of the SDK-85, interface and peripheral electronics was of the order of $1200.. This is a very modest outlay, considering the advantages of totally automated operation. The microcomputer may be adapted easily, if desired, to operate more elaborate control valves for various types of proportional temperature control.

APPLICATIONS FOR THE CYCLE LIFE TESTER

The apparatus was tested with stable reference materials such as paraffin wax or Wood's metal. PCMs for examination are to be chosen from published selective lists [4b, 5, 6] and various new suggested substances can be tried: H_2O + $Na_2SO_4.10H_2O$ [10] or $Mg(NO_3, Cl).6H_2O$ [11]. With suitable modification of the sample container, systems involving a combination of latent and sensible heats could be investigated [12, 13]. It could also be adapted to a systematic search for better nucleating agents [14, 15].

REFERENCES

1. W.A. Shurcliff, <u>Solar Heated Buildings: a Brief Survey</u>, 13th edition, 19 Appleton St., Cambridge, Mass., 1977.

2. H.G. Lorsch, "Thermal Energy Storage for Solar Heating", ASHRAE Journal,
    Nov. 1975, p. 47.

3. H.G. Lorsch, K.W. Kauffman and J.C. Denton, "Thermal energy storage for
    solar heating and off-peak air conditioning", Energy Conversion 15, 1
    (1975).

4. Proceedings of the Workshop on Solar Energy Storage Sub-systems for the
    Heating and Cooling of Buildings, Charlottesville, VA, ASHRAE, 1975.
    a) p. 169; b) p. 43; c) p. 56.

5. D.U. Hale et al., Phase Change Material Handbook, Report No. NASA
    CR-61363, Lockheed Missile and Space Co., Huntsville, AL, 1971.

6. G.J. Janz and co-editors, Physical Properties Data Compilations Relevant
    to Energy Storage. I. Molten Salts: Eutectic Data; Data on Single and
    Multi-component Systems, NSRDS-NBS 61, National Bureau of Standards,
    Washington, D.C., 1978/79.

7. J.W. Hodgins and T.W. Hoffman, "The storage and transfer of low potential
    heat", Can. J. Technol. 33, 293 (1955).

8. T.L. Etherington, "A dynamic heat storage system", Heating, Piping and
    Air Conditioning 29 (12), 147 (1979).

9. S. Cantor, "Applications of differential scanning calorimetry to the
    study of thermal energy storage", Thermochimica Acta 26, 39 (1978).

10. D.R. Biswas, "Thermal energy storage using sodium sulfate decahydrate and
    water", Solar Energy 19, 99 (1977).

11. N. Yoneda and S. Takanashi, "Eutectic mixtures for solar energy storage",
    Solar Energy 21, 61 (1978).

12. K.W. Kauffman, H.G. Lorsch and D.M. Kyllonen, "Thermal energy storage by
    means of saturated aqueous solutions", final report, TID-28330, The
    Franklin Institute, Philadelphia, PA, 1977.

13. L. Keller, "A new type of thermal phase-change storage", Solar Energy 21,
    449 (1978)

14. M. Telkes, "Nucleation of supersaturated inorganic salt solutions", Ind.
    Eng. Chem. 44, 1308 (1952).

15. Z. Stunic, V. Djurickovic and Z. Stunic, "Thermal storage : nucleation
    of melts of inorganic salt hydrates", J. Appl. Chem. Biotechnol. 28, 761
    (1978).

Sensitivity Analysis of a Community Solar System Using Annual Cycle Thermal Energy Storage

FRANK BAYLIN, ROSEMARY MONTE, and SANFORD SILLMAN
Solar Energy Research Institute
Golden, Colorado 80401, USA

ABSTRACT

The objective of this research is to assess the sensitivity of design parameters for a community solar heating system having annual thermal energy storage to factors including climate, building type, community size and collector type and inclination. The system under consideration uses a large, water-filled, concrete-constructed tank for providing space heating and domestic hot watter (DHW). This presentation outlines results and conclusions about system sizing; a system design study and economic analysis are underway.

1. INTRODUCTION

Incorporating an annual cycle of thermal energy storage (ACTES) into a solar system may permit more effective utilization of solar energy than do conventional designs that are based on diurnal storages. Although a number of ACTES solar systems have been constructed in Canada [1,2] and Sweden [3] and designs for others are being developed in Canada, Sweden [4] and France [5], no systematic study has been performed to assess critical determinants of system component sizing and economic competitiveness. (Note that a good deal of research which examines ACTES is in progress [1-7].

The objective of ACTES is to store heat collected in the summer for winter use, when the load is greatest. The need for seasonal storage is demonstrated in Fig. 1, which shows month-to-month variation in load and insolation for both a northern city (Madison, Wisconsin) and a southern city (Phoenix, Arizona).

Figs. 2a and 2b outline the simulated operation of a seasonal storage system for the two cities (Madison and Phoenix) and building types (single family and apartment). The top graphs show monthly load and collector gain. The difference between the two(shaded area) indicates the amount of heat to be provided by storage during the winter. The bottom graphs show the storage temperature and the collector efficiency. Storage temperature follows a similar pattern for both cities, rising to the $\simeq 75^\circ C$ by early autumn and then dropping through the winter as heat is drawn to satisfy the winter load. The collector efficiency drops sharply in the winter months. This effect is much more severe for Madison, where efficiency drops below 10%, than for Phoenix when efficiency remains above 20%. The low winter collector efficiency is another important reason for investigating seasonal storage, especially in northern cities.

In addition to collecting summer heat for use in winter, ACTES system provide other advantages. Collector stagnation during the summer months is all but eliminated. Collector field size is substantially reduced. ACTES systems can provide close to 100% solar heat, reducing or even eliminating the costs of a backup system and avoiding the burden of an increased winter peak load for electric utilities.

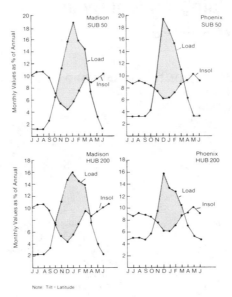

Figure 1. Month-by-Month Insolation and Heat Loads (Including Hot Water)

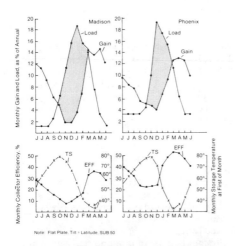

Figure 2a. Month-by-Month Collector Gain, Heat Load, Collector
Efficiency (EFF), and Storage Temperature (TS)

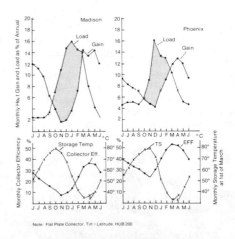

Figure 2b. Month-by-Month Collector Gain, Heat Loads, Collector
Efficiency (EFF), and Storage Temperature (TS)

   Use of a community-wide system rather than individual systems provides other
advantages. As storage size increases, the unit cost decreases. As well, the unit heat
losses decrease and storage efficiency increases because the surface-area-to-volume
ratio of the container decreases proportionately to increase in the radius. Statistical
averaging of demand in a community increases system reliability and may decrease
overall power requirements. Operation and maintenance is a shared expense. Finally, a
community system may have financing and tax advantages of a public utility.
   Disadvantages include the extra cost and energy losses in the storage and distribution system, increased problems of water freezing in the distribution system in winter,

possible overheating of buildings in summer, and the need for ongoing system management by trained personnel.

The trade-off between collector and storage size is being investigated further. The 440 systems in this study were designed to supply 100% of the annual space heating load requirement and about 85% of the domestic hot water. Storage volume was minimized with the constraint that heat would not be dumped in the summer. A larger storage size with the chosen collector field would result in a lower maximum storage temperature.

2. METHODS

An analysis based on an hourly simulation of an ACTES system is used to (1) size systems in 10 locations, (2) identify critical design parameters, and (3) provide a basic conceptual approach for future studies and designs. The computer code was developed by Hooper and his associates at the University of Toronto [8]. This code was used because it is the only hourly simulation available in North America, because it has been validated and updated in one demonstration project, Provident House [2], and because it has been used to design a second, larger facility, the Alymer community.

Community size and housing type, geographic location, and collector type and tilt angle were varied. Discussion of these design parameters follows.

2.1 Community Size and Housing Type

Several community sizes and housing types were examined. Single family detached homes, 10-unit condominiums, and 200-unit apartment complexes provide a range of building types and are judged to be representative of present U.S. housing trends. Community sizes were varied from 50 through 200, 400, and 1,000 units. Thus, a total of 11 configurations (3 x 4 minus the excluded 50-unit apartment complex) were considered.

The choice of building configuration was based on those from the recent OTA report on solar energy [9]. Single family 2,000 ft^2 residences and 10-unit 3-bedroom 1,300 ft^2 condominiums were modelled. The 200-unit apartment complex had 10 stories; each consisted of 160 one-bedroom 850-ft^2 units and 40 two-bedroom 950-ft^2 units.

2.2 Geographic Location

Typical meteorological year weather data from 10 U.S. cities were used. An isoinsolation map, Fig. 3, shows the location of those cities chosen for investigation. A number of variables—including total yearly insolation per square meter, yearly degree days, or yearly community heating load—are used in the analysis as proxies for location.

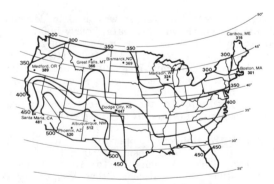

Fig. 3 - Geographic locations of Cities - Numbers indicate mean daily solar radiation in langleys (gram-calories/cm^2)

2.3 Collector Type and Tilt Angle

Two collector types are examined in this study: evacuated tube collectors (ETC) and a medium performance flat plate collector (FPC).
The collector parameters are:

	$F_r (\tau \alpha)$	$F_r U_L$ [w/m^2 °C]
FPC	0.711	6.1041
ETC	0.447	1.1697

Two collector tilt angles were chosen: tilt equal to latitude and tilt equal to latitude plus 10°. No procedure, has been devised, as yet, to determine the optimal tilt angle for annual storage. Such a procedure would be based upon the relative magnitudes of energy gains and losses and building loads over an entire season. In total, four configurations of collectors were used for this study.

A variety of parameters were chosen to have either a fixed value or a value which changed somewhat across the unconstrained variables. These were transmission losses; heat load factor; domestic hot water (DHW) delivery temperature; maximum design tank top temperatures; inlet temperature to the DHW System; DHW demand rate; thermostat setting; design ambient temperature; insulation thickness; and soil density, thermal capacity, and conductivity.

Soil Conductivity. The following values were used to describe the soil conditions:

- soil thermal conductivity—1.7307 w/m° C,

- soil density—1762.0 Kg/M^3, and

- soil thermal capacity—1.0 KJ/Kg° C.

The values are considered to be representative of soil conditions in North America.

Transmission Losses. The University of Toronto simulation was designed to model an ACTES system which would provide heating and DHW for only one building. Losses resulting from transmission of thermal energy among the storage facility, the load, and the collectors was, therefore, considered negligible. In order to estimate conservatively the effect of transmission losses in the piping, we assumed that single unit, multifamily, and apartment complexes had losses of 10%, 5%, and 0% added to the community heat load factor (see below). The ACTES was assumed to be either integral to or adjacent to the apartment complex. The single–family community has substantially more piping than does the multifamily group.

Heat Load Factor. The heat load factor was used to determine building energy load. When coupled with the hourly weather data, hourly building load could be calculated. The heat load factor for a single–family residence of 2,000 ft^2 was chosen to be 500 Btu/degree hour, based on a recent SAI study [10] and advice from SERI researchers. The value for the multifamily condominium based on the OTA study [9] was 202 Btu/degree hour per unit. (This is an average since the units on the end of the building with more exposed surface area will have higher heat losses than the intermediate units.) The heat loss factor for the apartment complex was 25,748 Btu/degree hour or 130 Btu/degree hour per unit (also drawn from the OTA study).

Domestic Hot Water (DHW) Delivery Temperature. The DHW delivery temperature was chosen to be 120° F (48.0° C), lower than the normal 140° F (60° C) but still in a perfectly functional range. This was selected for two reasons. First, this allows attainment of more nearly 100% solar systems. Other designs—for example, having one ACTES tank plus multiple DHW tanks that would be charged first—would easily permit attainment of 100% solar systems. Second, the lower temperature is in keeping with the philosophy of this study—use of renewable energy sources and conservation of energy.

Maximum Design Tank Top Temperature. The maximum design tank top temperature was chosen to be 175°F (79.4°C). This temperature is well within present limits on
plastic liners for storage tanks. It places less stress on tank insulation than would higher
temperatures. It is also the maximum design temperature of the Lyngby's home in
Denmark, an ACTES design that is currently in operation [11]. It could be argued that
higher tank temperature would allow better utilization of the ETC, but this change in
design would result in greater heat losses from storage and the transmission system.

Inlet Temperature to DHW System. The water main temperature was taken to be
the average temperature of shallow groundwater and is, therefore, location dependent.

Thermostat Setting. The effective thermostat setting is the temperature requirement that is actually experienced by the space heating system. It is always a few
degrees lower than the actual thermostat setting. In this study, 68°F (20°C) was chosen
as the actual thermostat setting and 65°F (18.3°C) was used as the design thermostat
readings.

3. FINDINGS

Results are organized as follows. First, critical factors in sizing the collector field
area and storage tank volume are analyzed. Second, the sensitivity of design parameters
to community size are investigated. Third, a preliminary comparison is made of annual-
versus daily-cycle storage/solar energy systems. (Single, 10-, and 200-unit building sizes
are abbreviated by SUB, TUB, and HUB, respectively, in the following discussion.) It
should be noted that even if further analysis would result in the resizing of some
components, the consistency from design to design in this analysis allows us to have
confidence in searching for system to system variations. Similarly, although the sizes of
the collector fields and the storage tanks can be proportionately different within limits,
the choice of relative sizes was reasonable and consistent across all systems.

3.1 Sizing Components

It was found that as total annual insolation and average ambient temperatures
increased, then collector efficiency increased. However, each of these factors could
vary somewhat independently and the above relations were only approximately linear. A
parameter that combines both the effects of average ambient temperature and total
annual insolation on collector efficiency can be derived from the familiar relationship:

$$= \text{Heat gain less heat loss}$$

$$= F_r \, (\tau \, \alpha) - F_r \, U_L \, \frac{T_i - T}{I}$$

where T   = outdoor ambient temperature

    T_i = collector inlet temperature

and   I = insolation.

Collector efficiency was plotted versus a yearly average quantity $57 - \langle T_A \rangle$ (where
$\langle T_A \rangle$ is the average ambient temperature) divided by total yearly insolation per square
meter. (57°C is chosen here because it is approximately the average temperature of the
storage fluid and is, therefore, the yearly average inlet fluid temperature to collectors.)
As seen in Fig. 4, this relationship is approximately linear.

Using this relationship and knowing the space and DHW load requirements and the
storage and distribution loss estimates, the collector area can be calculated from:

$$\text{Collector area} = \frac{\text{Total yearly space plus DHW load} + \text{storage and distribution losses}}{\text{Yearly insolation per } m^2 \times \text{efficiency}}$$

An algorithm for estimating storage volume is presented below.

Figure 4 also shows the relationship between flat plate (FPC) and evacuated tube collectors (ETC). For selected performance parameters, FPCs perform better when the difference between operating temperature and ambient temperature is small; ETCs perform better when this difference is large. As a consequence, ETCs are more advantageous in relatively cold and cloudy locations while FPCs are more efficient in the warmer climates.

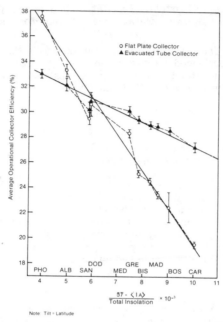

Note: Tilt = Latitude

Figure 4. Average Operational Collector Efficiency vs. $\dfrac{57-\langle TA \rangle}{\text{Total Insolation}}$

3.2 Storage Size

The difference between heat load and collected solar heat in winter (the shaded area in Fig. 2a and 2b) must be provided by storage. Storage size per unit versus this difference, the "winter net load" for November through February, is plotted in Fig. 5, 6, and 7. A linear relationship is obtained for all locations, community sizes, and collector types. This linearity makes the relationship useful for general system design.

Winter net load may be calculated by adding space heat and hot water loads, estimating storage loss, and subtracting collected solar heat for November through February. (The percentage of the DHW load must be estimated; it typically is 75% for the four winter months. Storage loss may be taken as one-third the annual loss.) Calculating collected solar heat presents some problems because collector efficiency for the winter months must be estimated. Table 1 gives representative average winter collector efficiencies for the 10 cities that are suitable for use in sizing storages.

Storage sizes were also plotted versus the total winter load without considering collector gain. The plot (see Fig. 8) is close to linear and also could be used for system sizing. However, this relationship is not as linear as the preceding plots. Some irregular patterns, such as the reduced storage size needed for evacuated tube collector and for warmer locations, are not accounted for and corrected in the storage size versus total winter load graphs.

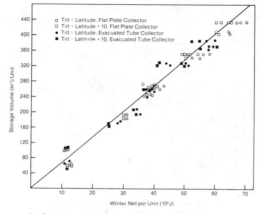

Note: The diagonal line is the same as the diagonal lines in the 10- and 200-unit graphs.
Winter net load is equal to load plus storage and transmission losses minus collector gain for the months of November through February.

Figure 5. Storage Volume per Unit vs. Winter Net Load per Unit: Single-Unit Building

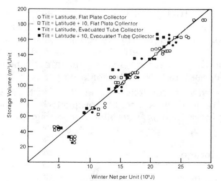

Note: The diagonal line is the same as the diagonal lines in the immediately preceding and following graphs, for single and 200-unit buildings.
Winter net load equals building load plus storage and transmission losses minus collector gain for the months of November through February.

Figure 6. Storage Volume per Unit vs. Winter Net Load per Unit: 10-Unit Building

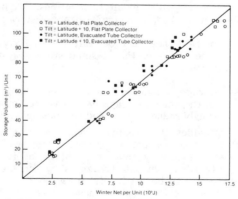

Note: The diagonal line is the same as the diagonal lines in the single and 10-unit graphs immediately preceding.
Winter net load is equal to load plus storage and transmission losses minus collector gain for the months of November through February.

Figure 7. Storage Volume per Unit vs. Winter Net Load per Unit: 200-Unit Building

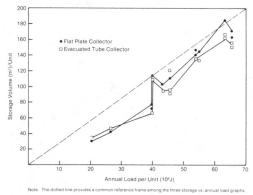

Figure 8. Storage Volume per Unit vs. Annual Load per Unit:
10-Unit Buildings (Space and Water Heat)

3.3 Sensitivity of Design Parameters to Community Size

Design variables were plotted against community size with building type and solar collector type as parameters. Results were as expected: collector area, solar energy collected, storage volume, and solar energy stored all increase linearly with community size and with building load. Collector efficiency remained constant for all community sizes within a given location. Storage losses became proportionately smaller with increasing community size, enabling the collector area per unit to drop slightly.

Storage size per unit was proportionately smaller for HUB and TUB than for SUB. This is because the DHW load is a greater proportion of the total load in larger buildings. As a result, winter load is a smaller percentage of the yearly load for the larger buildings and, therefore, storage size is reduced.

3.4 Annual Versus Daily Storage

A fundamental question in these considerations is how ACTES solar systems compare to conventional solar systems based on diurnal storages. Although a more thorough answer to this question, which examines the economics of collector-storage trade-offs, is presently under study; we can draw some preliminary conclusions here. To this end we compare those ACTES solar systems designed in this study with conventional solar systems for similar building types (SUB) in all 10 locations. Conventional systems are sized here by using the F-chart method, assuming 75 liters of storage per square meter of solar collection.

The percentage of solar heat that could be delivered by conventional systems with the same size collector are designed for the seasonal storage systems was calculated. Three observations were apparent.

- Without annual storage, about 65% of the heat load is provided by solar energy. Therefore, the annual storage can be viewed as adding 30% additional energy, correspondingly reducing the need for backup equipment.

- Annual storage provides the greatest advantage in cities with poor winter insolation, but this trend is not as pronounced as was expected. Medford, Oreg., which receives a very small percentage of its annual insolation in winter, is the city where annual storage is by far the most advantageous. Annual storage tended to be less useful in warmer climates (Phoenix, Albuquerque), but the difference between these cities and places like Boston and Madison was not very striking from this analysis.

- ETCs improve performance of an ACTES solar system as compared with FPCs because ETCs operate well over the relatively large temperature differences in seasonal storages. An annual cycle storage system can collect and store heat at 60-70°C, but conventional systems operate, on the average, at lower temperatures. Consequently, ETCs are more advantageous for ACTES solar systems. A counterbalancing trend occurs in cities with severe or cloudy winters. In such places, effective collection of winter insolation requires use of ETCs. Consequently, in Medford, which has a cold, cloudy winter, use of a diurnal storage system is less effective with FPCs than with ETCs.

The F-chart also was used to size daily storage systems that match the performance—96% solar—of the ACTES solar systems designed here. It was found that double to triple the collector area is required compared to the corresponding ACTES solar system.

4. CONCLUSIONS

Collector field area and storage volume have been sized for 440 community designs in 10 geographic locations. Analysis of the data has allowed identification of those parameters that have first order effects on component sizing. Storage size is determined by the difference between "winter net" and collected energy. Collector area then is sized to fully change storage.

Two linear relationships were derived which allow system sizing. The average ambient temperature is used to determine average yearly collector efficiency. This parameter combined with estimates of space/DHW loads, storage/distribution losses, and total yearly insolation per square meter allows estimation of collector area. Storage size can be estimated from the winter net load which is based on space and DHW loads, storage/distribution losses, and collector solar heat for the winter months.

The algorithms, which would be applicable to other types of annual storages such as aquifers, can be further refined as results from the operation of ACTES solar systems become available. Calculations also can be refined with more detailed knowledge of a particular community design.

In order to more accurately judge the relative merits of ACTES solar systems in different climates, a more detailed systems study and economic analysis is underway. Preliminary results indicate that as the DHW-to-space-heating-load ratio increases and as community size decreases, system economics become less favorable. Modifications to the design presented here such as incorporating a two-tank (annual storage for space heating; daily storage for DHW) storage systems or using multiple tanks for annual storage of both heat and cold, may be economically promising technologies.

REFERENCES

1. Hooper, F. C.; Attwater, C. R. Solar Space Heating Systems Using Annual Heat Storage. Progress Report. C00-2939-1. Washington, D.C.: U.S. Department of Energy; April 1977; pp. 57-61.

2. Hooper, F. C.; et al. Solar Space Heating Systems Using Annual Heat Storage. Progress Report. C00-2939-6. Washington, D.C.: U.S. Department of Energy; October 1978.

3. Margon, Peter; Roseen, Rutger. "Central Solar Heat Stations and the Studivih Demonstration Plant." The 13th Intersociety Energy Conversion Engineering Conference of the Society of Automotive Engineers. September 1978; pp. 1614-1619.

4. Swedish Energy Research and Development Commission. Directory of Research and Development Projects within the Swedish Governmental Programme for Energy Research and Development, 1975/76-1977/78. DFE Report #20. Swedish Energy Research and Development Commission; April 1979.

5. Torrenti, R. Seasonal Storage in Solar Heating Systems. International Solar Energy Society Conference; Atlanta, GA; May 1979.

6.   McGarity, A. E. "Optimum Collector-Storage Combinations Involving Annual Cycle Storage." Proceedings of Solar Energy Storage Options Conference, Volume 1. San Antonio, TX; March 19-20, 1979. CONF-790328-P1. Washington, D.C.: U.S. Department of Energy.

7.   Panel Report. "Annual Cycle Storage for Building Heating and Cooling." Proceedings of Solar Energy Storage Options Conference, Volume 2. San Antonio, TX; March 19-20, 1979. CONF-790328-P3. Washington, D.C.: U.S. Department of Energy.

8.   Hooper, F. C.; Attwater, C. R. "A Design Method for Heat Loss Calculations for In-Ground Heat Storage Tanks." Heat Transfer in Solar Energy Systems: Proceedings of ASME Winter Annual Meeting. Atlanta, GA; December 1979. Washington, D.C.: American Society of Mechanical Engineers; pp. 39-43.

9.   Office of Technology Assessment, U.S. Congress. Application of Solar Technology to Today's Energy Needs, Volume II. Washington, D.C.: OTA, U.S. Congress; September 1978.

10.   Hughes, P. J.; Morehouse, J. H. A Trnsys-Compatible, Standardized Load Model for Residential System Studies, Preliminary Draft of Final Report. SAI Report No. 80-915-WA. McLean, VA: Science Applications, Inc.; May 1979.

11.   Esbensen, Torben V.; Korsgaard, Vagn. "Dimensioning of the Solar Heating System in the Zero Energy House in Denmark." Solar Energy. Vol. 19 (No. 2): pp. 195-199.

12.   Baylin, Frank; Monte, Rosemary. Sensitivity Analysis of Community Annual Cycle Thermal Energy Storage Solar Systems, Volume 1. SERI. Golden, CO: Solar Energy Research Institute; forthcoming.

The Use of Concrete Block Directly under a Concrete Slab as a Heat Storage System in a Passive Solar Heated Building

ROBERT MITCHELL
Solar Systems Design, Inc.
Selkirk, New York 12158, USA

JOSEPH E. GIANSANTE
Schenectady, New York 12306, USA

ABSTRACT

In direct gain systems, concrete floors have their storage potential limited by their effective thickness and by the limited area that is irradiated at any one time. A system is described with the dwelling treated as a stratification tower whereby air heated at the irradiated surface of the floor storage slab rises to the building high point where it is ducted beneath the floor slab to an area of 6" concrete blocks. Performance of a specific building is given, and a thermal network analysis is developed for parametric studies of similar systems. A computer simulation of the thermal model is performed and results are given.

INTRODUCTION

Passively solar heated structures need an adequate amount of thermal mass to assure acceptable comfort levels. For our direct gain structure, we have chosen to use a system whereby 6" concrete blocks are laid on their sides below a 4" concrete slab that is the first floor of the building (See Figure 1). Trenches are constructed below the slab at each end of the parallel rows of concrete blocks. These trenches serve as supply and exhaust headers. A vertical duct connects the supply header with the highest point in the building, and a fan charges the system. Two inches of smooth extruded styrofoam and 6 mil. thickness of polyethylene separate the storage from the soil below.

We feel that there are certain advantages to this system, which we are in the precess of documenting. First, there is a reduction in required mass as compared to a totally passive direct gain system. This is due to forced convection within the blocks. Second, there is a reduced cost of construction compared to Trombe walls. This is due to both a reduction in required aperture (due to reduced operating temperature at the glazing) and the fact that the block storage system can be installed for about $2.70 per ft.^{2}, including fans and controls. Third, there is a high degree of comfort control. We have experienced no significant overheating (over 75°F) in our structures. Fourth, having an air handling system allows the use of electrostatic air filters and other devices that enhance comfort.

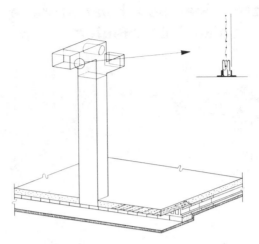

Fig. 1 - Heat Storage Schematic

We have monitored the performance of a particular building, the Lornell house, a 1750 sq. ft. building which was built for about $35 a sq. ft. The house has been occupied since May 1977.

1. BUILDING DESCRIPTION

Several features characterize this building (See Fig. 2).

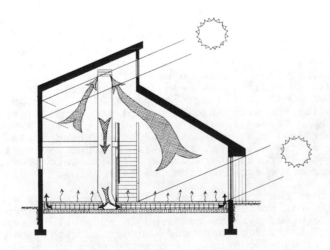

Fig. 2 - Lornell House Section

1.1 Apertures

Of the roughly 390 ft. gross aperture area, 325 ft. face south. All rooms except the bathrooms and utility room have direct south exposure. High clerestory windows provide back lighting that increases the overall light level and reduces glare.

1.2 Conservation

All walls are double studded 2' x 4's at 24" o.c. The
cathedral ceilings share two R-19 batts with a fully ventilated
space above them. The building sheathing is caulked to the frame.
Sill seal is used. One part foam is applied between the windows
and rough openings. Four inch perimeter insulation is used in
addition to 2" styro s.m. below the entire slab. Thermal
shuttering is provided throughout. As a result, the calculated
air change is .12/hr. Using 1/3 of an air change per hour the
building UA with shutters in place, is 297 BTU/hr °F or 7138 BTU/
degree days.

1.3 Stratification

Because of the juxtaposed cathedral ceilings (See Fig. 2)
a free convection path exists to the building high point. The
maximum heighth is 24' and daytime temperature differentials of
20°F are common even during storage fan operation.

1.4 System Control

A differential controller has sensors at the high point,
and in the storage exhaust trench. The controller is interfaced
with the furnace so that upon call for storage, dampers shift
and the furnace fan comes on. The system, although complex,
appears to work well.

1.5 Monitoring and Results

A low cost monitoring system was installed, incorporating
running time meters attached to the furnace gas solenoid and the
air handler. Daily readings were taken and correlated to the
gas meter. The result was that for the 1977-78 heating season,
natural gas usage was approximately 2/2 hundred cubic feet for
the cost of $86. Additionally, approximately 8 mil BTUs of heat
was provided by an airtight wood stove. This house is located
near Albany, New York, 6880 degree days and 134,700 BTU/ft.² total
cumulative horizontal radiation for October 1 through April 30.

2. SYSTEM SIZING

System sizing was originally done by a simple iterative
method for average and clear sky conditions. On an hourly basis,
space temperatures were calculated, mass temperature, and rate
of re-radiation and convection for the mass. This approach led
to oversized storage, with the result that the floor slab was
cold underfoot. To correct this, a thermal model of the floor
has been developed to facilitate computer simulation. The model
is still in a very early stage of development, and many gross
simplifications have been incorporated.

3.   THERMAL HOUSE MODEL

3.1   Node Diagram

Figure 3 is a schematic diagram of the thermal model of the
passive solar residence. The four basic components are divided
as follows:

Walls and roof. The exterior walls and roof are combined
together and are described by three nodes; 10, 11, 12.
Nodes 10 and 11 represent the exterior finish and sheathing,
while node 12 accounts for the studs, insulation, and interior
finish. It should be noted that this model does not account for
interior partitions.

Storage floor. The storage floor consists of a hardwood
floor (node 2), a poured slab (nodes 3, 4), the concrete block
(nodes 5, 6, 7, 8), and styrofoam ground insulation (node 9).
Node 14 represents the earth, which is assumed to have zero
thermal capacity.

Non-storage floor. This component consists of a rug, poured
slab, and ground insulation (nodes 17, 18). The insulation and
rug are assumed to have zero thermal capacity.

Indoor air. The indoor air is represented by node number
1, and does possess thermal capacity. A linear temperature
gradient with height is assumed to exist in the room. Studies
performed by Schutrum, et al [1] and the experience of the
author with the house described previously indicate that this
temperature gradient is approximately .8-1. degree Farenheit per
foot.

3.2   Energy Transfer Within the House

Figure 3 also indicates the relevant energy transfers within
the house. Both the storage and non-storage floor exchange energy
through convection with the air and radiation to the walls/roof.
There is no radiant energy exchange between them because they
lie in the same plane, and therefore do not "see" each other.
Energy is conducted through the non-storage floor to the earth
and through the walls/roof to the outside air. There is also
conduction heat transfer through the storage floor section.
Infiltration losses as well as conduction losses through the
windows are also considered. Thermal window shades cover the
windows during nighttime hours. Thermal capacity of the windows
is neglected. Supplementary heat and internally generated energy
are assumed to be convected to the air. All conduction is assumed
to be one-dimensional.

The solar gain is assumed to be absorbed by the walls/roof,
storage floor, and non-storage floor. It is assumed that none of
the solar gain is absorbed by the air. However, the walls/roof
and floors convect a portion of the solar gain back to the air,
the temperature of which, as stated earlier, is assumed to vary
linearly with height. If the air temperature in the ceiling
peak is sufficiently high the storage circuit fan is operated and
energy is convected from the air to the storage, thus charging

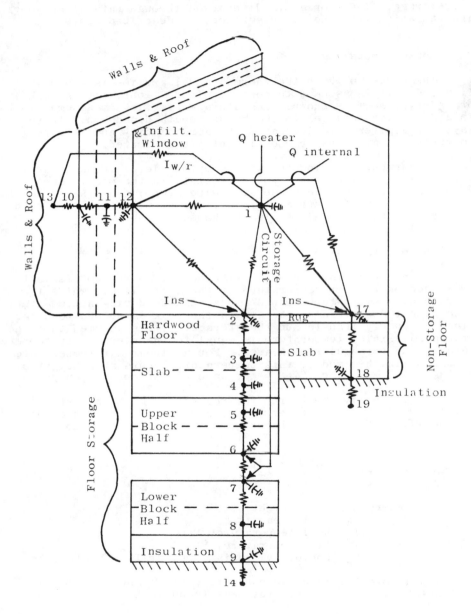

Fig. 3 - Passive Solar Storage Thermal Model

the storage. The storage is discharged through natural convection and radiation from the floor surface, as described above.

3.3 House Dimensions

The house to which the thermal model applies is shown in Figure 4. It is assumed to consist of one floor, with a peak or "A"-type roof as shown. The storage air flow is assumed to be in a direction parallel to the house length dimension. In this way, for a given air flow rate, the air velocity and resulting convective heat transfer coefficient are maximized.

A thermostat is located at the 5 foot level in the house. Mathematically, there are two thermostats in the model, one maintaining the desired house air temperature, the other used as a controller for the storage circuit. Presently these are both located at the 5-foot level in the house.

3.4 Thermal Parameters

In the analysis, the convective heat transfer coefficients are based on the air temperature at the 5 ft. level, while the average air temperature change is calculated during a time step. Also, the temperature of the air in the peak is needed to determine the energy transfer to the storage. It is assumed that the storage air inlet temperature is equal to the average air temperature in the triangular peak area. For a linear air temperature gradient, in this model, this temperature occurs at a height of c plus $1/3d$ (See Fig. 4 for an explanation of c and d).

The convective heat transfer resistance between vertical walls and air has been determined experimentally by Min et al. [2] in the ASHVE Environment Laboratory. This was found to be

$$R_W = 3.448 \frac{H^{.05}}{A_W} [T_w - T_a]^{-.32} \tag{1}$$

where H Wall height, ft.
      A_W Wall area, ft.^{2}
      T_w Wall temperature, ^oF
      T_a Air temperature, ^oF
      R_W Convective resistance, hr. ft.^{2} ^oF/BTU

Min also determined the convective resistance between horizontal floors and air. This was found to be

$$R_f = 2.564 \frac{D_e^{.08}}{A_f} [T_f - T_a]^{-.31} \quad \text{(Floor warmer than air)} \tag{2}$$

$$R_f = 24.39 \frac{D_e^{.25}}{A_f} [T_f - T_a]^{-.25} \quad \text{(Floor cooler than air)} \tag{3}$$

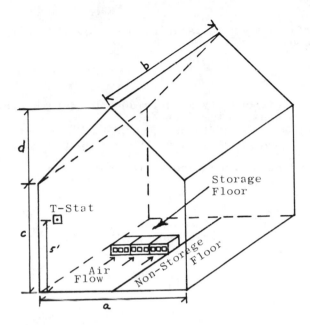

Fig. 4 - House Dimensions

where D_e Floor hydraulic diameter, ft.
      A_f Floor area, ft.^{2}
      T_f Floor temperature,$^{\circ}$F

The resistance is smaller when the floor is warmer than the air due to the fact that the heated air near the floor will naturally rise to the ceiling of the room.

The effective resistance for radiative heat transfer is

$$R_r = \frac{1}{h_r A_f} \tag{4}$$

where

$$h_r = \frac{4\sigma T_{avg}^3}{\frac{1}{F_{fw}} + \left[\frac{1}{e_f} - 1\right] + \frac{A_f}{A_w}\left[\frac{1}{e_w} - 1\right]} \tag{5}$$

σ Stefan Boltzman Constant
A_f Floor Area (Storage or Non-Storage), ft.^{2}
e_f emissivity of floor
e_w emissivity of walls
A_w Wall Area, ft.^{2}
F_{fw} Floor-Wall View Factor

T_{avg} Avg. of Wall and Floor Temperatures, $^{\circ}$R

For the case under consideration, $e_f = e_w = .9$, $F_{fw} = 1$, for both the storage and non-storage floors.

The infiltration resistance, based on the outside-inside air temperature difference is

$$R_i = \frac{1}{\rho V c p \; \#AC} \tag{6}$$

where: ρ air density at outside temperature, $lbm/ft.^3$
         V   house volume, $ft.^3$
         cp   air specific heat, $BTU/lbm-^\circ F$
         #AC   number of air changes per hour

The conduction resistances are found to be

$$R = \frac{1}{kA} \tag{7}$$

where: l   distance between nodes, ft.
         k   conductivity of node, $BTU/hr.ft.^\circ F$
         A   area of node, $ft.^2$

The thermal capacity of a node is

$$c = \rho x A c_v \tag{8}$$

where

     x   node depth, ft.
     p   node density, $lbm/ft.^3$
     c_v specific heat of node, $BTU/lbm^\circ F$

The concrete block was modelled as two slabs, each slab consisting of one-half of the block mass; a six inch block is shown in Figure 5. The leakage resistance is assumed to be

$$R_L = \frac{L}{kr_1 A_f} \tag{9}$$

where: L   leakage length, ft.
         R   block conductivity, $BTU/hr.-ft.^\circ F$
         A_f storage floor area, $ft.^2$
         r_1 ratio of web area to block area

The storage convective heat transfer coefficient was calculated, using the air flow velocity, from

$$h = \frac{k_a}{d_h} \left(1023 \; Re_{d_h} \right)^{.8} Pr^{0.4} \tag{10}$$

where: k_a air conductivity, $BTU/hr.ft.^\circ F$
         d_h hydraulic diameter of flow passage, ft.
         Re_{d_h} Reynolds number
         Pr   Prandtl number

The heat transfer area is equal to the flow passage surface area. The heat transfer to storage (when the storage circuit is on) is calculated to be:

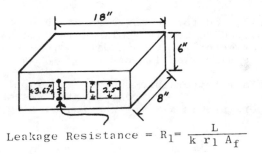

$$\text{Leakage Resistance} = R_1 = \frac{L}{k\, r_1\, A_f}$$

Fig. 5 - Concrete Block Model

$$\dot{Q}_{u\,(L)} = \frac{1}{2} m_a cp(T_1 - T_s) \left[1 - \exp\left(\frac{-2hA_s}{m_a cp}\right)\right]$$

$$+ hA_s\left(T_s - T_{u\,(L)}\right)$$

where $\dot{Q}_{u\,(L)}$ = heat transfer to upper (lower) half concrete block, BTU/hr.

m_a air mass flow rate, lbm/hr.

cp air specific heat BTU/lbm^{o}F

h convective coefficient, BTU

A_s surface area for half-block

T_s Average of upper and lower half-block temperatures

$T_{u\,(L)}$ temperature of upper(lower) half-block

The earth resistance is assumed to be

$$R_e = \frac{\Pi a}{4 k_e A_f}$$

where   a   width of house ft.

k_e earth conductivity, BTU/hr.ft.^{o}F

A_f floor area, ft.^{2}

This assumes that the conduction path is a semicircle centered at the intersection of the slab edge and earth, and that the radius of the semicircle is equal to one quarter of the width of the house. This method is suggested by ASHRAE. The temperature difference to be used is that between the outside air and the insulation - earth interface, for both the storage and non-storage floors.

The outside wind heat transfer coefficient is assumed to be equal to 4 BTU/hr.ft.^{2} ^{o}F, for a wind velocity of 7.5 mph.

3.5   Solar Gains and Outdoor Air Temperature

The solar gain to the house is assumed to be equal to that through a vertical south-facing window. Ninety percent of the gain is assumed to remain in the house, while the remainder is

assumed to be reflected to the environment through the windows.
The model is designed to receive solar gain data externally, or
if desired, it can generate its own data. This was done for the
example case to be discussed later.

The outdoor air temperature can also be supplied as an external input to the model. The current version assumes that the
outside air temperature is a ramp function of time, as shown in
Figure 6.

3.6   Computing Procedure

The computer model calculates the node temperatures and the
relevant energy transfers after the node capacitances and resistances are determined, using the following procedure:

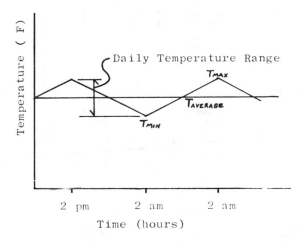

Fig. 6 - Outdoor Air Temperatuer vs
         Time

1)   If the air temperature at the thermostat level is
greater than the storage set point, and if the ceiling peak air
temperature is sufficiently large to cause a transfer of energy
into the upper concrete block half, equation (11) is used to
calculate the heat transfer rates into the upper and lower block
during the upcoming time step. Otherwise, the heat transfer to
storage is set equal to zero for the time step.

2)   The node equations are then solved, using the Euler
method, to determine the node temperatures at the end of the
time step.

3)   If the air temperature at thermostat level is less than
that desired, the amount of supplemental heat required to bring
the air temperature to the desired value is determined. If the
air temperature is greater than or equal to that desired, the
supplemental heat is zero for the time step.

4)   The relevant heat transfers are summed for the time step, and the cycle is repeated.

4.   SAMPLE SIMULATION

Figure 7 shows the performance of a typical floor storage system in Schenectady, New York, for a clear day in March.   The indoor and outdoor air temperatures (indoor at thermostat level), solar gain, and supplemental heat input are plotted as a function of time for a 24 hour period.   The following assumptions were made in the determination of the system performance:

```
Area weighted wall/roof=R-27
South-facing window area=310 ft.²
Total window area=485 ft.²
Double-glazed windows
Night window shade=R-10
Infiltration Rate=1/2 air change per hour
Storage Air Flow Rate=900 CFM
Storage Floor Area=450 ft.² (25% of total)
House Dimensions
        Length=60 ft.
        Width=30 ft.
        1st Floor Height=8 ft.
        Peak Ceiling Height=6 ft.
Indoor Air Vertical Temperature Gradient=1°F/ft.
```

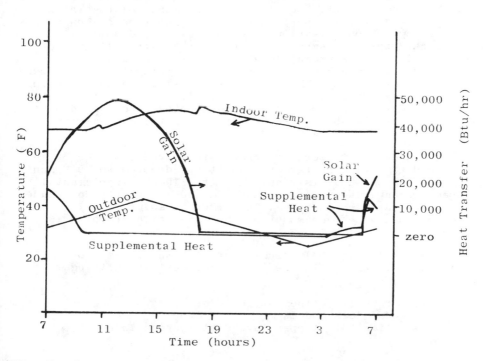

Fig. 7 - Passive System Performance, Schenectady, New York -
        March Clear Sky

Desired House Air Temperature=68°F
5' Air Temperature at which storage circuit is operated=
     70°F
Average Daily Outdoor Temperature=33.4°F
Daily Temperature Range=18.4°F
Storage Floor Construction
        3/8" Hardwood Floor (parquet)
        4" Concrete Slab
        6" Concrete Block
        R-11 Insulation
        6 Mil Poly
Non-Storage Floor Construction
        Rug
        4" Concrete Slab
        R-11 Insulation
        6 Mil Poly
Horizontal Daily Solar Insolation=1715 BTU/ft.^{2}-day
                (clear skies, March)

The solar gain to the house was computed assuming that the diffuse radiation was equal to 18% of the total daily horizontal radiation, as suggested by clear sky daily data of Liu and Jordan [3]. The ground area directly south of the window was assumed to possess a reflectivity of .7 (snow cover).

Returning to Figure 7, it can be seen that from 7:00 to 9:50, the house air temperature is 68°F, hence the non-zero supplemental heat input. Due to the increasing solar gain and outdoor temperature, the supplemental heat input decreases with time to zero at 9:50, at which time the integrated solar gain has increased the air temperature above 68°F. The slight drop in indoor air temperature at 10:45 is due to the fact that the storage floor was charged slightly, thus dropping the air temperature (Note that the air temperature at the start of the charging of storage was 70°F). The local peak in indoor air temperature at 18:00 is due to the lowering of the evening window shades at that time (sunset), which resulted in a dramatic decrease in house load.

The indoor air temperature drops steadily during the evening and nighttime hours; at 3:00 it has dropped to 68°F and again supplemental heat is required. At 6:10 the supplemental heat requirement rises sharply, because the evening window shades are withdrawn as the sun rises. At 6:20 the solar gain is sufficient to reduce the amount of supplemental heat which is required.

These results are only consistant with the assumptions outlined above. We are in the process of further refining the model.

Initial areas of refinement include:
        ·ground loss model
        ·vertical air temperature distribution in house

REFERENCES

1. Schutrum, L.F., Parmelee, G.U., and Humphreys, C.M.
    "Heat Exchanges in a Ceiling Panel-Heated Room", ASHVE
    Research Report #1473, ASHVE Trans., Vol. 59, 1953, p. 197.

2. Min, T.C., Schutrum, L.F., Parmelee, G.U., and Vouris, J.D.
    "Natural Convection and Radiation in a Panel-Heated Room",
    ASHVE Research Report #1576. ASHVE Trans., Vol. 62, 1956,
    p. 337.

3. Liu, B.Y.H., and Jordan, R.C., Solar Energy, 4, No. 3, (1960)
    "The Interrelationship and Characteristic Distribution of
    Direct, Diffuse, and Total Solar Radiation".

Membrane-Lined Thermal Storage Systems

R.C. BOURNE
University of Nebraska-Lincoln
Lincoln, Nebraska 68588, USA

ABSTRACT

Background and advantages of membrane-lined liquid thermal storage are presented. Surveys of existing projects and potential liner materials are reviewed. Alternate structural enclosure designs are discussed for basement, crawl space, and slab-on-grade foundation types, and optimal-cost designs are identified. Improved heat transfer methods are investigated. Design and laboratory test results are presented for several concepts providing forced air heating via an air jacket around the storage surface. Test results and cost-effectiveness studies are presented for 4 types of free-convection domestic water preheaters immersed in the storage container. Preferred storage inlet/outlet designs, and techniques for auxiliary heating of storage, are also discussed.

INTRODUCTION

Membrane-lined liquid thermal storage containers promise lower costs and easier installation than conventional steel or fiberglass tanks. A project to investigate and further develop membrane storage systems was begun in May, 1978, at the University of Nebraska-Lincoln, with funding from the Research and Development Branch, Office of Conservation and Solar Applications, United States Department of Energy.

Steel tanks are currently the most common storage containers for solar heated water. The membrane-lined storage alternative being investigated and further developed in this project has the following potential advantages compared to steel tanks:

<u>Reduced Corrosion Problems</u> Without a steel storage tank, no dissimilar metals need be used in liquid based solar heating systems.

<u>More Convenient Construction Scheduling</u> The storage tank components easily fit through doorways, and may be installed at any time in the construction process, with excellent applicability to "retrofit" projects.

Improved Heat Transfer Configurations Easy access to the tank
interior permits location of free-convection heat exchangers
within the container, eliminating pumps, external heat exchangers,
and some parasitic power consumption. Some membrane-lined designs
facilitate forced air heating from the storage surface area.

Cost Reduction For residential solar space heating systems,
membrane-lined storage containers may generally be completed for
less than one-third the cost of steel tanks. The savings result
partially from integration of the storage container into the
building foundation. For example, the storage structure may be
constructed in a basement corner, where only two new walls need
be added.

The project is designed to assess the current state of development of membrane-lined storage, develop optimized enclosure designs,
and develop designs which improve storage heat transfer and temperature stratification. These objectives are being pursued via
the following three project activity lines:

Activity 1 includes a survey of completed membrane-lined storage
projects, a review of alternative membrane materials, and identification of the most cost-effective membrane enclosures for a
range of building foundation types, for storage volumes of 250,
1500, and 5000 gallons.

Activity 2 involves development work aimed at improving heat
transfer to and from storage, including:

2a) Techniques for direct heat transfer from the storage sur-
     face to forced air, with laboratory testing of several
     promising design concepts and cost-effectiveness studies
     based on test results.

2b) Techniques and details for water inlet to and outlet from
     membrane-lined storage containers.

2c) Comparison of alternative free-convection heat exchangers
     for pre-heating of domestic water, with laboratory testing
     of the preferred designs and cost-effectiveness studies
     based on the test results.

2d) Concepts for input of auxiliary heat to storage, for
     application where utilities offer rate incentives for off-
     peak energy consumption.

Activity 3 involves designs which promote storage temperature
stratification to improve overall system performance, including
multiple and baffled tank designs, development of cost models,
and small-scale testing of promising designs.

     Results of Activities 1 and 2 are discussed in this paper.
Activity 3 will not be completed until Dec. 31, 1979, and
results will be reported at a later time.

Activity 1A SURVEY OF COMPLETED PROJECTS

The purpose of the completed project survey was to assess the current state of development of membrane-lined solar thermal storage. A multiple-choice survey form was prepared to solicit information on liner, structure, and insulation materials, heat exchange techniques, problem areas, and costs.

Agencies, firms, and publications provided sources for information on completed membrane-lined projects. Ultimately, consulting architects and engineers, identified from magazine articles and lists of completed solar buildings, became the primary information source. 22 firms or individuals were provided with survey forms by the end of October, 1978.

By early February 1979, several "follow up" telephone campaigns resulted in returned survey forms describing 61 completed projects. Of these, 35 use a "storage kit" produced by Acorn Structures of Concord, Mass. Some of the most significant results are summarized below:

. Only one manufactured membrane-lined storage system (Acorn) is now in widespread use.

. The earliest project reported was completed in 1975.

. 85% of the installations are residential; 97% provide space heating, 90% provide space and domestic water heating.

. 74% are located in basements, but 46% of non-Acorn systems are buried outside.

. Excluding Acorn (cylindrical), 88% of the installations are square or rectangular in plan.

. While most installations have insulated concrete floors, 29% of non-Acorn projects are placed on tamped sand or gravel. Polystyrene and urethane are equally popular as floor insulation materials, with thickness varying from 2 to 6 inches.

. Excluding Acorn (plywood and steel), most wall construction is of concrete (either poured or block), using polystyrene or urethane insulation.

. PVC (vinyl) is the most common lining material (77%), but 35% of non-Acorn projects use Hypalon. 93% of projects use a material 20 mils or more in thickness.

. 82% use a liner preseamed to shape, but 35% of non-Acorn projects use a flat sheet with folded corners.

. Only 4% are coupled to antifreeze collectors; all but one installation provide domestic water preheating via a free convection loop in storage.

. 77% contain 1000-2500 gallons, but 27% of non-Acorn projects
  contain 10,000 gallons or more.

. Excluding Acorn ($.60/gal), 35% had installed costs below
  $.60/gal, 31% were above $1/gal, including piping in
  storage.

. 13% reported storage problems, including leaks (4),
  material flaws (2), insulation(1) and structural (1).

In general, membrane-lined storage systems have performed
well compared to the more conventional alternatives. In a
recently published survey of 154 solar heating installations (1),
14% reported storage system problems, about the same frequency as
reported here.

Activity 1B MEMBRANE MATERIAL SURVEY

The goal of the membrane material survey was to identify
the most cost-effective liner materials for expected service
conditions in solar storage configurations developed in other
project activities. No laboratory testing was planned, from the
expectation that complete engineering data would be available
from manufacturers and fabricators of candidate membrane materials.

The following tasks comprised the membrane material survey:

(1)   Familiarization with the synthetic membrane industry
      in terms of material categories, production methods,
      industry organization, and comparable applications.

(2)   Identification of desirable membrane material prop-
      erties and characteristics.

(3)   Identification and description of candidate membrane
      materials, including combinations and composites.

(4)   Study of the economic implications of limited membrane
      life.

(5)   Recommendation of preferred materials for the range
      of storage system designs described in other project
      activities.

Both thermoplastic and elastomeric membranes are available.
Thermoplastic materials soften and melt at high temperatures,
and will "cold flow" when stressed. The elastomers are synthetic
rubber materials, and are usually "vulcanized" to achieve thermo-
setting characteristics. They have higher service temperatures
and do not cold flow, remaining "elastic" under load.

The elastomers are generally more expensive than thermoplastics, and are not available in sheets less than .020" thick.

Synthetic membranes are produced by either calendering or blow extrusion. Calendering is a hot rolling process, in which sheets 4 - 8' wide are produced. Greater widths are difficult because of the high pressures involved and the thickness variations which result. All the elastomers and most of the thicker thermoplastics are formed into sheets by calendering.

In the blow extrusion process, a hot plastic tube extruded through a die is inflated with high pressure air and "blown" into a large cylinder, which may be kept in its closed form (for products like plastic bags) or slit to form sheets as wide as 40'. Polyethylene sheets produced by this process are widely available through hardware and building materials stores in thicknesses up to .008" (8 mils).

Basic materials research and development, and chemicals production, are carried out by chemical firms such as DuPont, Dow, Monsanto, and several of the major oil companies. Membranes are usually produced by manufacturers who buy chemicals from the major firms, and use a variety of additives, filler materials, and plasticizers to generate the desired end product. (Some chemical firms do produce sheet products as well as chemical ingredients.) The chemical firms often provide technical expertise to assist in producing the best material for the application. Wide sheets and/or finished goods are produced by fabricators, and are seamed together from calendered rolls purchased from manufacturers.

No application for membrane materials directly comparable to solar thermal storage linings was identified. The three most similar applications are swimming pools, pond liners, and water beds. Virtually all water bed mattresses and swimming pool liners are made of 20 mil vinyl (PVC), with typical material life of 3 - 5 years for water beds and 7 - 10 years for pools. Failure is usually due to embrittlement resulting from leachout of the plasticizer (added originally to make the material flexible). Both heat and sunlight accelerate the leachout rate. Planned obsolescence may be a factor in both of these applications. A variety of pond liner membranes are used, usually 20 mils or thicker. Some are fabric reinforced, to aid in puncture resistance.

Many ASTM test methods are available for rating the physical properties of plastic sheets and films. With regard to strength properties, the relevance of the various methods depends on the quality of the support structure for the membrane. If the structure is well made, and caulked or filled where necessary to ensure that the membrane cannot be forced into a hole or crevice, very little membrane strength is required. Elongation (the ability of the material to stretch if necessary to reach the support structure) is important, as is puncture resistance. Aging resistance under high temperature "wet" conditions is obviously necessary to achieve the desired 20 year minimum life. Other desirable membrane material characteristics are:

. low water vapor permeability

. simple and reliable field patchability

. flexibility

. light weight

. low cost

A list of candidate materials was compiled through contacts with industry experts. For each material considered, a survey form was completed which covers (where available), properties and performance ratings for the desired membrane characteristics. Brief summaries of candidate materials are provided below.

a. Thermoplastics

PVC (polyvinyl chloride) is now most frequently used due to high strength, low cost, ease of patching and sealing; does age (and more rapidly at high temperature), but 20 year life for temperatures up to 150°F may be possible. High water vapor permeability is a potential problem.

CPE (chlorinated polyethylene) is more expensive than PVC, and difficult to produce thinner than 30 mils. Has low water vapor permeability, and does not have plasticizer, hence it does not embrittle as much as PVC with age. CPE can be solvent patched, but is weaker and more subject to material flaws than PVC. Produced only by calendering.

PE (polyethylene) is widely available in 2 - 8 mil thickness, but often has pinhole flaws. Now also available "cross-laminated" (inseparable layers) in a range of densities. Low in cost and aging (up to 150°F), available in wide sheets (blow extruded), and has low water vapor permeability. May require heat patching.

3110 (elasticized polyolefin) has characteristics similar to PE, but is more flexible and resistant to UV degradation. Blow extruded in 20 mil thickness, now used for pond liners. 158°F maximum recommended temperature, and not recommended for vertical wall applications.

Surlyn (DuPont tradename -a modified polyethylene) has good elongation characteristics, difficult to puncture. Used for product wrapping. Not currently available in wide sheets. Lower heat resistance than PE, and more expensive.

EVA (ethylene-vinyl acetate) copolymers - are softer and stretchier than PE, but more expensive, with lower heat resistance. More frequently used for film coating than for membrane sheets.

b. Elastomers

EPDM (ethylene propylene dieno monomer, a hydrocarbon rubber) -
retains desirable characteristics to +300°F (now being marketed as
a solar absorber material). Least expensive of the elastomers,
low moisture permeability. Current minimum thickness 30 mils.
Like other elastomers, must be adhesive or "hot melt" seamed.

Neoprene - lower temperature limit than EPDM, but better resistance to acids, oils, and solvents (not particularly valuable in
this application). More expensive than EPDM.

Hypalon (chlorosulfonated polyethylene) better ozone and sunlight resistance than Neoprene - and lower moisture permeability
(but not as low as EPDM). More expensive than EPDM. Field cured,
hence easier to seam. Not recommended above 158°F.

Butyl Rubber - lowest moisture permeability of the elastomers.
Good strength and elongation. More expensive than EPDM, without
performance advantages.

A study of the economics of liner replacement was carried
out based on a replacement cost estimate of $300 (current) for the
baseline 1500 gallon storage design. Most of the replacement cost
is labor. The results indicate the wisdom of investing up to
$1.17/ft^2 of material (vs. PVC at $0.25/ft^2) to double liner life,
if necessary to reach the desired 20 yr system life. However, it
is not certain that any currently available membrane material, at
any price, will regularly last 20 years at elevated temperatures.

Fabric - supported membranes have not been considered as
candidates because they are stiffer, do not elongate, may separate
at high temperatures, and are considerably more expensive than
simple membranes. Among the materials considered, EPDM, PVC and
cross - laminated, high density PE are considered to have the
greatest potential for achieving the desired life under expected
service conditions. Surprisingly, the three vary widely in cost.
The PE, which may be reliably produced in 3-4 mil thickness, can
be purchased for less than $.05/ft^2 in widths to 40'. 20 mil PVC,
in 20' widths, will cost $0.15-20/ft^2, while 20' wide EPDM runs
$40-0.50/ft^2.

Each of the three has its own peculiar question mark(s). The
EPDM material will almost certainly last 20 years containing water
at 212°F, but the seams are an uncertainty at that temperature.
Cross-laminated, high density polyethylene may be capable of withstanding prolonged high temperatures, but it is the most vulnerable
of the four to field damage. PVC seams and patches most easily
and reliably of the three, but it does degrade with time, at a rate
which accelerates with temperature. And, a 1500 gallon container
could lose up to 50 gallons/year by water vapor transmission
through a 20 mil PVC membrane, causing increased heat loss and a
need for regular replacement.

Experimental work is recommended to resolve the remaining
question marks for the three preferred candidate materials.

Activity 1C ENCLOSURE ALTERNATIVES

The purpose of activity 1C was to evaluate alternative membrane support structures for a range of storage sizes and locations. The 1500 gallon container size discussed in this paper represents a range appropriate for many residential combined solar space and domestic water heating systems. Designs are evaluated for installation with basement, slab-on-grade, and crawl space foundation types. The value of space occupied by storage is considered to be zero for storage location in a crawl space or beneath occupied floor levels. When storage occupies space above or in a "walkout" type basement which is otherwise entirely finished, the full cost of enclosing the space, including floor, walls, and ceiling or roof, must be charged to the storage system. A $12.75/ft² "cost of space" charge was used in the studies for above-ground, indoor storage.

Various degrees of prefabrication may be considered for a thermal storage system. For the greatest degree of prefabrication, a three dimensional storage container would be delivered and placed, permitting very little structural integration with the building in which it is located. For storage sizes typical for space heating, this approach requires container placement before the building enclosure is completed, since the container would be too large to move through doorways or down stairs.

A more versatile approach uses partial prefabrication, based on a kit of components designed for rapid installation in the field. All of the enclosure types evaluated here may appropriately be marketed as kits, although some (masonry and wood frame, in particular) require more onsite construction than others.

The four basic enclosure configurations considered are rectangular, cylindrical, hopper, and quadrant. For each configuration, an optimum set of dimensions may be identified at which the enclosure cost is minimized. Since horizontal and vertical structural component unit costs differ, the optimum dimensions are not typically those resulting in minimum surface area.

Fig. 1 shows isometric views of the four basic configurational alternatives. The annual report (2) discusses materials and construction details.

Rectangular enclosure designs were evaluated for masonry, wood, and "sandwich" wall structures. For basement applications, a corner location was assumed to take advantage of existing walls. Added walls may easily be finished to become interior surfaces for adjacent spaces. As in all configurations, the floor membrane rests on rigid foam insulation. Fiberglass blanket insulation rests on the top membrane which floats on the water surface.

The major advantages of the cylindrical enclosure configurations are structural efficiency and light weight. Cylinders are particularly appropriate for retrofit installations, where the quality of existing basement walls may be low, discouraging rectangular corner storage locations. Since only tensile forces are

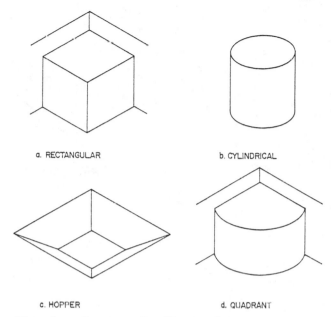

a. RECTANGULAR   b. CYLINDRICAL

c. HOPPER   d. QUADRANT

Fig. 1 - Storage Configuration

exerted on the cylinder walls, a thin sheet metal cylinder may be
used without requiring a structural frame. The cylindrical tank
designs were not evaluated for crawl space and buried slab-on-
grade applications, since they cannot withstand earth backfill
against their side walls.

The quadrant configuration takes advantage both of existing
structure, with its corner location, and of efficient tension-
stressed sheet metal walls for the remaining structural enclosure.
A rectangular framework fastened at the floor and ceiling eliminates any reliance on structural connections to the existing
walls, and permits rapid installation.

In the hopper configuration, sloping side walls are used to
eliminate the need for wall structure. Because it must rely on
earth for support, the hopper is considered for "below slab"
applications only. For "slab-on-grade" buildings, a floor must
be constructed above the hopper to avoid the "cost-of-space"
penalty. Even for basement applications, a floor is necessary
for protection.

Cost models were developed for each of the configurational
alternatives. Material and labor costs for early 1979 were used
and cost vs depth was evaluated for the 1500 gallon containers.
The cost studies include the complete insulated container,
installed, without heat exchangers or internal piping. A 20 mil
PVC liner was assumed.

Figs. 2, 3, and 4 show results for the basement, slab-on-
grade, and crawl space foundation types, respectively.

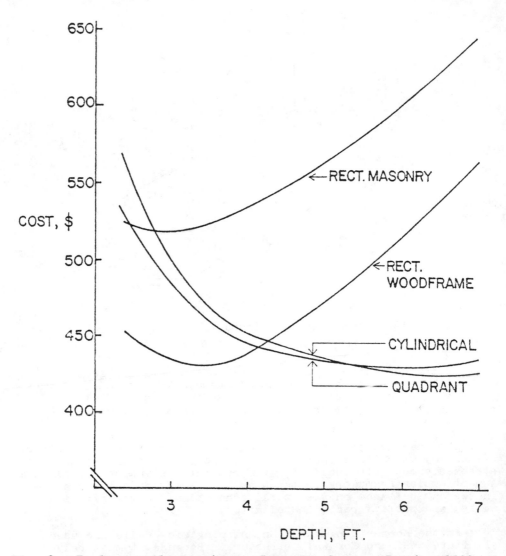

Fig. 2 - Enclosure Alternatives - Basement Cost vs Depth - 1500
        Gallon Container

    For the basement location, the rectangular types are square
in plan, with depths less than half the square dimension, in the
optimal configuration. The cylindrical and quadrant types have
optimal depths near 6' due to their higher ratio of horizontal
element to vertical element costs. The lower profile rectangular
designs require slightly more floor area, but allow easier access
for installation and service, and space above the units may be
used for storage, if support racks are constructed.

    At their optimal depths, three of the alternatives are
projected to be within a 2% range for installed cost, at less
than $.30/gallon. The rectangular masonry alternative is approx-
imately 20% higher, and is not as well insulated.

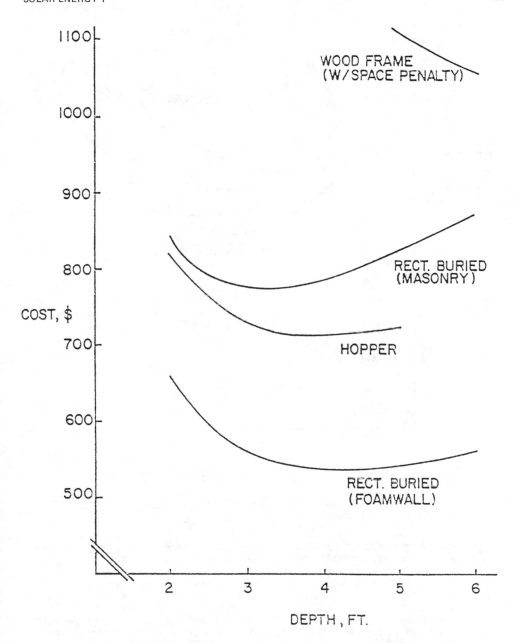

Fig. 3 - Enclosure Alternatives - Slab on Grade Cost vs Depth -
1500 Gallon Container

For the slab-on-grade foundation type, four alternatives
are shown, of which only the woodframe design is located above
the floor. The "cost-of-space" penalty prevents the above-slab

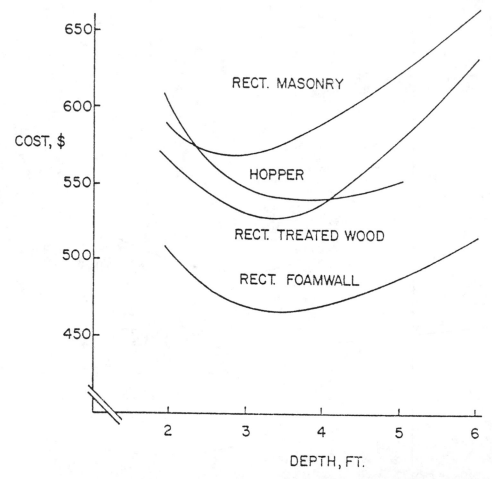

Fig. 4 - Enclosure Alternatives - Crawl Space Cost vs Depth - 1500
        Gallon Container

designs from competing with designs where storage is placed below
the floor. The "foamwall" design is experimental, using 8" polystyrene walls without masonry structural support. In addition
to its projected cost advantage, the foamwall enclosure has more
than twice the insulting value of the other below-slab designs.
The surprising cost advantage of the foamwall rectangular configuration compared to the hopper design is due partly to the larger
floor required above the hopper. The precision required in pre-
paring the sand bed and cutting the insulation boards also
increases projected hopper configuration costs.

        The crawl space systems are generally 10 - 15% more expensive
than the basement systems, due to excavating costs and the need
for an access door from above. However, they are 20 - 30% less
than the slab-on-grade enclosures, since no charge is necessary
for the floor above the storage container. Again, the foamwall
enclosure shows a significant cost advantage.

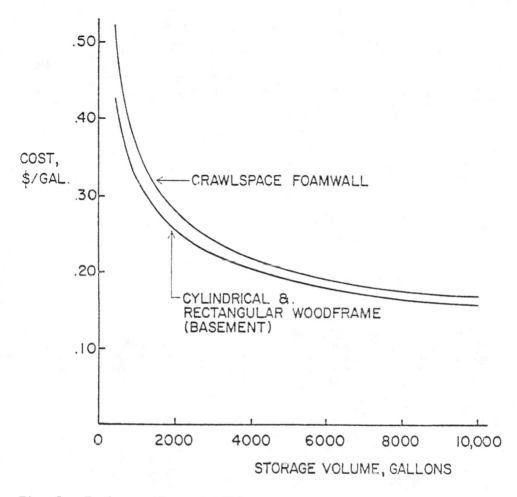

Fig. 5 - Enclosure Cost vs. Volume

The cost equations were also used to evaluate relationships between storage volume and cost. As storage volume is increased, the greater container structural loads were considered. Figure 5 shows cost vs volume projections for two of the most promising storage configurations. For the cylindrical design, maximum depth is 7', so that beyond 2000 gallons, non-optimal configurations are assumed. For the crawl space foamwall design, depth is held constant at 4' to permit backfilling against the 8" bead-board prior to filling the water container. As the curves show, the membrane-lined containers promise economics of scale which have not been realized with conventional steel and fiberglass storage containers.

Activity 2A   DIRECT AIR HEATING

The goal of Activity 2A was system simplification via incorporation of the load heat exchanger into the storage sub-

system, using the storage surface area for direct heating of
building return air. In the direct heating concept, return air
flows through a metal jacket in contact with the membrane liner
to withdraw heat from the storage water, and then proceeds to the
auxiliary heat unit, for additional heating if necessary. The
storage unit must be bypassed for summer cooling operation. The
heating coil, pump, piping, and pump energy consumption of the
conventional solar space heating system are eliminated with the
direct air heating approach.

A previous paper (3) describes performance requirements,
design alternatives, initial test results, and cost-effectiveness studies for several promising direct heating concepts. That
work is summarized as follows:

The evaluation strategy was to establish heat exchanger
performance requirements for a typical conventional system,
then design direct heat exchangers with equal performance, and
compare costs.

- Performance Requirements

A residence heat loss characteristic of 335 BTU/hr-^{o}F was
assumed (fairly typical for a 1200 ft^2 home with basement). For
this heating load, a recommended heat exchange coil characteristic
of 670 BTU/hr-^{o}F was assumed, where the temperature parameter is
the difference between storage water and return air. Maximum
allowable air flow pressure drip was assumed to be 0.20" water
at the assumed flow rate of 1200 cfm.

- Design Alternatives

Direct air heating concepts were evaluated for both rectangular and cylindrical enclosure designs. Of the four concepts considered, two were considered sufficiently promising to merit laboratory testing. These were the jacketed cylinder and the floating finned plates.

In the jacketed cylinder concept (Figure 6), a galvanized
steel cylinder supports the membrane liner, with a second metal
layer whose inwardly projecting fins contact the structural
cylinder. An annular air flow zone is thus defined, with heat
transfer improved by the finned design. The finned plate concept, shown in Figure 7, uses a metal plate with upward-projecting
fins. The assembly is floated on the top membrane, and has a
rigid, insulated top sheet to enclose air flow paths between fins.

- Cost Projections

Conventional system component installed costs were estimated
to be $714, and the pump energy requirement was estimated to
justify an additional $120 initial investment, for a total target
cost of $834. Estimated incremental costs for the direct heating
components (installed) were estimated to be $384 for the jacketed
cylinder and $563 for the finned plate. This initial analysis
suggested good market potential for both designs.

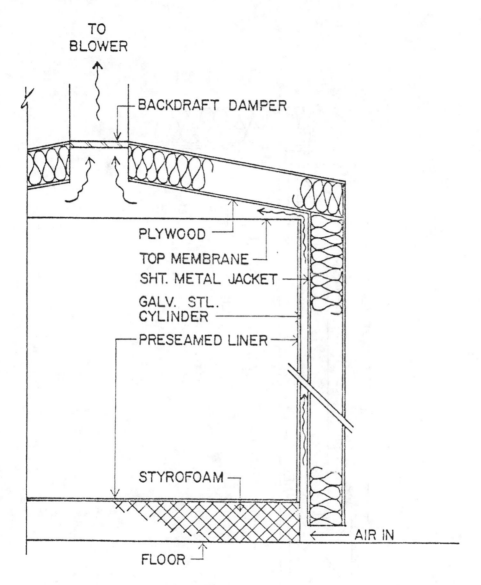

Fig. 6 - Section View - Jacketed Cylinder Concept

Since the previous paper, test installations have been constructed and preliminary testing completed for both jacketed cylinder and finned plate designs.

The test objectives were:

1. to assess the accuracy of the pressure drop and heat transfer calculations

2. to develop and evaluate assembly details and procedures.

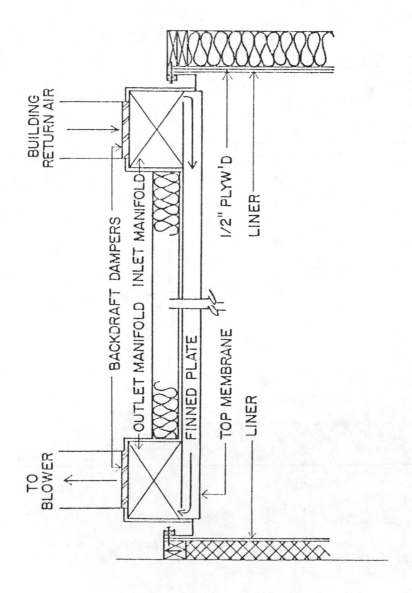

Fig. 7 - Finned Plate Concept

BUILDING RETURN AIR

BACKDRAFT DAMPERS

INLET MANIFOLD

OUTLET MANIFOLD

TO BLOWER

FINNED PLATE

TOP MEMBRANE

LINER

1/2" PLYW'D

LINER

Fig. 8 shows the jacketed cylinder test unit. The 320 gallon tank is 4' in diameter with 3'-9" water depth. The .019" aluminum jacket is formed with 5/8" fins spaced 4" apart. A top-mounted blower draws air into the annular flow zone around the entire lower circumference of the cylinder. Upon exiting the flow channels at the top, the air passes through a 3" deep plenum on the tank top surface before entering the blower.

a) Finned Aluminum Cylinder

b) Galvanized Steel Shell

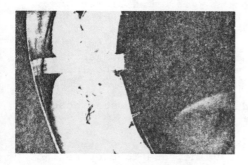

c) Wall Section and Support

d) Blower Inlet

Fig. 8 - Cylinder Test Unit

Thermocouple probes were placed at strategic locations within the water, in the airflow zones, and in the surrounding space. Static pressures were measured using an oil-type manometer, at 5 locations around the circumference, at top and bottom of the air flow zones. Air flow rate was measured using a pitot tube in the duct downstream from the blower.

For the tests, tank water temperature was raised to 110°F, and a water pump was operated to maintain a fully mixed condition in the tank.

Pressure drop and heat transfer measurements were taken at two flow rates for comparison to predicted results. Figure 9 shows the predicted and actual performance curves, with the latter extended from the two measured values. The pressure drop comparison is for the vertical wall portion only, where the test results are approximately 20% lower than predicted. However, the inlet pressure drop averaged 0.34" at 1000 CFM, indicating that redesign is necessary to eliminate turbulence at the lower entrance to the air flow zones.

The heat exchange characteristic was found to be approximately 80% of the calculated value, which translates to a U-value approximately 25% lower than predicted. Both reduced liner thickness and improved fin-to-cylinder contact might improve heat transfer. Without improvement, the performance objective could be achieved by reducing fin spacing to 2½". This would increase system cost to $418, 50% of the conventional system cost. Thus, the concept remains very attractive if inlet and outlet pressure drops can be reduced.

Procedures for the finned plate tests were similar to those for the jacketed cylinder. For the tests, a narrow section of the plate designed for the optimum rectangular enclosure was fabricated. Figure 10 shows the finned heat transfer plate design. The plate was fabricated from .019" aluminum sheet, with ten 3/4" wide air flow zones with 2" high fins. A 3/8" plywood top was used, and a small centrifugal blower was mounted in a housing at one end to draw air through the air passages. The blower was sized to develop flow velocities between 14 and 18 feet per second in the passages, as required in the full scale unit.

Pressure drop through the finned plate flow passages was approximately 50% of the predicted value. However, the inlet pressure drop again exceeded that through the straight passages. For full scale tests contemplated in the next project phase, improved inlet and blower housing designs should reduce those pressure drops to acceptable levels.

In the initial heat transfer tests, the measured average U-value was only 50% of the predicted value, due to poor contact between membrane and uneven bottom surface of the finned plate. Subsequent adjustments improved heat transfer to 85% of the predicted value. Based on these results, a selection chart (Fig. 11) was developed, which predicts that the performance objectives may be achieved using a 98# finned plate. This results

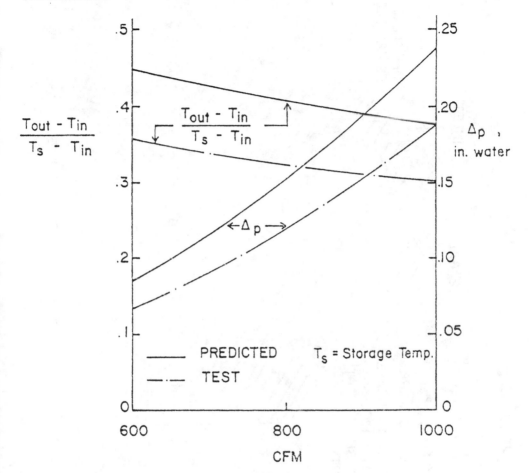

Fig. 9 - Cylindrical Unit - Predicted vs. Test Performance

in a $66 savings compared to the original estimate. The $497
component cost is 60% of the conventional system estimate.

Activity 2B INLET/OUTLET DESIGNS

        The goal of the water inlet/outlet studies was to identify
and design preferred techniques for extracting and returning water
to the membrane-lined storage container, and for penetrating the
membrane liner with heat exchange tubing. When the direct air
heating techniques of section 2a are not used, a maximum of six
tubing connections to storage are required:

        domestic water preheater inlet and outlet (2)

        collector loop suction and discharge (2)

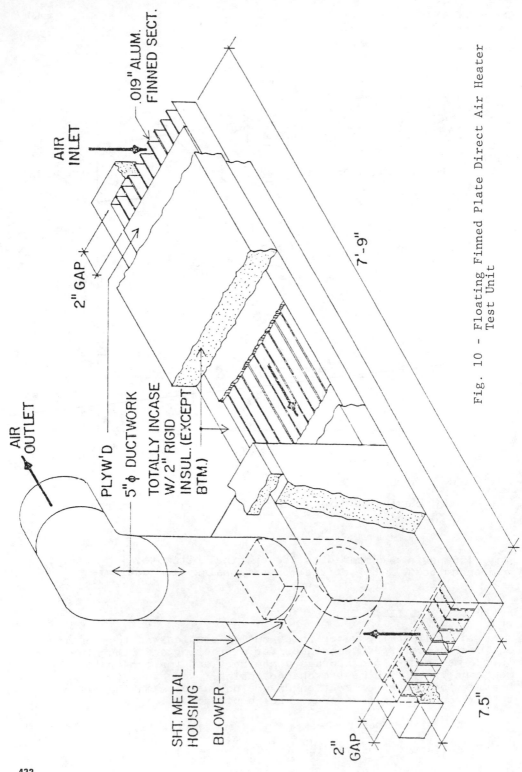

AIR INLET

.019" ALUM. FINNED SECT.

2" GAP

AIR OUTLET

PLYW'D

5" φ DUCTWORK

TOTALLY INCASE W/2" RIGID INSUL. (EXCEPT BTM.)

SHT. METAL HOUSING

BLOWER

2" GAP

7.5"

7'-9"

Fig. 10 – Floating Finned Plate Direct Air Heater
Test Unit

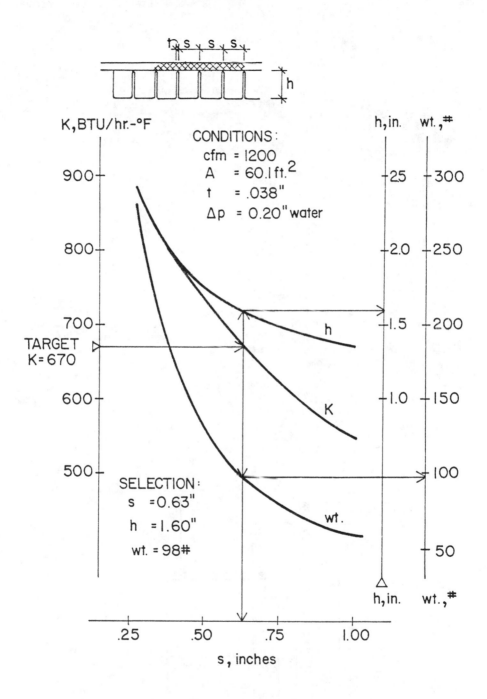

Fig. 11 - Finned Plate Selection Chart

space heating loop suction and discharge (2)

If an antifreeze liquid is circulated through the collection
loop, it must transfer heat to storage water via a heat exchanger.
For membrane-lined storage, the heat exchanger is preferably of
the free convection type, immersed in storage. Thus, the results
of project Activity 2C, which evaluates alternative storage-
immersed domestic water heaters, may also be applied to antifreeze-
to-storage heat exchange, although additional studies are required
to determine optimum heat exchanger size.

When storage water is used as the collection fluid, the
collection loop must, for most of the U.S., be of draining design
to prevent collector freeze damage. Since most collector-loop
circulating pumps are not of the selfpriming type, it is desirable
to locate the pump below water level to prevent air-lock of the
pump during collector drainage.

If piping connections are made through the side wall of the
storage vessel to a collector loop pump below the storage liquid
level, air-lock will not occur. However, side wall connections
through the membrane are not recommended. Fig. 12 shows four
potential solutions to the problem of preventing collector pump
air lock in draining systems with penetrations through the top
membrane only. Design (a) uses an accumulator (which may simply
by a long section of pipe) allowing air to escape upward before
being drawn into the pump. The pump must be placed well below
the water level, which requires additional piping if collectors
are roof-mounted. Design (b) uses a "foot valve" below the pump,
with a small, angled drainback line above the pump. This
configuration maintains water in the pump, (even though it is
above water level) while allowing collector drainage. The
angled inlet uses the "jet pump" principle to prevent short-
circuiting of the pump discharge flow through the drainage bypass.

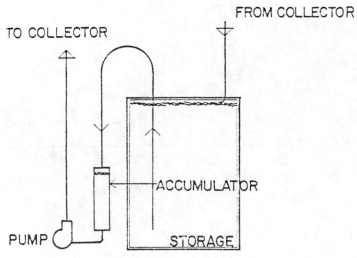

Fig. 12a - Preventing Collector Pump Air Lock - Accumulator

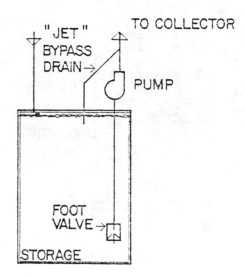

Fig. 12b - Preventing Collector Pump Air Lock - ByPass Drain &
          Foot Valve

      In design (c), the pump is located on the storage top membrane
and is forced down below water level after the connections are
made. At the tank top, where water pressure is low, the membrane
may easily be deformed to accomodate the pump. However, the
design is recommended for use only with prefabricated liners with
sleeves, or with elastomeric top membranes.

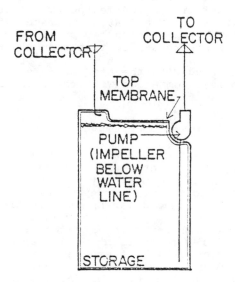

Fig. 12c - Preventing Collector Pump Air Lock - Floating Pump

In design (d), a submersible pump is used within the storage container. This approach eliminates any worry of air lock, but makes pump service difficult, and means that a watertight electric wire and motor are located in the storage water. All pump motor heat is transferred to the storage water.

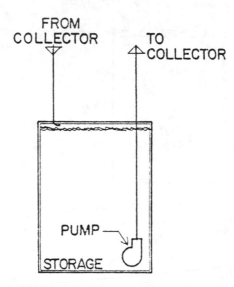

Fig. 12d - Preventing Collector Pump Air Lock - Submersible Pump

The preferred choice among these alternatives depends on the storage configuration and materials. Of the designs discussed in the enclosure studies, the rectangular and hopper configurations use folded-sheet liners, while the cylindrical and quadrant configurations use prefabricated membrane liners. Details of inlet/outlet and storage piping designs for these two configurational categories are presented in the annual report (2).

Concept (C) is used with the prefabricated liners, while a submersible pump is recommended with the folded membrane liners.

Activity 2C DOMESTIC WATER PREHEATERS

The purpose of the domestic water preheater is to increase water temperature prior to entering a conventional hot water heater, to reduce auxiliary fuel costs. Domestic water preheating may be accomplished in several ways. Figure 13a shows one common approach for combined solar space/domestic water heating. Stored solar heated water is pumped through a heat exchanger in a modified water heater, whenever domestic water flow occurs. Figure 13b shows a "free convection" preheater concept, in which supply water flows through a heat exchanger in the solar storage tank prior to entering a conventional hot water heater. This approach eliminates the pump, controller, and two tank connections,

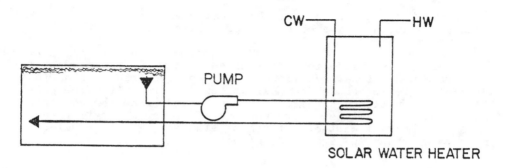

Fig. 13a - Domestic Water Preheating Alternatives - Conventional

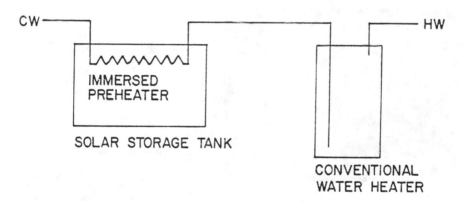

Fig. 13b - Domestic Water Preheating Alternatives - Free Convection

and requires no parasitic energy. It appears particularly viable for membrane-lined storage tank designs, where easy access may be provided to the tank interior, and where potential galvanic corrosion of tank walls is eliminated.

Many solar heating systems have been installed using the free-convection preheater concept. Various preheater types have been used including tanks, serpentine copper loops, soft copper spirals, and Rollbond panels. The objective of this project activity was to investigate the cost-effectiveness of alternative free-convection preheaters.

Four tank-type and ten tankless preheater design concepts were identified for consideration. The tank preheaters can perform efficiently with only modest surface area due to their storage capacity. During periods of sustained flow, however, the tank preheaters suffer considerable performance reduction. A demand simulation was developed to put designs of both types to a fair and equal test.

Preliminary evaluations determined that many of the preheater concepts could not be made to withstand the required combination of supply water pressure and maximum temperature. From the preliminary evaluations, three copper "tankless" preheater concepts and the steel tank-type design were selected for laboratory testing and subsequent cost-effectiveness studies. The four selected preheaters, shown in Figure 14, were:

1. steel tank

2. ½" type M copper tubing

3. copper Rollbond panels

4. 5/8" Wolverine WH Trufin (extruded copper fintube)

a) steel tank

b) 1/2" type "M" copper tubing

Fig. 14 - Domestic Water Preheaters

c) copper Rollbond panel

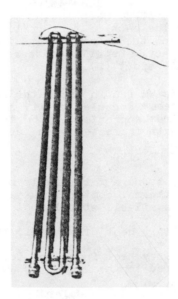

d) 5/8" Wolverine WH Trufin
    (copper fintube)

Fig. 14 - Domestic Water Preheaters

A description of the demand simulation may be found in the annual report (2). The computerized technique calculates an average weekly preheater efficiency using a load pattern averaging 95 gallons per day. Nine demand types are used, with both long and short load durations, and with both high and low flow rates.

Heat transfer coefficients for use in the cost-effectiveness studies were determined in laboratory tests. Sample preheaters were placed in a rectangular tank containing 1100 gallons of water heated to approximately 110°F. Supply water entered the preheaters at approximately 60°F. Storage, inlet, and exit temperatures were recorded via thermocouples connected to a multipoint potentimeter. Water flow rates were varied from approximately 1 -4.5 gpm and measured with a Hersey flow meter. The tank water heating flow was maintained during the tests to ensure mixing, and to minimize tank temperature drop during the test. Figure 15 shows the overall heat transfer coefficients determined from the tests, based on external preheater surface area.

The cost-effectiveness studies were designed to identify optimum preheater size for each type. Costs at optimum size could then be compared to select a preferred preheater type. Fig. 16 shows the results, based on 4¢/kwh electricity. The annual cost parameter ("ACP") used is the total of preheater amortization cost and the cost of that portion of the domestic water heating load which the solar storage fails to supply due to preheater inefficiency. Detailed cost assumptions are provided in the annual report (2).

The Wolverine Type WH Trufin preheater shows a clear advantage over the other three. For electricity at 4¢/kwh, the optimum preheater design requires approximately 50 feet of the 5/8" finned tube. The optimum length increases with fuel cost.

Activity 2D AUXILIARY HEATING OF STORAGE

When solar heated storage water falls below the temperature necessary to maintain thermal comfort, auxiliary heat is required. Two basic options are available at this point:

1) provide auxiliary heat directly to the load

2) provide auxiliary heat input to storage, from which it may be delivered to the load.

With most currently marketed solar heating systems, auxiliary heat is delivered directly to the load, in response to 2nd stage demand by the building thermostat. This approach maximizes collector performance, since average storage temperature is not raised by the addition of auxiliary heat.

Despite some reduction in collector efficiency, the auxiliary heated storage (AHS) approach has several attractive features, including "off-peak" auxiliary energy consumption and system simplification by elimination of conventional hot water and space

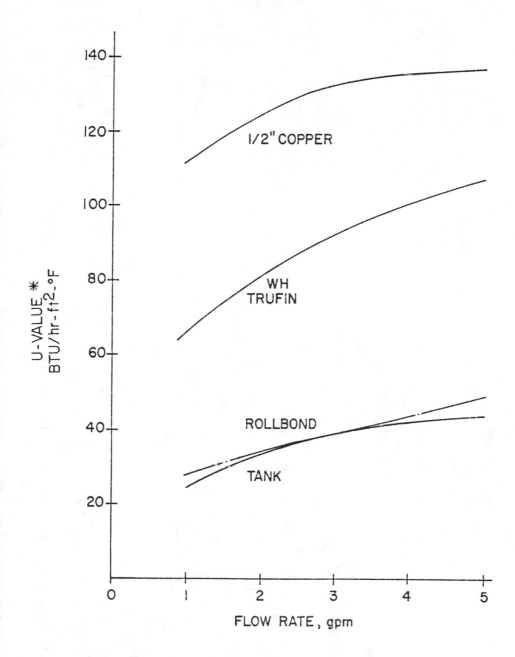

* BASED ON EXTERIOR SURFACE AREA

Fig. 15 - Preheater Test Results

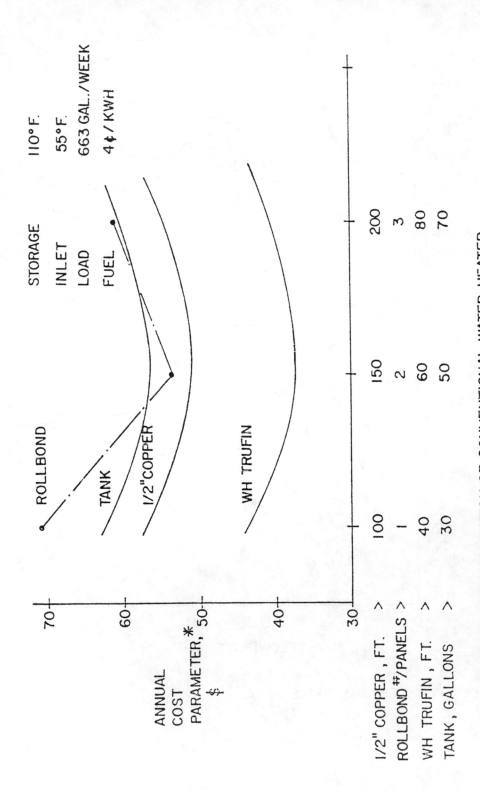

Fig. 16 – Preheater Cost Effectiveness

heating components. And, with a well-stratified storage container of adequate size, collector performance may not be not-
iceably diminished by AHS design, if auxiliary heat is delivered
only to the high temperature portion of storage. Tank bottom
water, unaffected by auxiliary heat, would be delivered to the
collector, resulting in high collector efficiency.

The objective of Activity 2d was to evaluate AHS alternatives
and identify a preferred approach. Concepts were considered for
both electric and combustion-type heat sources. However, fewer
alternatives were generated for the latter due to problems in-
herent in the requirement for combustion air and exhaust gas
ducting. The combustion-type systems must have a separate
heating unit, since immersion of the entire auxiliary heater in
the storage tank does not appear feasible. Of the alternatives
considered, only two appear to merit further development. These
are:

1) immersion electric heating elements

2) use of a pumped hot water loop through the domestic
   water preheater.

For immersion coils, designs were developed in which the
heating coil and the domestic water preheater might be delivered
as one unit, to reduce installation costs and promote separation
of the electric coil from the membrane liner. Figure 17 shows
the schematic of the preferred version of the combined preheater/
auxiliary heater unit, which locates the electric heating coil
inside the preheater. This design could eliminate the conventional
furnace, the hot water heater, and the heat losses from their
surfaces.

For combusion-type auxiliary and non-immersion electric
auxiliary, the preferred AHS approach is shown in Figure 18. A
slightly oversized domestic water heater provides auxiliary heat
to storage via the immersion preheater loop identified under
Activity 2C. This design also eliminates the conventional furnace
and its associated wiring and flue.

Calculations indicate that the optimum preheater is of
adequate capacity for auxiliary heating of storage.

CONCLUSIONS

Membrane-lined liquid thermal storage systems have been shown
to have significant advantages over conventional tanks in terms of
both cost and versatility. The following specific conclusions may
be drawn from the project results to date:

1) A significant number of membrane solar storage systems
   are completed and operating successfully.

2) EPDM appears to be the longest life membrane material

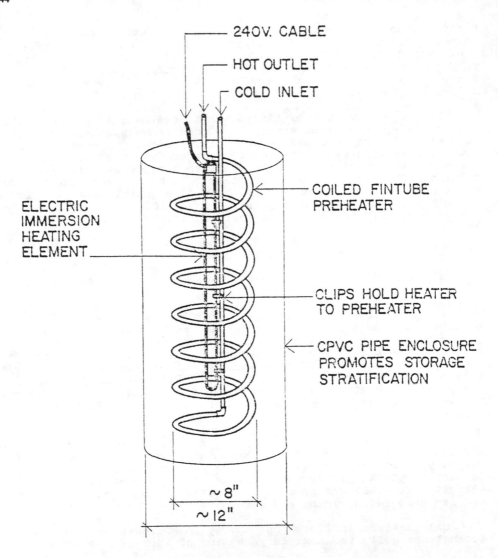

- ENTIRE UNIT IMMERSES IN STORAGE

- IMMERSION UNIT HEATS STORAGE AND DOMESTIC WATER

- ELIMINATES FURNACE/BOILER AND HOT WATER HEATER

Fig. 17 - Combined Preheater & Immersion Auxiliary Heater

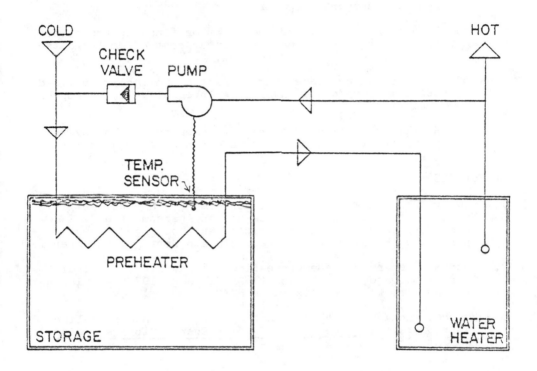

WHEN STORAGE FALLS BELOW MINIMUM HEATING TEMP., THE PUMP RECIRCULATES WATER HEATER OUTPUT THROUGH THE PREHEATER, HEATING STORAGE.

USING OVERSIZED WATER HEATER, FURNACE IS ELIMINATED.

Fig. 18 - Preferred Auxiliary Heated Storage Concept

under expected temperature conditions, but PVC and cross-laminated PE are promising lower cost materials.

3) Site-built rectangular and kit-type cylindrical enclosures appear to provide the most cost-effective membrane support structures.

4) Both jacketed cylinder and floating finned plate direct air heating design concepts appear cost-effective after small-scale laboratory testing.

5) Domestic water preheating is most economically provided using a commercially available integrally-finned copper tube.

6) Auxiliary heating of storage appears to promise system simplification in addition to off-peak energy utilization.

REFERENCES

1. Sawyer, Stephen W., "The Cost, Performance, and Reliability Patterns of Solar Heating Systems: An Assessment of 177 Owners", presented at Solar Heating and Cooling Systems Operational Results Conference, Colorado Springs, November 1978.

2. Bourne, Richard C., "Membrane-Lined Foundations for Liquid Thermal Storage, Annual Report (Draft) Aug. 1979" available from author, to be published by NTIS.

3. Bourne, Richard C., "Direct Air Heating from Membrane-Lined Water Storage", Proceedings of Solar Energy Storage Options, San Antonio, March 1979, pp. 417-427.

Solar-Powered Saline Sorbent-Solution Heat Pump/Storage System

HARRY ROBISON and SAMUEL HOUSTON
University of South Carolina
Conway, South Carolina 29526, USA

ABSTRACT

Coastal Energy Laboratory Chemical HEAt Pump (CEL-CHEAP) is a redesigned open-cycle liquid desiccant air conditioner. Heat is discharged to shallow-well water by dehumidification-humidification for cooling and extracted by humidification-dehumidification for heating. Direct solar radiation concentrates the desiccant. For continuous operation, a small uninsulated tank stores concentrated solution. This chemical heat pump needs no mechanical compressor, condenser, vacuum system, or pressure system. The collector-regenerators are inexpensive. The refrigerant is water and the desiccant is calcium chloride. First cost and operating expenses are very low.

INTRODUCTION

The U.S.C. open-cycle desiccant air conditioner [1] has been redesigned to include a heating cycle by replacing the triethlene glycol sorbent solution with calcium chloride solution.

CEL-CHEAP (Coastal Energy Laboratory - Chemical HEAt Pump) uses, well-water, salt and the sun for heating or cooling, as well as humidification or dehumidification. This chemical heat pump needs no mechanical compressor. Instead, a saline sorbent-solution is used to pump the refrigerant. No vacuum or pressure systems are required as the conditioned air stream acts as the transfer medium. No condenser is necessary for an open cycle is utilized. The refrigerant, which is water, is evaporated directly into the process air stream and need not be recovered. No generator is needed as direct solar radiation removes the excess moisture from the dilute desiccant solution as it flows over the roof of the building. Thus, the simple trickle collector acts as both collector and regenerator. No insulated sensible heat storage is required. Energy is stored for both heating and cooling in the form of a concentrated salt solution.

Concentrated salt solutions contain, as potential energy, that work that was expended in order to concentrate the solution. Figure (1) shows an engine designed to operate on the difference in vapor pressure between solutions contained in two tanks that are maintained at the same temperature. One tank contains a concentrated salt solution with a low vapor pressure and the other tank contains water with a much higher vapor pressure. This is a concentration difference potential engine [2].

Figure (2) shows the same two tanks separated from one another. A large pipe connects the tanks and maintains the same pressure in each tank. Because evaporation is taking place in the water tank, this tank becomes cold and, because absorption is taking place in the saline tank, this tank becomes hot.

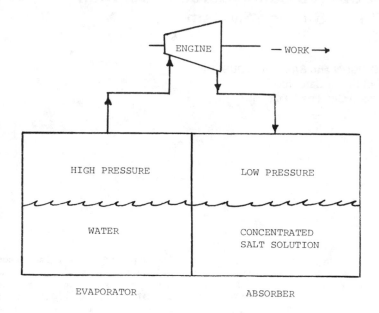

Fig. 1 - Concentration Difference Engine

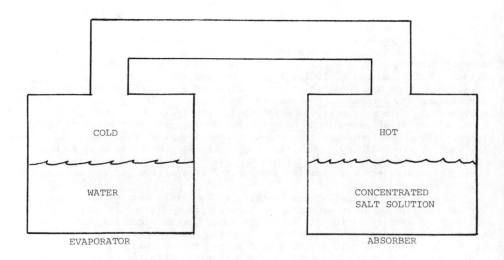

Fig. 2 - Chemical Heat Pump

This is a chemical heat pump [3].

    If the tops of the tanks in the chemical heat pump are cut off and the
resulting pans of water and salt solution are placed in a duct through which
air is blown as in figure (3), an open-cycle desiccant air conditioner results
[4]. To increase the rate of mass transfer, a spray can be used to recirculate
the desiccant. Well-water removes both sensible heat and the heat of absorption.

    Figure (4) shows CEL-CHEAP operating in the heating mode. The evaporator
has become the absorber and the absorber has become the evaporator. Well-water
now furnishes the heat of vaporization and the air is adiabatically heated by
the heat of absorption to a temperature higher than the well-water temperature.

OPERATING CYCLES

    During the cooling cycle, warm, humid, outside air is continuously cooled
by shallow well-water and dehumidified by a concentrated saline sorbent solution [5]. See Figure (5). Both sensible and latent heat are removed. The dry
process-air is then further cooled to a temperature lower than the well-water
temperature by dehumidification with cool water. Sensible heat is converted to
latent heat. After picking up the room heat-load, the process air flows out
through the collector-regenerator (area = 10 m^2/3.5 KW = 100 ft^2/ton) and
carries away the moisture from the hot, dilute desiccant solution, which has
been heated by direct solar radiation as the solution flows through the collector-regenerator. Sensible heat absorbed by the liquid and not used for mass
transfer is recovered by a heat exchanger.

    During the heating cycle, cool, room air is heated and humidified as it
flows through the collector-regenerator across the hot, dilute sorbent solution.

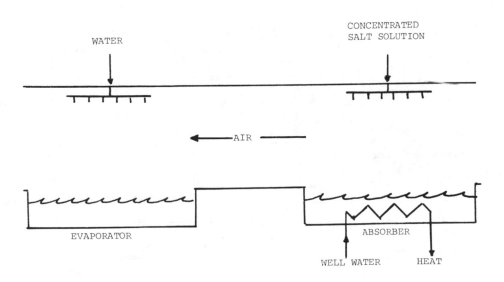

Fig. 3 - CEL-CHEAP(Coastal Energy Laboratory-Chemical Heat Pump)
         Cooling Mode

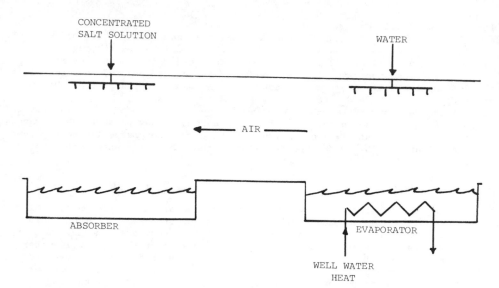

Fig. 4 - CEL-CHEAP (Coastal Energy Laboratory-Chemical Heat Pump)
         Heating Mode

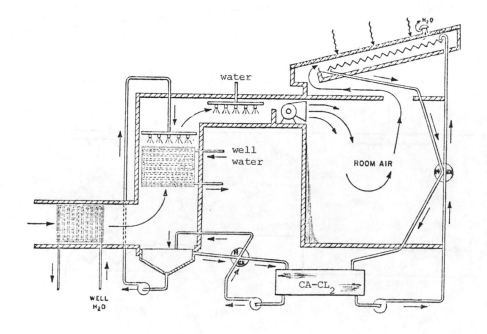

Fig. 5 - CEL-CHEAP

The air gains both sensible and latent heat (latent heat/sensible heat = 2.5).
The warm, wet air then flows through the conditioner. With the cooling water
cut off, dehumidification by the warm concentrated salt solution further heats
the air to a temperature higher than the collector temperature. Latent heat is
converted to sensible heat. This process allows efficient operation of the
collector at a low temperature.

Storage of the concentrated solution [6] is at ambient temperature during
the cooling cycle. During the heating cycle, storage is at room temperature.
Diurnal or seasonal energy storage is possible with energy density as high as
600 kJ/L (2,500 BTU/gal) of saline solution.

ACKNOWLEDGEMENT

This work is now supported by the Coastal Educational Foundation, Inc.,
the Horry County Higher Education Commission, and the Research and Productive
Scholarship Committee of the University of South Carolina.

REFERENCES

1. Robison, H.I. 1978. Liquid Sorbent Solar Air Conditioner. Alternative
    Energy Sources: An International Compendium. Veziroglu, T.N., ed.
    Hemisphere Publishing, Washington, D.C. Vol. II.

2. Isshiki, N., Maekawa, Y., Takeuchi, M., Nichai, I., Akuta, T., and
    Kamoshida, J. 1977. Energy Conversion and Storage by CDE (Concentration
    Difference Energy) Engine and System. Proceedings of the Twelfth Inter-
    society Energy Conversion Engineering Conference, Washington, D.C. American
    Nuclear Society, La Grange Park, Ill.

3. Offenhartz, P.O.D. 1978. Chemically Driven Heat Pumps for Solar Thermal
    Storage. Sun: Mankind's Future Source of Energy, Proceedings of Interna-
    tional Solar Energy Society Congress, New Delhi, India. de Winter, F. and
    Cox, M., eds. Pergamon Press, Elmsford, N.Y.

4. Robison, H.I., Houston, S.H. 1979. Passive Solar Heat Pump. Proceedings
    of the 4th National Passive Solar Conference, Kansas City, MO. Franta, G.E.,
    ed. American Section of the International Solar Energy Society. Newark,
    Delaware.

5. Robison, H.I. and Houston, S.H. 1979. Open-Cycle Solar-Powered Chemical
    Heat Pump/Storage System. Proceedings of Fourteeth Intersociety Energy
    Conversion Engineering Conference, Boston, Massachusetts. American Chemical,
    Society, Washington, D.C.

6. Robison, H.I. and Houston, S.H. 1979. Thermo-Chemical Energy Storage for
    Heating and Cooling. Solar Energy Storage Options, San Antonio, Texas.
    McCarthy, M.B., ed. Trinity University, San Antonio, Texas.

High Temperature Storage for a Wind Energy System

RAYMOND RAMSHAW and DONALD BOWMAN
University of Waterloo
Waterloo, Ontario, Canada, N2L 3GL

ABSTRACT

For given environmental conditions and a specified "top-up" load requirement the sizing of the storage unit, the wind turbine and the generator are compared.

1. INTRODUCTION

A wind energy system design procedure was developed for the purpose of supplying supplemental "top-up" energy to a space heating load. Since the availability of wind energy and load demand do not always coincide, storage is necessary. High temperature high energy density storage is well suited when the energy is transferred by electric generation. This storage technology has been developed for the design of off-peak heat-storage furnaces and may be applied to the wind energy system.

2. SYSTEM DESCRIPTION

The system consists of a vertical axis wind turbine, a d.c. generator and a high temperature storage unit. The energy from the storage unit is used for "top-up" heating for a load that is proportional to the ambient temperature. The generator is field modulated to deliver the maximum possible energy at any given wind speed. This energy is fed directly to the storage unit that acts as a load buffer and converts the electrical energy to thermal energy. The non top-up portion of the load energy is supplied by resistance heating.

2.1 High Temperature Storage

The high temperature storage facilitates the use of high energy density storage. The thermal energy is maintained as high grade heat enabling it to be used as a power or space heating source.

Conventional units (see Fig. 1) such as European storage furnaces consist of a rectangular shaped refractory core placed in a high temperature insulation jacket. The core consists of horizontal passages in which the elements are placed and vertical air passages for the extraction of heat. These air passages form a series of inverted u tubes when the top of the core is sealed off and the air inlet and outlet are located at the core base. This method of construction prevents natural convection losses during idle periods.

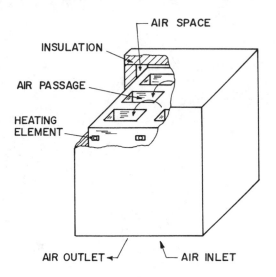

Fig. 1 – Conventional Storage Unit.

However, it complicates the design of the base insulation. Vermiculite is used for the base insulation where compressive strength and rigidity are requirements.

The volume of the core comprises two parts; the refractory material and the air passage space. The required air passage space is dependent on the air temperature, the output power, and air velocity. A linear dependence is determined for the output power and air velocity with respect to the air passage volume. The mass flow rate depends exponentially on the air temperature. This results in maximizing the air temperature to minimize the core size. The tube diameter and height are calculated to give sufficient heat transfer.

It was found that the core material provided the limiting design factor. The thermal conductivity of most refractory materials is low enough to result in temperature gradients in the material and hence low surface temperatures and poor heat transfer performance. Materials such as aluminum have superior thermal properties but suffer from high cost and oxidation. A material called Feolite, a high density refractory of 250 lb/ft^3 compared to 160 lb/ft^3, has a superior thermal conductivity and exhibits no problems when exposed to high temperatures.

The electrical conductivity of refractories increases exponentially with temperatures in excess of 1400°F. This is a limitation for the maximum core temperature. The lower core temperature is limited by the required size of the air passage which increases exponentially as the minimum air temperature decreases. The available power output decreases as the core temperature decreases and thus the heat transfer surface must be designed for the lower core operating temperature.

The heating elements are of 80/20 nickel chrome alloy wire which is formed into an open wire spiral and placed in the element passages. The dominant heat transfer mechanism is by radiation. Due to the typical operating temperatures of 1400°F for the core and 1600°F for the element wire, a high grade alloy is required. To prevent over heating, each element has a safety thermal link.

The insulation jacket thickness and cost will increase exponentially with temperature. This is a result of the thermal conductivity increasing exponentially with temperature. The insulation jacket cost is also a limiting factor in the choice of the maximum core operating temperature. The bulk of the insulation jacket can be reduced by use of Microtherm, a material which has a thermal conductivity less than that of still air.

An alternative high temperature storage unit is a packed bed. See Fig. 2. This storage consists of a container filled with small pebbles of refractory material. The bed is charged and discharged by the passage of air through the bed. Rapid heat transfer is assured by the large ratio of pebble surface area to volume.

During the charging process air is heated to a high temperature by electric elements and forced downwards through the bed. The air being at a high temperature results in rapid heat transfer to the pebbles. A temperature wave forms with near instantaneous heating of the core material as the wave

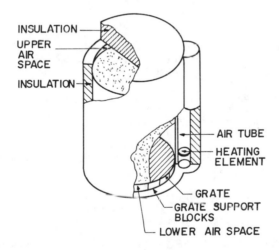

INSULATION
UPPER AIR SPACE
INSULATION
AIR TUBE
HEATING ELEMENT
GRATE
GRATE SUPPORT BLOCKS
LOWER AIR SPACE

Fig. 2 - Packed Bed Storage Unit.

progresses from the top to the bottom of the bed. During the discharging cycle air is forced through the bed from the bottom to the top. Since the air temperature at the top of the bed is that of the maximum core temperature a very large output power may be realized.

The design procedure involves a compromise between small pebbles for good heat transfer and large pebbles for a low air pressure drop. Once a size has been determined it becomes standard for units of all energy and power ratings. A minimum air pressure drop of 0.2 inches of water is required to give a good air flow distribution in the bed.

The packed bed cores are economically suitable for short term storage where the numerical value of stored energy in kWh is less than or of the same magnitude as the output power in kW. In contrast, the conventional brick cores are more suitable for long term storage from an economical standpoint [1].

2.2 Wind Turbine and Generator

The characteristics of vertical axis wind turbines were used in the simulations. For the several diameters the same power coefficient curve was used. The maximum power coefficient was 0.6 at a velocity ratio of 2 and the cut-out speed was at a velocity ratio of 6.8. The curve represented experimental results [2] published by the Canadian National Research Council.

The generator is a controlled field, direct current machine. It is assumed that the resistance of the generator load is constant and the field is excited proportionally to produce a mechanical load on the turbine to maintain its operating speed at the peak velocity ratio of 0.6. The efficiency curve of the generator at reduced loading was taken from particular specifications and used to calculate the available output power.

The models of turbine and generator were lumped together as one unit. A transfer function was tabulated. This expressed the electrical power output for each value of wind speed for a given wind turbine diameter and generator size.

2.3 Load Model

The load for the storage unit was the space heating of a residence. Thus the load is proportional to the ambient temperature. The heating demand of a structure can be calculated by the following equation.

$$\text{Demand} = (65 - \beta) \alpha,$$
where   Demand is in kW,
         β is the ambient temperature $^\circ F$ and
         α is the heat loss factor $kW/^\circ F$.

The value of 65 is defined by A.S.H.R.E. as a suitable space temperature to use. It reflects the heat produced by internal gains.

For the purpose of sizing equipment a maximum design value of β is used. This will determine the maximum load. This load will then be proportioned between the electrical auxiliary heating and the "top-up" from wind energy. Three general types of "top-up" systems may be considered. The wind power may be used to supply the load and be limited to a certain maximum value. Thus when the demand exceeds this value the auxiliary source is utilized. When the auxiliary source is electric heating an attempt may be made to limit the wind generated supply and to limit the maximum demand on the auxilary power source. This reduces the electric utilities peak loading and provides a better distribution of energy over the heating season. The auxiliary demand may also be kept constant and the wind supplied power allowed to vary to take up the required load.

3. OPTIMIZATION

Using particular wind data alone, the energy output of the turbine generator may be computed. Since the output power varies as the square of the turbine diameter and the power coefficient varies as the first power of the diameter there is no convenient way of using a per unit diameter system of representation. The generator efficiency curve is also nonlinear. Thus several values of turbine diameter and generator size were studied. As shown in Fig. 3, there is an optimum generator size for each turbine diameter.

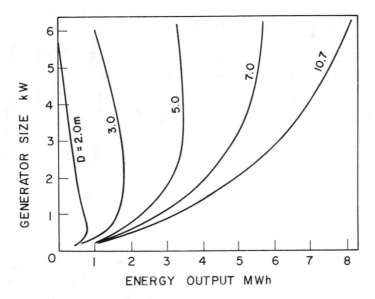

Fig. 3 - Optimized Energy Output

However, the maximum generation of energy may not coincide with the demand of energy and thus a sub-optimization is required where the supply and demand are matched with the storage acting as a buffer.

4. SUB-OPTIMIZATION

To supply the demand, that is determined by the ambient temperature, from a supply, that is controlled by the wind speed, requires the optimal selection of the combined turbine generator and storage units.

Scaling of the results may not be used because of the nonlinear functions involved. Thus several combinations of the values must be used and an optimum point selected by interpolation.

For each generator turbine combination a characteristic curve may be developed that presents information concerning the percentage of demand met by various degrees of storage. As the storage capacity increases the percentage of demand met will tend to one hundred. If the overall generated energy is greater than the required demand energy, this will result in a one hundred percent case. However, when the generated energy is less than the demand energy the percentage of daily demand will saturate at some value. Fig. 4 presents a characteristic curve where percentage daily demand is plotted against percentage of time that demand is supplied. The integration of the area to the left of the curve represents the percentage of the system energy supplied by the wind system.

The sub-optimization process requires that for a chosen point on Fig. 3 several characteristic curves of the form of Fig. 4 be generated for points in the area of the one chosen. Next a criterion for the allowable percentage of time that less than 100% of the demand will be met must be established. In Fig. 2 a value of 10% was used. This then establishes a required storage size

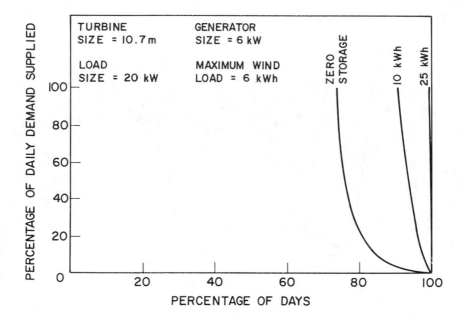

Fig. 4 - System Characteristics.

for the given turbine generator size.

For each of the chosen points the storage size may be plotted as a function of the turbine generator size. To determine the optimum cost effective system the axes may be superimposed with the cost of the units and a total cost plotted for each case.

5. CONCLUSIONS

Energy from the wind and thermal energy demand in a residence do not necessarily coincide. Therefore, there is need for energy storage if a wind turbine is to be used to provide a specified proportion of a load on demand. The design of a wind turbine and generator for optimum energy output does not provide an economic solution. Instead, the interaction of all four units, the wind turbine, the generator, the high temperature, high density energy storage unit and load demand has to be simulated for overall economic efficiency. The criteria are that the storage unit must provide a particular power on demand and that throughout the season there will be times when the storage unit is depleted.

6. REFERENCES

1. D. Bowman, R.S. Ramshaw and S. Stricker, "Electric Thermal, High Temp-
    erature Storage Heater Design Procedures". UMR, DMR Conference on Energy,
    University of Missouri 1978, P672 - 678.

2. R.S. Rangi, P. South and R.J. Templin, "Wind Power and The Vertical-Axis
    Wind Turbine Developed at The National Research Council". Canada.

Assessment of High-Head Turbomachinery for Underground Pumped Hydroelectric Storage Plants

S.W. TAM, A.A. FRIGO, and C.A. BLOMQUIST
Argonne National Laboratory
Argonne, Illinois 60439, USA

ABSTRACT

Underground Pumped Hydroelectric Storage (UPHS) plant costs for plants equipped with advanced reversible pumped turbines have been considered. Equipment used includes single- and two-stage reversible pump turbines for operating heads from 500 to 1500 m. The effects of machinery costs, operating heads, plant configurations and sizes have been taken into account. The results indicate that the use of advanced machinery seems to push the minimum UPHS plant cost to heads greater than 1500 m. The employment of advanced, reversible pump turbines seems to be economically attractive. It is shown that pump-turbine efficiencies and the so-called charge/discharge ratio are very important design parameters for UPHS applications and the interactive effects of these parameters have been analyzed. The results show that under certain conditions a pump-turbine option with a higher charge/discharge ratio at the expense of somewhat lower operating efficiency can be desirable. Careful integration of pump-turbine performance characteristics with the overall system consideration (i.e., UPHS plant design and utility grid requirement) bears significantly on the direct cost savings due to the differing capital costs of the various machines. More importantly, such considerations have a significant impact on the storage cost, the most important element in UPHS plant cost considerations.

1. INTRODUCTION

Base load power plants (thermal and nuclear) generally have idle capacity during low load periods (e.g., during the night). If these idle capacities are utilized and the energy generated stored in some way, the energy can be utilized to supply power during peak demand hours (usually during daytime and early evening). This serves both to displace the reliance of peak power generation upon premium fuels (oil and gas) to coal or nuclear-based plants as well as better utilization of those base load plants.

One way to utilize energy storage for utility peaking application is via the concept of underground pumped hydroelectric storage (UPHS) schemes. Conceptually UPHS is very similar to surface pumped hydro storage (SPHS). SPHS utilizing reversible pump-turbine has been in the U.S. since 1953 when the U.S. Bureau of Reclamation installed a 9 MW unit at Flatiron. Today many systems are operational However, a SPHS system requires a suitable site for the upper-level reservoir located at sufficient elevation in the vicinity of the lower reservoir and also at a reasonable distance from the load center concerned. Sites with such characteristics are being rapidly depleted today within the U.S. The concept of UPHS is designed to minimize these siting and transmission difficulties while still fulfilling the primary mission of gener- ating peak power for the electric utility grid.

Conceptually, UPHS (Fig. 1) is very similar to SPHS, and, similarly, does not require premium fuel. The major difference from SPHS is that the upper reservoir for UPHS is at ground-level while the lower reservoir (an excavated cavern in general) is located underground. This important feature results in the advantages of siting flexibility and reduced transmission costs and retains the good system reliability and availability characteristics of SPHS.

A typical proposed UPHS plant will have a capacity of 1000-2000 MW with 6-10 hours of storage, and an overall efficiency close to 70%. Presently, there are two basic conceptual plant configurations -- the one-drop and the two-drop schemes (Figs. 2 and 3).

In the one-drop scheme, there is only underground reservoir. The turbo- machinery is housed in the powerhouse located below the underground reservoir to provide pump submergence. Access, cable, and equipment shafts connect the powerhouse to the aboveground service building. The powerhouse is linked to the upper reservoir through the penstock shaft, and the lower cavern is vented to the atmosphere through the vent shaft.

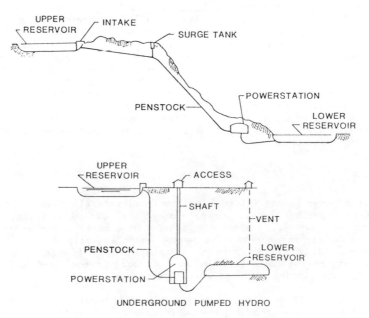

Fig. 1. Diagrammatic Sketch of Conventional
and Underground Pumped Hydro Scheme

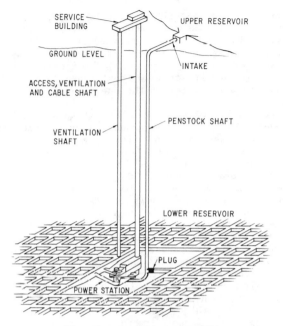

Fig. 2.   One-drop UPHS Scheme

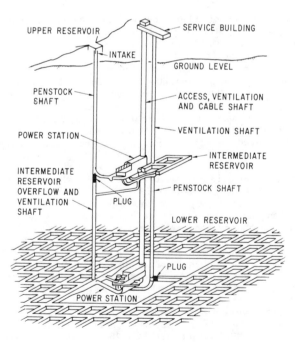

Fig. 3.   Two-drop UPHS Scheme

The two-drop configuration utilizes single-stage, reversible pump turbines (RPT) with an average gross head of about half the depth at which the lowest reservoir is situated. A smaller intermediate reservoir is located half way between ground level and the underground cavern. A powerhouse is connected to each of the two underground reservoirs. The intermediate cavern allows the two power plants to operate in series without synchronizing the turbomachinery. The shaft requirements are similar to the one-drop scheme.

The energy generation capability of a UPHS plant is proportional to the product of the volume of the underground reservior and the operating head. Since the underground reservior excavation represents the largest cost component (see discussion below and Table 1), a strong incentive exists to utilize higher heads to reduce the lower cavern volume requirement for a given plant power generation and storage capacity. On the otherhand the high head requirement ($\gtrsim 1000$ m) means care must be exercised in choosing the specific speed N_s of a pump-turbine. N_s is proportional to ([Rotational speed, RPM] x [Power, MW]/[Head,m]$^{5/4}$). This parameter is an essential factor characterizing the operating efficiency of the machine. In general, for high-head operation it is important to select a large N_s to maintain high efficiency,[2,9] by either increasing the RPM and/or the power (capacity). However, the RPM must be matched to the available synchronous speeds of the generator/motors and generator manufacturers are somewhat reluctant to fabricate high-speed machines. Therefore, increasing the capacity is the only reasonable alternative. Besides large capacity machines have a better accessibility to the water passages resulting in a smoother surface finish and fairness. A smooth surface is an extremely important factor in reducing cavitation, particularly for high-head operation.

Thus, the application of UPHS for electric utility peaking service is enhanced by the use of relatively high-head (>1000 m) and large-capacity (>350 MW) turbomachinery.[1,7] These constraints are presently not mutual. High-head but small-capacity (<150 MW) turbomachines are currently available as multistage, reversible pump turbines or tandem units (Pelton-type waterwheel and multistage pump). Large-capacity, Francis turbines are available for low-head operations, e.g., 700 MW @ 87-m head. Single-stage, reversible pump turbines are under construction for output powers of 315-MW at 625 m, and 508-MW at 384 m. Recent design studies[2] have shown the technical feasibility of advanced machinery such as 500-MW-capacity, single-stage, reversible pump turbines at a 1000-m head and two-stage, gated, reversible pump-turbines with an operating head of 1500 m.

The choice of turbomachinery for UPHS plants rests ultimately on economic considerations. The result of preliminary cost analyses are discussed in this report. The principal objective is the comparison and contrasting of the costs associated with the utilization of state-of-the-art pump turbines and those associated with advanced machinery.

UPHS plant costs have been estimated in a recent study conducted by Charles T. Main, Inc.[3] Although typical UPHS plant costs ($350-400/kW) are comparable to that of conventional pumped storage plants of the same capacity, the relative costs of the various plant components are quite different. For example, the two reservoirs in an SPHS scheme are comparable in cost. However, the lower reservoir in an UPHS plant is much more costly than that of the ground-level, upper reservoir. Typical cost percentages for the various components of a 2000-MW UPHS plant with a 1200-m operating head are shown in Table 1.[3] It can be seen that the underground cavern cost is over 30% of the total plant cost, and the pump-turbines are the second most expensive item. The turbomachinery options chosen in this recent study are either state-of-the-

Table 1. Percentage of Total Plant Cost for the Major UPHS Plant Components
          (2000 MW at 1200 m head)*

Description	Percentage Cost
LAND & LAND RIGHTS	0.4
POWER PLANT STRUCTURE & IMPROVEMENTS	
Main Equipment Shaft	3.5
Cable Shaft	2.8
Underground Powerhouse	4.9
Gate Gallery	0.2
Above Ground Structures	0.3
RESERVOIRS, DAMS & WATERWAYS	
Upper Reservoir, Dam & Reservoir	3.5
Upper Reservoir, Intake	0.3
Lower Reservoir:	
Excavation	27.3
Rockbolting & Shotcrete	5.6
Disposal	2.9
Water Conductors:	
Penstock Shaft	3.0
Manifold	3.0
Draft Tunnels	0.8
Lower Reservoir, Ventilation Shaft	2.1
Construction Adits	0.8
Reservoir Filling	0.4
PUMP TURBINES & GENERATOR MOTORS	
Pump Turbines, Governors, Valves & Installation	11.4
Generator Motors, Starting Equipment & Installation	7.8
ACCESSORY ELECTRICAL EQUIPMENT	
High-Voltage Buses	1.6
All Other Electrical Equipment	3.4
MISCELLANEOUS MECHANICAL EQUIPMENT	
Cranes & Hoists	1.4
All Other Mechanical Equipment	1.0
ROADS & BRIDGES	0.9
ENGINEERING, SUPERVISION & OVERHEAD	10.7
TOTAL	100

*Calculated from data in Ref. 3.

art or a slight extension of currently available technology.[3] A total of
seven different UPHS plant configurations were considered. The major plant
characteristics together are listed in Table 2. Major conclusions are:

  1. For a fixed plant size and type of machinery, the minimum
      plant cost ($/kW) seems to occur in the head range of
      1200-1500 m.

   2. Multistage, reversible units appear to be the most economical
      machinery option.

Table 2.   UPHS Plant Configurations

Schemes	Capacity (MW)	Average Gross Head (m)	Storage (Hours)	Machinery (No. of Units)
I	2000	1200	10	Multistage (6)
II	2000	2 x 600 (two-drop)	10	Single-Stage (2 x 4)
III	2000	1200	10	Tandem (6)
IV	2000	1500	10	Multistage (6)
V	2000	900	10	Multistage (6)
VI	2700	1200	10	Multistage (8)
VII	1300	1200	10	Multistage (4)

3. A plant capacity of 2000 MW represents the most economical plant size, when interest during construction and escalation are added.

In the present work, advanced turbomachinery is considered for UPHS plant configurations similar to those shown in Table 2, thus providing a common basis for comparison with the Charles T. Main study. Civil engineering costs from Reference 3 (updated to Sept. 78 dollars) have been utilized. The advanced machinery considered includes single-stage, reversible pump turbines and two-stage, reversible pump turbines. These machines represent considerable extension to the state-of-the-art, reversible pump-turbine technology. At present, single-stage, reversible, Francis-type pump turbines have been designed for operating heads as high as 625 m (Bajina Basta Station in Yugoslavia) and for power outputs up to about 500 MW (Bath County Plant in Virginia, U.S.A.).

Preliminary engineering design and cost studies of higher-head and greater-output, single-stage, reversible, Francis-type pump turbines (SSRPT) have been conducted by Allis-Chalmers Hydro-Turbine Division for Argonne National Laboratory. Three, high-head machines were investigated. The three units are designed to produce a power output of 500 MW while operating under heads of 500, 750, 1000 m respectively.

A similar design analysis has been carried out on two-stage reversible pump-turbines (TSRPT) with operating heads of 1000, 1250 and 1500 m. At present, there are no known operating gated TSRPT. Performance characteristics [2,4] of these pump-turbines have been utilized to investigate UPHS plant costs for plants equipped with such advanced turbomachinery. Major factors considered were machinery costs, operating heads, plant configurations and sizes. The pump-turbine efficiencies as well as the so-called charge/discharge ratios of the machines are very important elements, and the interacting effects of these factors will be analyzed.

2. TURBOMACHINERY ECONOMICS FOR UPHS APPLICATION

Capital costs (as functions of operating head) for single-stage, reversible pump-turbines and two-stage, reversible pump-turbines are shown in Figs. 4 and 5. Single-stage machines have a lower unit cost than two-stage machines for the same head. At the same time, one notes that both types of machines have decreasing unit cost for increasing operating head. This is basically due to the reduction in the machine size as the head increases. Thus, from the point of view of turbomachinery, there are benefits in going to higher head.

The two major conceptual UPHS plant configurations -- the one-drop and the two-drop schemes have been discussed in the previous section. The effect of plant configuration on UPHS plant cost have been investigated in the recent Charles T. Main study,[3] which indicates that the one-drop scheme with multistage units is the most economical choice. An estimate was made to determine the possible cost savings if two-stage, gated, reversible pump-turbines are used for the various UPHS plant configurations all with a 1000-MW plant capacity at a 1200-m head. The results are listed in Table 3 and indicate that the two-stage unit becomes increasingly favorable when compared to multistage, single-stage (two-drop scheme), and tandem units. Consequently, most of the subsequent cost comparisons will be made between two-stage and multistage reversible pump-turbines with a one-drop plant configuration.

The estimate below has been made at one particular head and plant capacity. The important question is whether the previous conclusion (i.e., that two-stage, reversible pump-turbines offer an economically attractive option) remains valid when different operating heads and capacity are considered. Cost savings were found to be somewhat insensitive to plant size, which means that the selection of multistage or two-stage units does not affect the optimal size of a UPHS plant. Subsequently, all discussions are based on a UPHS plant capacity of 2000 MW, which is within the optimal capacity range as indicated by the Charles T. Main study.[3]

The effect of different operating heads between 900 to 1500 m on UPHS plant cost has also been considered. The cost ($/kW) for a 2000-MW UPHS plant similar to Scheme T in Table 2, but employing four, 500-MW, two-stage, reversible pump-turbines, as a function of operating head is shown in Fig. 6. For comparison purposes, the plant cost[3] utilizing multistage units is also shown. It can be seen that the conclusion that two-stage machines give rise to lower plant cost holds over the head range of 1000-1500 m. The minimum plant

Table 3.   Percentage Saving in Plant Cost When Two-Stage
           Reversible Pump Turbines Are Used Instead of
           the Turbomachinery Listed

Machinery Type	Savings ($/kW)[a]	Percentage Savings (%)
Multistage, Reversible Pump Turbine	16	4.4
Single-Stage, Reversible Pump Turbine (Two-Drop)	28	7.5
Tandem Unit	48	12.2

[a]1978 dollars.

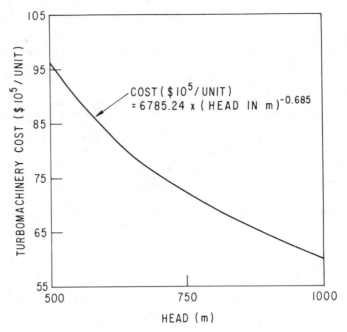

Fig. 4. Single-stage, Reversible Pump-Turbine
         Cost (500 MW size)

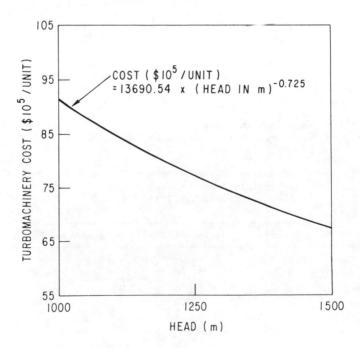

Fig. 5. Two-stage, Reversible Pump-Turbine
         Cost (500 MW Size)

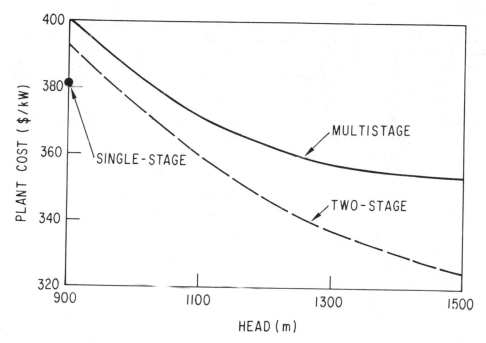

Fig. 6. UPHS Plant (2000 MW) Cost ($/kW) as a
Function of Operating Head

cost is probably at a head beyond that of plants employing conventional pump turbines.

For heads above 1000 m, the two-stage, reversible pump-turbine is an economically attractive option. The point denoted by single-stage is the cost for a similar UPHS plant utilizing 900-m-head, single-stage, reversible pump-turbines. The cost data shows that it is economically the best machine choice for operating heads less than 1000 m.

The advanced machinery concepts considered have different capacities and larger submergence requirements than the multistage reversible pump-turbines considered in the Main Study. These factors, in principle, will lead to somewhat different shaft and draft tube lengths and powerhouse size differences. Estimates of the differences were made by utilizing existing civil engineering cost information[3] to assess the cost corrections and to determine whether these perturbations alter the basic conclusions reached in the preceding analysis. The results indicate that the total cost correction due to these factors amounts to $2-$3/kW, which represents only a minor perturbation on the basic attractiveness of the two-stage reversible pump-turbines considered.

2.1 Pumping Energy Cost and Pump-Turbine Efficiency

The overall UPHS plant efficiency η_u is affected by the operating efficiency of the turbomachinery which in turn depends on factors such as pump-

turbine type and/or operating head. The plant efficiency reflects on the UPHS
plant economics through change in the pumping energy cost. Since an UPHS
plant lifetime is on the order of 50 years, the corresponding capitalized cost
penalty/savings[5] must be taken into account in evaluating turbomachinery
options. The effect has been addressed in a parametric manner.

Pertinent factors that must be considered are:

1. type of base load plants (nuclear or coal-fired) supplying
    the pumping energy to the UPHS plant,

2. fuel cost for each type of baseload plant, and

3. number of hours of annual operation for the UPHS plant.

The baseline UPHS plant considered has a 50-year life and a 70% overall
efficiency. The effective UPHS plant efficiency η_e is given by

$$\eta_e = 0.7(1 - \delta)$$

where δ is a measure of the change in plant efficiency due to a change in
the product of the turbine and pump efficiencies, $\eta_t \eta_p$. If the original
efficiency product is $(\eta_t \eta_p)_o$, then the new product $\eta_t \eta_p$ is related to δ by

$$1 - \delta = \frac{\eta_t \eta_p}{(\eta_t \eta_p)_o}$$

The resulting cost penalty P ($/kW) due to different values of δ (1 and
4%) has been computed for both coal-fired, and nuclear, base-load plants. The
results are shown in Figs. 7-9. Pertinent conclusions are:

1. For small differences in pump-turbine efficiencies
    ($\delta \sim 1\%$, Fig. 7), 2000 hours of annual operation and typical
    coal cost ($\sim$ $1.5/MBtu), P is about $5/kW.

2. For large differences in pump-turbine efficiencies
    ($\delta \sim 4\%$, Fig. 8), the corresponding P is about $20/kW or about
    the same order as the capital cost differences between
    various turbomachinery options.

3. A corresponding result for using nuclear base-load plants
    is shown in Fig. 9. The effective cost penalty, P, result-
    ing from the lower pump-turbine efficiency is less for
    nuclear baseload plants than for coal-fired ones.

From these results, one can conclude that consideration of efficiency is an
important element in choosing turbomachinery options.

2.2 Charge/Discharge Ratio Considerations

The so-called charge/discharge ratio, r, must also be taken into account
in pump-turbine design considerations,[6,8] because this factor may have an
effect on the pump-turbine efficiency. This factor is defined as the ratio of
the average pumping load to the rated generation capacity. The averaging
should be performed over the designed cyclic operating period of the pumped-
storage plant. For example, plants can be designed to operate on a daily,
weekly, or even longer cycle.

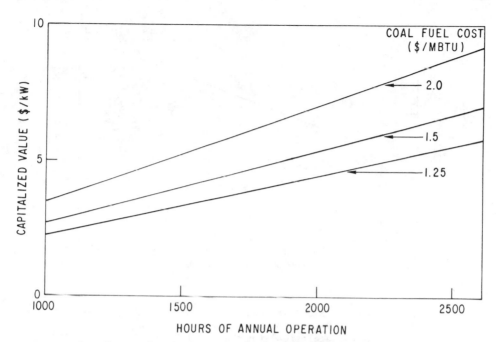

Fig. 7. Capitalized Cost Penalty P ($/kW) for 1% Difference
(δ) in Pump-Turbine Efficiency (50 years lifetime
at 8% Escalation) for Coal Plants

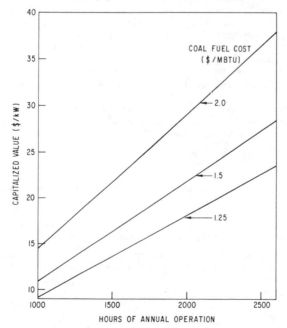

Fig. 8. Capitalized Cost Penalty P ($/kW) for 4% Difference
(δ) in Pump-Turbine Efficiency (50 years lifetime
at 8% Escalation) for Coal Plants

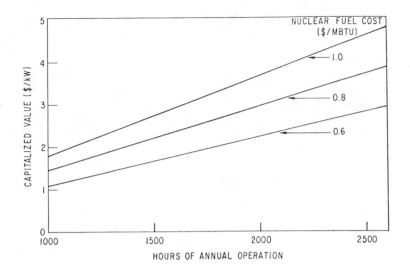

Fig. 9.   Capitalized Cost Penalty P ($/kW) for 1% Difference
          (δ) in Pump-Turbine Efficiency (50 years lifetime
          at 8% Escalation) for Nuclear Plants

A pump-turbine that has a charge/discharge ratio of close to 1 is said
to have an electrically balanced design, and the output is roughly the same as
the pumping load. On the other hand, a high r value (1.3 or 1.4) means the
generation time, T_G, is comparable to the pumping time, T_p. Therefore, the
pump discharge rate, Q_p, is about the same as the turbine discharge rate, Q_T.
The pump-turbine is then in a hydraulically balanced mode. This so-called
charge/discharge ratio can be adjusted within reasonable limits via pump-
turbine design (e.g., oversizing) or wicket-gate position adjustment.[8,9]

The significance of r is that the larger this ratio, the longer the
available generation time for a given amount of pumping time. The generation
time T_G is related to the pumping time T_p by the equation

$$T_G = r \; \eta_u \, T_p$$

The advantages as well as disadvantages of an increased charge/discharge
ratio are as follows:

1. Possible reduction in the additional required storage cost
    due to insufficient daily pumping time.

2. Increase in the balance-of-plant cost including larger
    hydraulic, electrical, and mechanical equipment plus a
    larger powerhouse.

3. Increases in the pumping energy cost due to a possible
    reduction in the pump-turbine efficiency.

Assuming a daily pumping time of 8 hours during the weekday, the cost
penalties due to the last two factors are shown in Fig. 10. Both factors are
important. Depending on the required generation time (Fig. 10) the total cost
penalty amounts to $30-40/kW. However, the cost savings from the reduction in
the required additional storage cost due to insufficient pumping time during
the weekdays can amount to about $80/kW if the required generation time is
about 8 hours or more (Fig. 11). Note that α is the product of the charge
discharge ratio r and the UPHS plant efficiency η_u. In these cases it is
advantageous to increase the charge/discharge ratio even at the expense of some
losses in efficiency. On the other hand, for generation of about 6 hours, then
the cost penalty from 2) and 3) outweighs the benefit from 1).

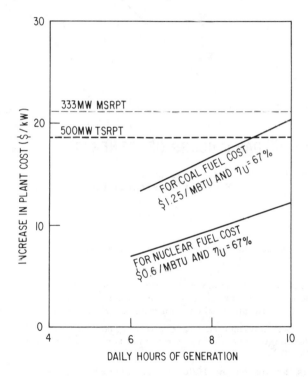

Fig. 10. Increase in Balance-of-Plant Cost and Pumping
          Energy Cost vs Daily Hours of Generation

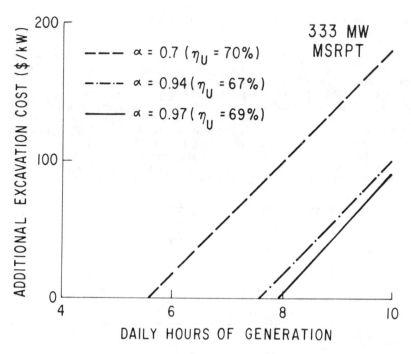

Fig. 11.   Additional Excavation Cost ($/kW) vs Daily
           Hours of Generation for 333 MW MSRPT

CONCLUSION

From the results of the system studies presented, the following conclusions can be drawn:

1. The utilization of advanced turbomachinery seems to push
    the minimum UPHS plant cost at a head range beyond that for
    UPHS plants equipped with conventional turbomachinery.

2. The percentage savings in plant costs using advanced machin-
    ery is somewhat insensitive to variation in plant capacity
    (1300-2700 MW).

3. For heads at or below 1000 m, single-stage, reversible
    pump turbines appears to be a logical choice. For heads
    above 1000 m, the present work indicates that two-stage,
    reversible pump-turbines represent an economically attrac-
    tive option. A significant cost reduction results when a
    plant with two-stage reversible units is compared with
    two-drop configuration plants or plants equipped with tandem
    units. However, this cost reduction may be diminished to
    some extent when differences in pump turbine efficiencies
    are considered. Because of the long plant and machinery
    lifetime, seemingly small differences in performance can
    result in nonnegligible cost penalties or benefits. Correc-

tions due to submergence and powerhouse size differences tend to have opposing effects. As a result, the net correction is not large enough to alter significantly the basic attractiveness of the advanced turbomachinery options considered here.

4. Increasing the charge/discharge ratio will reduce the relative storage cost. This reduction is accompanied by increases in the balance-of-plant cost and cost penalties resulting from possible reduction in turbomachinery efficiency. The cost benefit will depend on the amount of generation capacity required. If the generation time is large enough, a significant cost saving from reduction in storage cost will outweigh the other associated cost increases. In these cases, some sacrifices in efficiency from increasing the charge/discharge ratio may be acceptable. This factor illustrates the point that evaluation of both capital cost and performance are essential in making a judicious choice of turbomachinery options for UPHS application. Maximizing operating efficiency is only one among several important aspects in pump-turbine selection considerations. Careful integration of pump-turbine performance characteristics with the overall system consideration (i.e., UPHS plant design and utility grid requirement) bear significantly on the direct cost savings due to the differing capital costs of the various machines. More importantly, such considerations have a significant impact on the storage cost, the most important element in UPHS plant cost considerations.

ACKNOWLEDGMENTS

The research activities in underground pumped hydro storage, of which this paper is a part, were funded by the Division of Energy Storage Systems, U.S. Department of Energy.

REFERENCES

1. Blomquist, C.A., S.W. Tam, and A.A. Frigo, Turbomachinery Options for an Underground Pumped Hydro Storage Plant, Proceedings of the 14th Intersociety Energy Conversion Engineering Conference, Boston (Aug. 5-10, 1979).

2. Allis-Chalmers Corp., unpublished studies for Argonne National Laboratory (Oct. 1978).

3. Underground Pumped Hydroelectric Storage. An Evaluation of the Concept, Final Report. Solicitation No. 6-07-DR-50100, Charles T. Main, Inc. (Nov. 1978).

4. Blomquist, C.A., A.A. Frigo, and S.W. Tam, Underground Pumped Hydroelectric Storage (UPHS), mid-year program report ANL/EES/TM-60 (April 1979).

5. Newman, D.G., Engineering Economic Analysis: Engineering Press, pp. 50-53 (1976).

6. PSE&G, An Assessment of Energy Storage Systems Suitable for Use by Electric Utilities, Vol. III EM-264 EPRI project 225 ERDA E (11-1)-2501 (July 1976).

7. Rodrique, P., The Selection of High-head Pump-turbine Equipment for Underground Pumped Hydro Energy Storage Application, Pump Turbine Schemes: Planning, Design, and Operation. Proceedings of Joint ASME -CSME Applied Mechanics, Fluids Engineering and Bio-engineering Conference, Niagara Falls, N.Y., June 18-20, 1979, pp. 1-9.

8. Mitchell, W.S., Underground Pumped Hydro Storage, presented at the Engineering Foundation Energy Storage Conference, Pacific Grove, California (Feb. 8-13, 1976).

9. Stelzer, R.S., Estimating Reversible Pump Turbine Characteristics, Pump Turbine Schemes, see Ref. 7, pp. 139-149.